ELEMENTS OF PHYSICAL GEOGRAPHY

ELEMENTS OF PHYSICAL GEOGRAPHY

ARTHUR N. STRAHLER   ALAN H. STRAHLER

John Wiley & Sons, Inc.

New York   London   Sydney   Toronto

Cover art by Nita Engle.

Library of Congress Cataloging in Publication Data:

Strahler, Arthur Newell, 1918—
  Elements of Physical Geography.

  Includes index.
  1. Physical Geography — Text-books — 1945–
I. Strahler, Alan H., joint author. II. Title

GB55.S83   910′.02   75-25592
ISBN 0-471-83140-9

Printed in the United States of America

10 9 8 7 6 5 4 3 2

Throughout this book, illustrations with credit reading *Based on Goode Base Map* use as the base map Goode Map. No. 201HC World Homolosine, copyrighted by the University of Chicago. Used by permission of the Geography Department, University of Chicago.

Preface

Elements of Physical Geography is written for use in one-semester and one-quarter survey courses. Both content and structure are designed for students who are following general education programs in nonscience fields and who need only an overview of global physical environments. In keeping with this objective, we stress important concepts and basic facts, treating them descriptively and minimizing the use of technical terms. Emphasis is on ways in which the environment influences Man, both directly and indirectly. These influences act through climate, soils, vegetation, and landforms.

If we are to make physical geography more than a mere collection of science topics, and if we are to give students an insight into the nature and goals of professional geography, we must emphasize spatial distribution of physical environmental variables and their interactions. We stress the global patterns of climate, soils, vegetation, and landforms and explain these patterns as simply as possible in terms of natural processes.

Two contemporary trends in physical geography are emphasized. One trend is toward a climatology geared more closely to the soil-water balance and the availability of soil water to plants. A second trend is toward linking physical and organic processes through the concept of cycling of energy and materials within ecosystems. To analyze problems of world food resources, the geographer must integrate his knowledge of physical systems with the ecologist's knowledge of ecosystem dynamics.

In two subject areas we faced difficult problems of decision. These deal with classification systems of climates and soils, perennially debated by professional geographers. In the area of climate classification, we emphasize the basic environmental qualities of a dozen or so distinctive climate types, without introducing any elaborate code or a set of arbitrary boundary criteria. The Köppen system is offered separately for those instructors who require it. In the area of soils we de-emphasize classification in favor of descriptions of a few basic soil types having fairly clear climatic and vegetative associations. A plunge headlong into the new Soil Taxonomy with its 10 orders and 47 suborders, all bearing newly coined names, seems inappropriate to the aims of a general explanatory descriptive treatment of physical geography.

Physical geography is involved more and more with analysis of Man's impact on the environment. A trend toward deeper involvement is understandable, since physical geography integrates most of the diverse factors contributing to environmental changes. Wherever relevant, we include material on the environmental impacts of Man's activities on the store of finite natural earth resources and on resources that are renewable through inorganic and organic cycling of energy and matter.

Review questions and a glossary are available in our *Study Guide,* published separately. This guide not only gives a complete abstract of all important concepts, facts, and explanations found in the textbook, but it has a complete glossary as well, defining all italicized terms underlined in color in the text. The *Study Guide* also includes a complete set of self-testing review questions covering facts, concepts, and terms, and a set of sample objective tests. An *Instructor's Manual,* written expressly for our textbook, includes discussions of chapter content with background material on various topics and many suggestions for lecture preparation. There are many reading suggestions in the *Instructor's Manual;* these can be selected to fill the need for outside readings in any desired number and depth.

To enhance the enjoyment of a course in physical geography, our book uses an attractive and open design, for which we thank Suzanne Bennett of the Design Department of John Wiley and Sons. A number of color plates are introduced to illustrate features in which natural color is important — soils, vegetation types, and landforms. Aside from the NASA materials, most of these color photographs are reproduced from slides contributed by professional geographers and geologists.

Arthur N. Strahler
Alan H. Strahler

Biographical Note

ARTHUR N. STRAHLER (b. 1918) received his B.A. degree in 1938 from the College of Wooster, Ohio, and his Ph.D. degree in geology from Columbia University in 1944. He is a fellow of the Geological Society of America and the Association of American Geographers. He was appointed to the Columbia University faculty in 1941, serving as Professor of Geomorphology from 1958 to 1967 and as Chairman of the Department of Geology from 1959 to 1962. His published research deals mostly with quantitative and regional geomorphology. He is the author of several widely used textbooks of physical geography, environmental science, and the earth sciences. He resides in Santa Barbara, California.

ALAN H. STRAHLER (b. 1943) received his B.A. degree in 1964 and his Ph.D. degree in 1969 from The Johns Hopkins University, Department of Geography and Environmental Engineering. His published research lies in the fields of quantitative plant geography and forest ecology. He is Assistant Professor in the Department of Geography of the University of California at Santa Barbara. From 1973 to 1974 he held an appointment as Bullard Forest Research Fellow at Harvard University. He is the coauthor of two textbooks on environmental science.

Contents

ELEMENTS OF PHYSICAL GEOGRAPHY

Introduction

PHYSICAL GEOGRAPHY BRINGS together and interrelates the important elements of Man's physical environment. As in all branches of geography, emphasis is on spatial relationships — the systematic arrangements of environmental elements into regions over the earth's surface and the causes behind those patterns.

Physical geography draws on several natural sciences for its subject matter, among them sciences of the atmosphere (meteorology, climatology), oceans (oceanography), solid earth (geology), soils (soil science), vegetation (plant ecology), and landforms (geomorphology). Yet physical geography is much more than a collection of topics drawn from other sciences; it weaves that information into patterns of interaction with Man in a way that we do not find expressed within each of the contributing sciences.

The focus of physical geography is on the life layer, a shallow zone of the lands and oceans containing most of the world of organic life, or biosphere. Quality of the life layer is a major concern of physical geography. By quality we mean the sum of the physical factors that make the life layer habitable for all forms of plants and animals, but most particularly for Man. We use "Man," the first letter capitalized, as the English equivalent of genus *Homo,* thus referring collectively to all individuals of the human race. Man's full latin name is *Homo sapiens;* Man is the only surviving species of that genus.

The surface of the lands is the home of Man, because this animal requires a terrestrial rather than an aquatic environment. The quality of the physical environment of the lands is established by factors, forces, and inputs coming both from the atmosphere above and the solid earth below. The atmosphere dictates climate, which governs the exchange of heat and water between atmosphere and ground. The atmosphere also supplies vital elements — carbon, hydrogen, oxygen, and nitrogen — needed to sustain all life of the lands. The solid earth forms the stable platform for the life layer and is also shaped into landforms. These relief features — mountains, hills, and plains — bring another dimension to the physical environment and provide varied habitats for plants. The solid earth is also the basic source of many nutrient elements, without which plants and animals cannot live. These elements pass from rock into the shallow soil layer, where they are held in forms available to organisms.

Acting together, the inputs of energy and materials into the life layer from atmosphere and solid earth determine the quality of the environment and the richness or poverty of organic life it can support. In this way we come to recognize environmental regions, each with its particular qualities for life support. A given environmental region usually has certain definite locations on the globe in terms of latitude and continental position. It has a characteristic combination of soil type and native plant cover and offers a certain set of opportunities to Man to derive vital supplies of fresh water and food. Some environmental regions are richly endowed with water and food; others are very poorly endowed. The poorly endowed environments are too cold, too dry, or too rocky to support much life. A major goal of physical geography is to evaluate each environmental region in terms of its life-support capacity.

Thus an understanding of physical geography is vital to planning for survival of the earth's rapidly expanding human population. Survival will not only depend on how much fresh water and food is available, but also on protecting the environment from forms of pollution and destruction that will reduce the capacity of the land to furnish those necessities. Here we enter another of the important goals of physical geography: to evaluate the impact of Man on the natural environment. Environmental science, the study of interaction between Man and the environment, is gaining wide recognition today; many persons think of it as a new discipline. Actually, geographers have been investigating environmental science for many decades. Physical geography has always been at the heart of environmental studies, because physical geography is strongly oriented toward the interaction between Man and environment.

Besides natural resources needed for the support of life, the earth provides resources needed to support Man's industrial society: fresh water, minerals, and hydrocarbon fuels. Geographers are interested in the distributional patterns and interchanges of these industrial resources. Physical geography includes a study of the way in which these resources occur on and within the earth. A knowledge of resource origin and distribution is essential to planning for their future use and in evaluating the impact that use will have on the environment.

Our plan of study of physical geography is to begin with the atmosphere and the ways in which it furnishes light, heat, and water to the life layer. An evaluation of climate follows; here we define global environmental regions and assess their life-support capabilities. Next, we turn to the solid earth, reviewing geological principles essential to understanding major forms of the earth's crust and its mineral resources. Finally, we turn to the configuration of the earth's land surface, investigating landforms and the processes that shape them. Throughout all of these topics we shall stress the way in which Man interacts with the physical environment.

Our Spherical Habitat

Chapter 1

OUR SEARCH FOR understanding of Man's physical environment on Planet Earth starts with an inquiry into things astronomical. The most basic environmental controls relate to two facts: first, the earth is approximately spherical in form; second, the spherical earth is in motion, spinning on an axis and, at the same time, traveling in a nearly circular path around the sun. Let us look into these facts in order to derive significant interpretations relating to the quality of the life layer.

A spherical earth

In this age of orbiting satellites, the spherical form of the earth is such an obvious fact that we may have difficulty in imagining ourselves living at a time when the extent and configuration of the earth were entirely unknown. To a sailor of the Mediterranean Sea on his ship and out of sight of land, the sea surface looked perfectly flat and seemed to be terminated by a circular horizon. Based on this perception, the sailor might well have inferred that the earth has the form of a flat disk, and that if he traveled to its edge, he would fall off. Yet even this man sensed optical phenomena that could have suggested that the sea surface is not flat, but curved so as to be upwardly convex like a small part of the surface of a sphere. Given sufficiently acute eyesight and very clear air, the sailor could observe that as another vessel moved away from his, it seemed to pass below the horizon, so that the upper sails and rigging might remain visible even though the hull was out of sight. This phenomenon is easily observed today with the use of a telescope (Figure 1.1) and is a crude proof of the earth's curvature.

A second phenomenon that might have been put to use in reasoning about the earth's form is one we have all observed. After the sun has set below our horizon, its rays continue to illuminate clouds above us, or the peaks of high mountains, should any be near us. This effect could be explained by a curved earth surface and is perhaps the reason that scholars among the ancient Greeks believed the earth to be spherical. Pythagoras (540 B.C.) and associates of Aristotle (384-322 B.C.) held this view. They had also speculated on the length of the earth's circumference, but with highly erroneous guesses.

It remained until about 200 B.C. for Eratosthenes, librarian at Alexandria, Egypt, to perform a direct measurement of earth circumference based on a sound principle of geometry. He observed that on a particular date of the year (close to summer solstice, June 21), at Syene, a place located on the upper Nile River far to the south, the sun's rays at noon shone directly on the floor of a deep vertical well. In other words, the sun at noon was in the zenith point in the sky, and its rays were perpendicular to the earth's surface at that point on the globe (Figure 1.2). At Alexandria, on the same date, the rays of the noon sun made an angle with respect to the vertical. The magnitude of this angle was

figure 1.1
Seen through a telescope, a distant ship seems to be partly submerged.

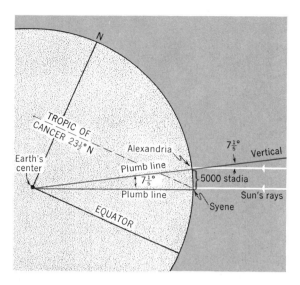

figure 1.2
Eratosthenes' method of measuring the earth's circumference.

one fiftieth of a complete circle, or $7^1/_5$ degrees.

Eratosthenes needed only to know the north-south distance between Syene and Alexandria to calculate the earth's circumference; he would simply multiply the ground distance by fifty to obtain the circumference. In those days distances between cities were only crude estimates, based on travelers' reports. Eratosthenes took the distance to be 5000 stadia, and so evaluated the earth's circumference as 250,000 stadia. It is not easy to rate his results in terms of accuracy, since we are not sure of the length equivalent of the distance unit (the stadium) that he used. If it were the Attic stadium, equivalent to 607 ft in modern English units, the circumference would come to about 26,700 English (statute) miles. Considering that the true circumference is close to 25,000 mi (40,000 km), Eratosthenes' results seem offhand to be remarkably good.

From Eratosthenes' classic experiment, it is an easy step to design an astronomical method of measuring the earth's figure using star positions

instead of the sun. We need only to select a north-south line, whose length can be measured directly on level ground by surveying means. This line should be several tens of miles long. At the ends of the line, the angular position of any selected star can be measured at its highest point above the horizon or with respect to the vertical, using a level bubble or a plumb bob as a means of establishing a true horizontal or vertical reference. The difference in angular positions of the star will be equal to the arc of the earth's circumference lying between the ends of the measured line. This very procedure is believed to have been followed by ninth-century Arabs. Their measurements were probably much more accurate than those of Eratosthenes but, because the units of measure are not known in modern equivalents, their work cannot be checked.

In the eight centuries following Eratosthenes' work Western science lay in stagnation. Then, about 1615, Willebrord Snell, a professor of mathmatics at the University of Leyden, developed methods of precise distance and angular measurement that he applied to the problem of the earth's circumference. His work presaged the era of scientific geodesy ("geodesy" comes from the Greek word meaning "to divide the earth") and led to remarkably accurate measurements of the earth's figure about a century later.

Gravity in the environment

What is significant, from the standpoint of life on earth, about the fact that the earth's form very closely approximates that of a true sphere? One answer is, of course, "gravity." Gravity is the force acting on a small unit of mass at the earth's surface, tending to draw that mass toward the earth's center. Gravity is a special case of the phenomenon of gravitation, the mutual attraction between any two masses — special in the sense that gravity refers to a unit of mass so small in comparison with the earth's mass that the attraction of the unit mass for the earth can be disregarded. Gravitational attraction varies inversely as the square of the distance separating the centers of two masses. So gravity depends on the distance of any particle of matter from the earth's center of mass. That center lies close to the geometrical center of the sphere.

Perhaps you recall a principle of geometry to the effect that a sphere is defined as a solid on which all surface points lie equidistant from a common point: the center of the sphere. Because of this principle gravity turns out to be an almost constant value at all points at sea level over the entire globe. This is a fact of fundamental importance to all life forms on earth. Life has evolved through geologic time under the influence of a value of gravity uniform over the earth, and probably little changed during the major evolutionary period of a billion years or so. Gravity thus represents the lowest common denominator of the planetary environment.

The force of gravity acts as an environmental factor in many ways. It separates substances of differing density into a layered arrangement, the least dense being at the top and the most dense at the bottom. Air, liquid water, and rock are arranged as we find them in order of density because of their response to gravity. As a result, the life layer is defined as an interface between atmosphere and ocean and between atmosphere and the solid land surface.

Trees, animals, cliffs of rock, and man-made structures must have the strength to withstand the force of gravity, which tends to collapse them and crush them. Under a weaker planetary gravity such structures could rise higher or be constructed of weaker materials to attain a given height. Gravity supplies power for important physical systems of the life layer, particularly streams and glaciers that erode the land. To appreciate the importance of gravity as an environmental factor, we need only speculate as to what would happen if its influence should be cancelled out and a condition of weightlessness takes its place. Environmental destruction would be total within a short time!

There are very small systematic differences in the value of gravity from place to place over the earth. The value at the equator is slightly less than at the two poles; there is also a very slight decrease in gravity as we rise to higher elevations above sea level. But, for practical purposes, gravity can be taken as a constant the world over.

The constancy of gravity over the earth's surface might be used in an experiment to prove the earth's sphericity. If we first assume that Newton's law of gravitation is valid, it follows that a given object should register the same weight at all places over the earth's surface. Using a spring balance as the scales, we might travel widely over the earth, repeatedly weighing a small mass of iron and recording the values. If they proved to be unvarying, we could conclude that we have taken our measurements at points all equidistant from the earth's center of mass, and so we are on a spherical surface. Actually, this same experiment, carried out with great precision and using highly refined instruments, has shown that the earth's true figure departs slightly from that of a true sphere. We will refer to this subject again.

Earth rotation and the geographic grid

The earth spins like a top on an axis; this is a phenomenon called *rotation*. One earth-turn with respect to the sun defines the *solar day*. Man has arbitrarily assigned a value of 24 hours to the mean duration of the solar day, and so all cultural aspects of his life can be regulated and coordinated.

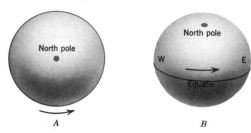

figure 1.3
The direction of rotation of the earth can be thought of as (A) counterclockwise at the north pole, or (B) eastward at the equator.

The direction of earth rotation can be determined by using one of the following rules. (1) Imagine yourself to be looking down on the north pole of the earth; the direction of turning is counterclockwise. (2) Place your finger on a point on a globe near the equator and push eastward. You will cause the globe to rotate in the correct direction (Figure 1.3). This explains the common expression "the eastward rotation of the earth." (3) The direction of rotation of the earth is opposite that of the apparent motion of the sun, moon, and stars. Because these bodies appear to travel westward across the sky, the earth must be turning in an eastward direction. Earth rotation affects Man's environment in two quite different ways: one is cultural; the other is physical.

Taking first the cultural influences of earth rotation, the spin of the earth on an axis allows us to set up a system designating the location of points and the directions of lines on the sphere. Without such a system human society would, indeed, be lost in the literal sense of the word. First, the axis of spin determines the *north pole* and *south pole,* which are fixed points of reference. Second, as any given point on the earth surface (except the pole points) moves with the earth's spin, the point generates a curved line. With the completion of one earth-turn with respect to fixed stars the line becomes a full circle, known as a *parallel of latitude* (Figure 1.4).

The largest parallel of latitude lies midway between the two poles and is designated the *equator.* As a unique circle on the sphere, the equator is a fundamental reference line for measuring position of points on the globe. Parallels of latitude also define the cardinal directions east and west on the globe.

If the earth is imagined to be sliced in two by a plane passing through the poles, a circle oriented at right angles to the plane of the equator is generated. The half-part of the circle, connecting one pole to the other, is known as a *meridian of longitude* (Figure 1.4). Meridians can be produced in any desired number, and a meridian can be found to pass

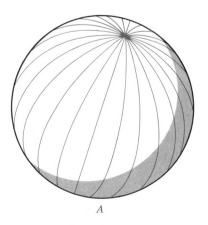 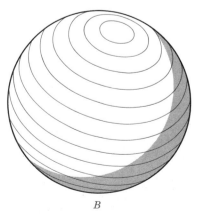

figure 1.4
Meridians and parallels.

through any point on the earth's surface. Meridians define the cardinal directions north and south.

The numbers of parallels and meridians are infinite, just as the number of points on the sphere is infinite. Every point on the globe has its unique combination of one parallel and one meridian; their intersection defines the position of the point. The total system of parallels and meridians forms a network of intersecting circles, and this is called the *geographic grid*.

Latitude and longitude

To communicate effectively about any phase of geography — whether it be physical or cultural — a numerical system must be fitted to the geographic grid. Master it well, because it will appear a thousand times in this book. *Latitude* is a measure of the position of a given point in terms of the angular distance between the equator and the poles. Latitude is an indicator of how far north or south of the equator a given point is situated. Latitude is measured in degrees of arc from the equator (0°) toward either pole, where the value reaches 90°. All points north of the equator — in the northern hemisphere, that is — are designated as north latitude; all points south of the equator — in the southern hemisphere — are designated as south latitude. Figure 1.5 shows how latitude is measured. The point *P* lies at latitude 50 degrees north, which we can abbreviate to lat. 50° N. Notice that latitude is actually measured along a meridian; it is an arc of that meridian.

Longitude is a measure of the position of a point eastward or westward with respect to a chosen reference meridian, called the *prime meridian*. As Figure 1.5 shows, longitude is the arc of a parallel of latitude between a given point and the prime meridian. The point *P* lies at longitude 60 degrees west (long. 60° W). The prime meridian is almost universally accepted as that which passes through the

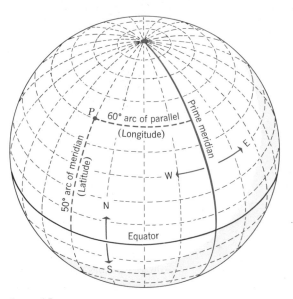

figure 1.5
The point P has a latitude of 50° N, a longitude of 60° W.

old location of the Royal Observatory at Greenwich, near London, England, and is referred to as the Greenwich meridian. This meridian has the value long. 0°. The longitude of any given point on the globe is measured eastward or westward from this meridian, whichever is the shorter arc. Longitude may thus range from 0° to 180°, either east or west. When both the latitude and longitude of a place are given, it is accurately and precisely located with respect to the geographic grid.

Statements of latitude and longitude tell nothing about distances in miles. However, you can easily make rough conversions from degrees of latitude into miles. One degree of latitude is approximately equivalent to 69 mi (111 km) of surface distance in the north-south direction. This value can be rounded off to 70 mi for multiplying in your head. For example, if you live at lat. 40° N (on the 40th parallel north), you are located about 40 × 70 = 2800 mi north of the equator. East-west distances cannot be converted so easily from degrees of longitude into miles because of the convergence of meridians in the poleward direction. Only at the equator is a degree of longitude equivalent to 69 mi (111 km). At lat. 60° N or S, one degree of longitude is reduced to half its equatorial value, or about 35 mi (56 km).

Global time

Besides generating the geographic grid, the rotation of a spherical earth has a cultural impact on Man through the phenomenon of global time. Our clocks run by the sun's schedule, but the sun sends us only a single set of parallel rays. This means that what is noon for persons on one side of the globe is midnight for persons on the opposite side.

To get a grasp of the global time picture, return to the days of the astronomer, Claudius Ptolemy (about 250 B.C.), and suppose, as he did, that the sun revolves about the earth. Where the sun's rays strike the earth most directly, _solar noon_ is occurring. A single meridian, which we shall call the _noon_

meridian, marks the north-south line on which it is noon (Figure 1.6). Under the Ptolemaic system of astronomy, the noon meridian sweeps westward around the globe, always staying directly beneath the sun. Directly opposite on the globe is an imaginary _midnight meridian,_ also sweeping westward and maintaining a separation of 180° of longitude from the noon meridian. Whereas the noon meridian separates the forenoon and afternoon of the same calendar day, the midnight meridian is the dividing line between one calendar day and the next.

Because the noon meridian sweeps over 360° of longitude every 24 hours, it must cover 15° of longitude every hour, or 1° of longitude every 4 minutes. We therefore find it convenient to state that 1 hour of time is the equivalent of 15° of longitude. This equality forms the basis for all calculations concerning time belts of the globe. For example, if the noon meridian reaches one place on the globe 4 hours after it leaves another place, the two places are separated by 60° of longitude.

Enlarging this concept of time meridians still further, imagine that in addition to the noon and midnight meridians there are 22 hour circles, each one separated by 15° of longitude from its neighbor. The hour circles are equidistantly spaced between the noon and midnight meridians (Figure 1.6). Each hour circle will then represent a given hour of the day and can be labeled with a specific hour number, which it keeps permanently. Together with the noon and midnight meridians, the hour circles can be imagined to form a birdcagelike net enclosing the globe and attached only at the north and south poles.

A working model of global time relations is illustrated in Figure 1.7. Two discs of different radius are attached at their centers in such a way that one disc can be turned while the other remains still. On the inner disc radii are drawn to represent 15° meridians of a globe

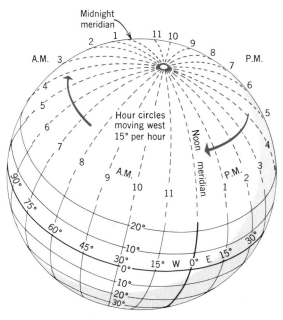

figure 1.6
Imaginary hour circles move ceaselessly westward around the globe.

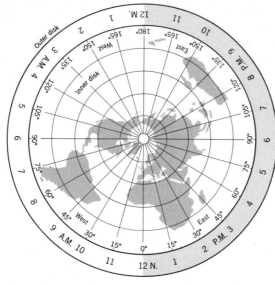

figure 1.7

This working model of global time zones uses a movable inner disc that can be set to any given hour at any selected standard meridian.

seen from a point above the north pole. On the outer disc similar radii are marked in hours to represent the time net. As a further refinement, the inner disc may be a hemispherical world map.

As set in Figure 1.7, it is noon on the Greenwich meridian (0° longitude), while it is midnight on the 180° meridian in mid-Pacific.

Figure 1.7 also illustrates global *standard time,* based on *standard meridians* spaced 15° apart around the globe. With reference to the Greenwich meridian, which is the world standard for reckoning time, 12 *time zones* lie in the eastern hemisphere and 12 in the western hemisphere. However, the 12th zone is shared by the two hemispheres. That half of the zone lying on the Asiatic side of the 180th meridian is exactly one full calendar day later than the half on the American side. For this reason, the 180th meridian defines the position of the International Date Line.

The United States encompasses six time zones.

Their names and standard meridians of west longitude are:

Eastern	75°
Central	90°
Mountain	105°
Pacific	120°
Alaska-Hawaii	150°
Bering	165°

Had it been carried out precisely, the system would have resulted in belts extending exactly 7½° east and west of each standard meridian, but a glance at the map (Figure 1.8) shows that great liberties have been taken in locating the boundaries. Wherever the time-zone boundary could conveniently be located along some already existing and widely recognized line, this was done. Natural physiographic boundaries have been used. For example, the Eastern time-Central time boundary line follows Lake Michigan down its center, and the Mountain time-Pacific time boundary follows a ridge-crest line also used by the Idaho-Montana state boundary. Most frequently, the time-zone boundary follows state and county boundaries.

The time belts are by no means equally distributed on both sides of the standard meridians, as a glance at the map will show. An extreme example is found in western Texas, where the Central time-Mountain time boundary follows the western boundary of Texas, even crossing west of the standard meridian (105° W) of the Mountain time zone. Although such deviations may seem odd, they cause little difficulty when the boundaries are well established. The advantage gained by extreme deviations of the boundaries is in permitting entire states to operate under one kind of time.

Because many human activities, especially in urban areas, start well after sunrise but continue long after sunset, it is desirable to set forward the hours of daylight so as to utilize them to best advantage. A considerable saving

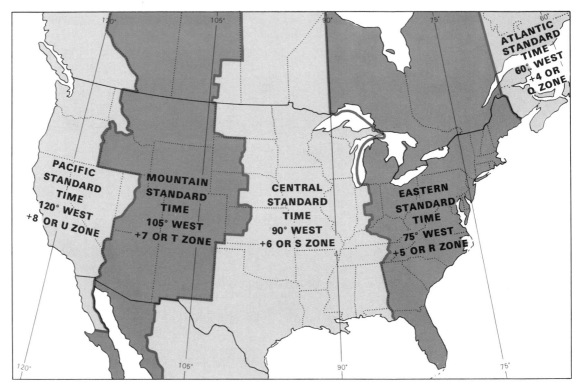

figure 1.8
A time-zone map of the contiguous United States and southern Canada. Figures with plus signs tell the time difference between the zone and Greenwich time. Zones are also designated on a global basis by letters of the alphabet.

in electric power can be made in summer when the early morning daylight period, wasted while schools, offices, and factories are closed, is transferred to the early evening when the large majority of persons are awake and busy. The adjusted time system is known as *daylight saving time* and is obtained by setting ahead all timepieces by 1 hour.

The spherical globe on flat maps

A cultural problem posed by the earth's spherical form arises when the geographer tries to depict the geographic grid on a flat sheet of paper, thereby producing a map. Much of the geographical information about world environments is best presented on a map showing the entire global surface. Unfortunately, a spherical surface cannot be transferred to a flat surface without considerable distortion.

Any orderly system of parallels and meridians drawn on a flat surface is a *map projection*. Many map projections have been devised, and each has its advantages as well as its shortcomings. Map projections used in this book have been very carefully chosen with respect to the kind of information they carry. For further information about these map projections, their good points and bad points, refer to the Appendix at the end of the book.

Physical effects of earth rotation

The physical effects of earth rotation are truly profound in terms of the environmental processes of the life layer. First, and perhaps most obvious, is that rotation imposes a daily, or diurnal, rhythm on many phenomena to which plants and animals respond. These phenomena include light, heat, air humidity, and air motion. Plants respond to the daily rhythm by storing energy during the day and releasing it at night. Animals adjust their activities to the daily rhythm, some preferring the day, others the night for food-gathering activities. The daily cycle of input of solar energy and a corresponding cycle of air temperature will be important topics for analysis in Chapters 3 and 4.

Second, as we shall find in the study of the earth's systems of winds and ocean currents, earth rotation causes the flow paths of both air and water to be consistently turned in direction: toward the right hand in the northern hemisphere and toward the left in the southern hemisphere. This phenomenon goes by the name of the Coriolis effect; we shall investigate it further in Chapter 5.

A third physical effect of the earth's rotation is important to Man. Because the moon exerts its gravitational attraction on the earth, while at the same time the earth is turning with respect to the moon, a set of forces known as the tide results. Tidal forces induce a rhythmic rise and fall of the ocean surface, and these motions in turn cause water currents of alternating direction to flow in shallow waters of the coastal zone. To a grain farmer in Kansas, the ocean tide has no significance whatsoever but, for the clam digger and charter boat captain on Cape Cod, the tidal cycle is a clock regulating their daily activities. For many kinds of plants and animals of salt-water estuaries tidal currents are essential to maintain a suitable life environment. The tide and its currents are discussed in Chapter 23.

Another effect of earth rotation is of great scientific importance but will not figure significantly in our study of Man's environment. Any mass in circular motion is affected by centrifugal force, a force tending to cause the mass to leave its circular path and fly off in a straight line, tangent to the circle. The earth itself is subjected to centrifugal force, and the value of that force increases from zero at either pole to a maximum value at the equator. In response to the force, the earth's form is distorted from a sphere into an *oblate ellipsoid*.

As illustrated in Figure 1.9, a cross section of the earth through the poles takes the form of an ellipse. The length of the polar axis is reduced by about 27 mi (43 km) as compared with the diameter of the equatorial circle, which has been somewhat increased. Lengths in miles of the polar axis and the equatorial diameter are given on the figure. The figure of 8000 mi (13,000 km) is a good round number for the earth's average diameter. The earth's oblateness must be taken into account in many

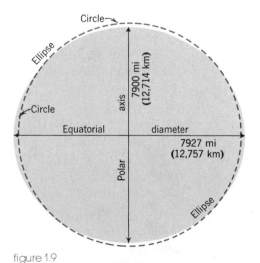

figure 1.9
The earth is slightly elliptical in a cross section passing through its polar axis.

forms of scientific activity, among them astronomy, precision navigation, and mapmaking.

The earth in orbit around the sun

Simultaneously with its rotation on an axis, the earth is in motion in an orbit around the sun. This motion is called *revolution*. As a first approximation, the orbit can be considered a circle with the sun lying at the center. To be exact, the orbit is an ellipse in which the sun occupies one focus. The elliptical form of the orbit does, in fact, have important astronomical consequences but produces only secondary environmental influences. By using the circular orbit as a model, we shall not miss out on any important environmental influences on processes of the life layer. The earth completes its circuit about the sun in about 365¼ days, a period known as the *tropical year*. The year is defined by astronomers in other ways as well, but these are not important here.

Every 4 years the extra one-fourth day difference between the tropical year and the calendar year of 365 days totals nearly one whole day. By inserting a 29th day in February, every leap year, Man is able to correct the calendar with respect to the tropical year. Further minor corrections are necessary to perfect this system.

In its orbit the earth revolves in such a direction that if we imagine ourselves in space, looking down on the earth and sun so as to see the north pole of the earth, the earth is traveling counterclockwise around the sun (Figure 1.10). It is further worth noting that nearly all planets and their satellites in our solar system have the same direction of rotation and revolution, suggesting that their motions were imparted to them at a time when they and the sun were originally formed.

The important point about the period of the

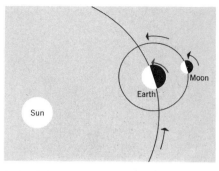

figure 1.10
Viewed as if from a point over the earth's north pole, the earth both rotates and revolves in a counterclockwise direction.

earth's revolution is that it sets the timing for climatic seasons that profoundly influence life on earth. To understand how seasons are generated, we must take another astronomical step in which we examine the relationship between the earth's axis and the plane of the orbit.

Tilt of the earth's axis

First, imagine the earth's axis to be exactly perpendicular with the plane in which the earth revolves about the sun. Astronomers call the plane containing the earth's orbit the *plane of the ecliptic*. Under these imagined conditions the earth's equator would lie exactly in the plane of the ecliptic. The sun's rays, which furnish all energy for life processes on earth, would always strike the earth most directly at a point on the equator. In fact, at the equator, the rays would be exactly perpendicular to the earth's surface at noon. The rays would always just graze the north and south poles. Conditions on any given day would be exactly the same as conditions on every other day of the year (assuming the orbit to be circular). In other words, there would be no seasons.

In reality, the earth's axis is not perpendicular to the plane of the ecliptic; it is inclined by a substantial angle of tilt, measuring almost exactly 23½° from the perpendicular. Figure 1.11 shows this axial tilt in a three-dimensional perspective drawing. The angle between axis and ecliptic plane is then 66½° (90° − 23½° = 66½°).

To proceed we must couple the fact of axial tilt with a second fact: the earth's axis, while always holding the angle 66½° with the plane of the ecliptic, maintains a fixed orientation with respect to the stars. The north end of the earth's axis points constantly toward Polaris, the North Star. To help in visualizing this movement, hold a globe so as to keep the axis tilted at 66½° with horizontal. Move the globe in a small horizontal circle, representing the orbit, at the same time keeping the axis pointed at the same point on the ceiling.

Solstice and equinox

What is the consequence of these two facts: (1) the earth's axis keeps a fixed angle with the plane of the ecliptic; and (2) the axis always

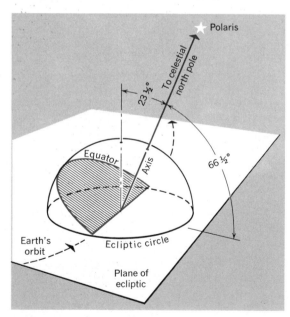

figure 1.11
The earth's axis keeps an angle of 66½° with the plane in which its orbit lies.

points to the same place among the stars? You will find that at one point in its orbit the earth's axis leans toward the sun; at an opposite point in the orbit the axis leans away from the sun. At the two intermediate points, the axis leans neither toward nor away from the sun (Figure 1.12). Next, consider the four critical positions in detail.

On June 21 or 22 the earth is so located in its orbit that the north polar end of its axis leans at the maximum angle 23½° toward the sun. The northern hemisphere is tipped toward the sun. This event is named the *summer solstice*. Six months later, on December 21 or 22, the earth is in an equivalent position on the opposite point in its orbit. At this time, known as the *winter solstice*, the axis again is at a maximum inclination with respect to a line drawn to the sun, but now it is the southern hemisphere that is tipped toward the sun.

Midway between the dates of the solstices occur the *equinoxes*, at which time the earth's axis makes a 90° angle with a line drawn to the sun, and neither the north nor south pole has any inclination toward the sun. The *vernal equinox* occurs on March 20 or 21; the *autumnal equinox* occurs on September 22 or 23. Conditions are identical on the two equinoxes as far as earth-sun relationships are concerned, whereas on the two solstices the conditions of one are the exact reverse of the other.

Consider first the conditions at the equinoxes, since this is the simplest case. Figure 1.13 shows that the earth is at all times divided into two hemispheres with respect to the sun's rays. One hemisphere is lighted by the sun; the other lies in darkness. Separating the hemispheres is a circle, the *circle of illumination*; it divides day from night. You will notice that at either equinox the circle of illumination cuts precisely through the north and south poles. Looking at the earth from the side, as in Figure 1.13, conditions at equinox

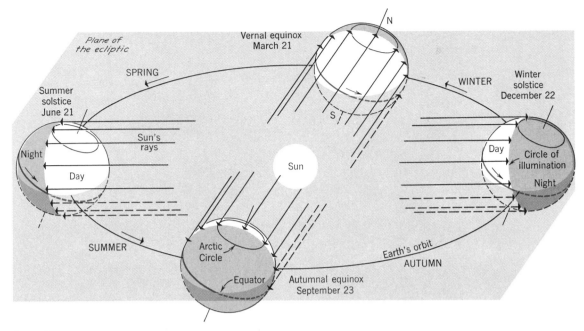

figure 1.12
The seasons result because the tilted earth's axis keeps a constant orientation in space as the earth revolves about the sun.

are essentially as follows. The point at which the sun's noon rays are perpendicular to the earth, or *subsolar point,* is located exactly at the equator. Here the angle between sun's rays and the earth surface is 90°. At both poles, the sun's rays graze the surface. As the earth turns,

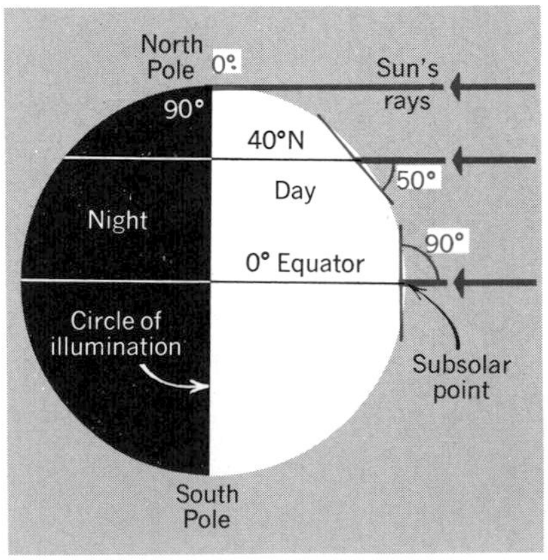

the equator receives the maximum intensity of solar energy; the poles receive none. At an intermediate latitude, such as 40° N, the rays of the sun at noon make an angle with the surface equal to 90° minus the latitude.

Next, examine winter solstice conditions, as shown in three dimensions in Figure 1.14. Because the maximum inclination of the axis is away from the sun, the entire area lying inside the *arctic circle,* lat. 66½° N, is on the dark side of the circle of illumination. Even though the earth rotates through a full circle during one day, this area poleward of the arctic circle remains in darkness. Conditions in the southern hemisphere during winter solstice are shown in Figure 1.15. All of the area lying south of the *antarctic circle,* lat. 66½° S, is under the sun's rays and enjoys 24 hours of day. The subsolar point has shifted to a point on the *tropic of capricorn,* lat. 23½° S.

At summer solstice, conditions are exactly reversed from those of winter solstice. As

figure 1.13
Equinox conditions. From this viewpoint the earth's axis appears to have no inclination.

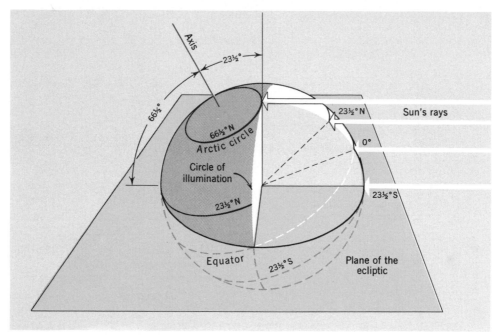

figure 1.14
At winter solstice the entire area between the arctic circle and the north pole is in darkness throughout the 24-hour period.

Figure 1.15 shows, the subsolar point is now on the *tropic of cancer*, lat. 23½° N. Now the region poleward of the arctic circle experiences a 24-hour day; that poleward of the antarctic circle experiences a 24-hour night. From one solstice to the next, the subsolar point has shifted over a latitude range of 47°.

The progression of changes from equinox to solstice to equinox and back to solstice initiates the *astronomical seasons:* spring, summer, autumn, and winter. These are labeled on Figure 1.12. The inflow of energy from the sun is thus varied in an annual cycle, and climatic seasons are generated. In Chapter 3 we shall study the seasonal changes in solar energy reaching various latitude zones of the earth. Make sure the astronomical facts are clear in your mind, because they lie behind the differentiation of the earth's surface into a wide spectrum of life environments.

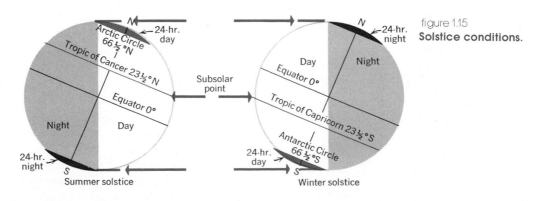

figure 1.15
Solstice conditions.

Man's spherical habitat

Looking back over what we have covered in this chapter, a number of salient concepts emerge.

Treating the earth as an ideal sphere, the uniformity of gravity over the entire planetary surface is the lowest common denominator of the environment; its influence on life forms during their evolution through geologic time has been profound.

Earth rotation on an axis divides the planet into two hemispheres, one northern and one southern, each one the mirror image of the other. To a first approximation, and disregarding surface inequalities, the two-hemispherical earth model requires that all physical phenomena of one hemisphere have a counterpart in the other, alike in all respects except for an inversion of the symmetry. Actually, this model is well displayed in the motions of atmosphere and oceans.

Earth rotation generates the diurnal rhythms of the environment, because solar energy is applied from a single point source and can be intercepted by only a hemisphere. Life processes show a strong response to the diurnal cycle of energy input. Earth rotation has the effect of turning the direction of motion of air and water. This Coriolis effect profoundly influences the surface environment. Earth rotation generates the ocean tide and its ceaseless rhythmic water motions essential to the life of the shallow coastal waters. Earth rotation causes a spherical earth to be deformed into an oblate ellipsoid.

The accident of a rotational axis inclined from the ecliptic plane generates a seasonal rhythm of changes in solar energy of the life layer and is a major factor in causing the environment to differ from one latitude zone to another.

In this study of the astronomical controls of environment our planet is visualized as a perfectly smooth, solid sphere of uniform surface quality, exposed to the near-vacuum of outer space. Now we must take a step toward realism and recognize that our planet has an envelope of air — the atmosphere — an extensive water layer — the oceans — and large patches of exposed solid rock — the continents. Variation in the global substance is to be found as we probe in both vertical and horizontal directions. Chapter 2 takes inventory of the properties and structures of the atmosphere and the oceans.

The Earth's Atmosphere and Oceans

Chapter 2

MAN LIVES AT the bottom of an ocean of air. Humans are air-breathers dependent on favorable conditions of pressure, temperature, and chemical composition of the atmosphere that surrounds them. Humans also live on the solid outer surface of the earth, which they depend on for food, clothing, shelter, and means of movement from place to place. But the air and the land are not two entirely separate realms; they constitute an interface across which there is a continual flux of matter and energy. Man's surface environment is the *life layer,* [a shallow but highly complex zone in which atmospheric conditions exert control on the land surface, while at the same time the surface of the land exerts an influence on the properties of the adjacent atmosphere.]

Essentially the same statements apply to the surface of the oceans and the atmospheric layer above it. Man utilizes the surface of the sea as a source of food and a means of transportation. There is a continual flow of energy and matter between the sea surface and the lower layer of the atmosphere. Here, again, we find an interface of vital concern to Man. The sea influences the atmosphere above it, while the atmosphere influences the sea beneath it.

Our objective in these early chapters is to examine the atmosphere and oceans with particular reference to the air-land and air-sea interfaces that are so vital to Man. To geographers, the distributions of physical properties of the ocean and atmosphere are matters of special interest, concerned as they are with spatial relationships on a global scale. Physical geographers describe and explain the ways in which the environmental ingredients of weather and climate change with latitude and season and with geographical position in relation to oceans and continents. They seek out the broad patterns of similar regions and attempt to define their boundaries and organize them into systems of classes. More important, geographers try to evaluate the environmental qualities of each region, emphasizing the opportunities as well as the limitations of each for future development of natural resources such as food, water, energy, and minerals.

States of matter

One of the most basic science concepts in physical geography is that of states of matter and their changes. Mostly, we are concerned with three common states: gas, liquid, and solid.

A *gas* is a substance that expands easily to fill any small empty container, and it is readily compressible. A gas is usually much less dense than liquids and solids of the same chemical composition. While the atmosphere is largely in the gaseous state, it also contains varying amounts of substances in the liquid and solid states.

A *liquid* is a substance that flows freely in response to unbalanced forces but maintains a free upper surface. Liquids are compressed in volume only slightly under strong pressures. Liquids have densities closely comparable with solids of the same composition. Although the world ocean is largely composed of water in the liquid state, it also contains substances in the gaseous and solid states.

Solids are substances that resist changes of shape and volume. Solids are typically capable of withstanding large unbalanced forces without yielding. When yielding does occur, it is usually by sudden breakage. Although the earth's crust is largely in the solid state, it also contains substances in gaseous and liquid states.

A change of state occurs frequently in the world of nature. Most important and wide-spread is the change of state of water from water vapor (a gas) to liquid water and vice versa, and from liquid water to ice (solid state) and vice versa (Chapter 6). Changes of state require either an input of heat energy or the disposal of heat energy, depending on the direction of the change.

These observations about states of matter will acquire more meaning as you apply them to various processes acting in the atmosphere and oceans.

Composition of the atmosphere

The principles we shall review in this chapter belong to two areas of natural science: *meteorology*, the science of the atmosphere, and *physical oceanography*, the physical-science aspect of the oceans.

The earth's *atmosphere* consists of a mixture of various gases surrounding the earth to a height of many miles. Held to the earth by gravitational attraction, this envelope of air is densest at sea level and thins rapidly upward. Although almost all of the atmosphere (97 percent) lies within 18 mi (29 km) of the earth's surface, the upper limit of the atmosphere can be drawn approximately at a height of 6000 mi (10,000 km), a distance approaching the diameter of the earth itself. From the earth's surface upward to an altitude of about 50 mi (80 km) the chemical composition of the atmosphere is highly uniform throughout in terms of the proportions of its gases.

Pure, dry air consists largely of nitrogen, about 78 percent by volume, and oxygen, about 21

percent (Figure 2.1). Nitrogen does not enter easily into chemical union with other substances and can be thought of as primarily a neutral substance. Very small amounts of nitrogen are extracted by soil bacteria and made available for use by plants. In contrast to nitrogen, oxygen is highly active chemically and combines readily with other elements in the process of oxidation. Combustion of fuels represents a rapid form of oxidation, whereas certain forms of rock decay (weathering) represent very slow forms of oxidation. Animals require oxygen to convert foods into energy.

The remaining 1 percent of the air is mostly argon, an inactive gas of little importance in natural processes. In addition, there is a very small amount of carbon dioxide, amounting to about 0.033 percent. This gas is of great importance in atmospheric processes because of its ability to absorb radiant heat and so allow the lower atmosphere to be warmed by heat rays coming from the sun and from the earth's surface. Green plants, in the process of photosynthesis, utilize carbon dioxide from the atmosphere, converting it with water into solid carbohydrate.

All of the component gases of the lower atmosphere are perfectly diffused among one another so as to give the pure, dry air a definite set of physical properties, just as if it were a single gas.

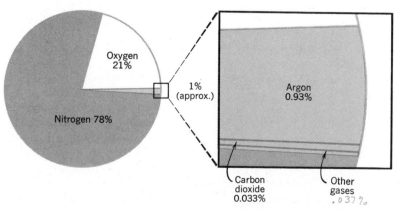

figure 2.1
Components of the lower atmosphere. Figures tell percentage by volume.

Oxygen 21%

1% (approx.)

Argon 0.93%

Nitrogen 78%

Carbon dioxide 0.033%

Other gases
.037%

Atmospheric pressure

Although we are not constantly aware of it, air is a tangible, material substance, exerting *atmospheric pressure* on every solid or liquid surface exposed to it. At sea level, this pressure is about 15 pounds per square inch (about 1 kg/sq cm). Because this pressure is exactly counterbalanced by the pressure of air within liquids, hollow objects, or porous substances, its ever-present weight goes unnoticed. The pressure on 1 sq in. of surface can be thought of as the actual weight of a column of air 1 in. in cross section extending upward to the outer limits of the atmosphere. Air is readily compressible. That which lies lowest is most greatly compressed and is, therefore, densest. In an upward direction, both density and pressure of the air fall off rapidly.

Atmospheric science uses another method of stating the pressure of the atmosphere, based on a classic experiment of physics first performed by Torricelli in the year 1643. A glass tube about 3 ft (1 m) long, sealed at one end, is completely filled with mercury. The open end is temporarily held closed. Then the tube is inverted and the end is immersed into a dish of mercury. When the opening is uncovered, the mercury in the tube falls a few inches, but then remains fixed at a level about 30 in. (76 cm) above the surface of the mercury in the dish (Figure 2.2). Atmospheric pressure now balances the weight of the mercury column. When the air pressure increases or decreases, the mercury level rises or falls correspondingly. Here, then, is a device for measuring air pressure and its variations.

Any instrument that measures atmospheric pressure is a *barometer*. The type devised by Torricelli is known as the *mercurial barometer*. With various refinements over the original simple device it has become the standard instrument. Pressure is read in inches or centimeters of mercury, the true measure of the height of the mercury column. Standard

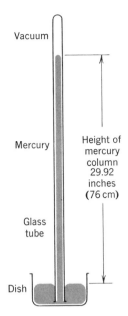

figure 2.2
The mercurial barometer of Torricelli.

sea-level pressure is 29.92 in. (76.0 cm) on this scale.

Another unit has been introduced by meteorologists. This is the *millibar* (mb). One inch of mercury is equivalent to about 34 mb, and each $1/10$ in. of mercury is equal to about 3.4 mb. Standard sea-level pressure is 1013 mb.

Figure 2.3 shows how pressure falls with increasing altitude. For every 900 ft (275 m) of rise in altitude, the mercury column falls $1/30$ of its height. As the graph shows, by a steepening of the curve, the rate of drop of the mercury becomes less and less with increasing altitude. Beyond a height of 30 mi (50 km), the decrease is extremely slight.

Atmospheric pressure is an important environmental factor affecting the life processes of plants and animals. Air-breathing animals depend on air pressure to force air into the lungs. They rely on the abundant supply of oxygen present in air of normal density for exchange with waste carbon dioxide through the lung tissues.

Temperature structure of the lower atmosphere

The atmosphere has been subdivided into layers according to temperatures and zones of temperature change. Of greatest importance to Man and other life forms is the lowermost layer, called the *troposphere*. If we sent up a sounding balloon carrying a recording thermometer and repeated this operation many times, we would obtain an average or representative profile of temperature. We would find that air temperature falls rather steadily with increasing altitude. The average rate of temperature decrease is about 3½ F°/1000 ft (6.4 C°/km) of ascent. This rate is known as the *environmental temperature lapse rate*. When used repeatedly, we shorten this ponderous term to *lapse rate*.

Figure 2.4 shows a typical air sounding into the troposphere at midlatitudes (45° N) on a

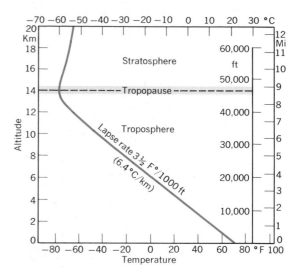

figure 2.4
A typical summer altitude-temperature curve in midlatitudes.

figure 2.3
Atmospheric pressure decreases with increasing altitude, but the rate of decrease falls off rapidly.

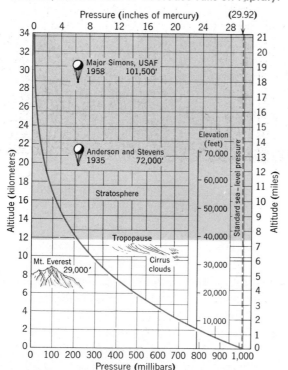

summer day. Altitude is plotted on the vertical axis, temperature on the horizontal axis. The resulting curve is a sloping line. Temperature drops uniformly with altitude to a height of about 8 mi (13 km). Don't be surprised if the captain of your jet aircraft, flying at 40,000 ft (12 km) altitude, announces that the outside temperature is −60°F (−50°C) — it's a fact!

The lapse rate changes abruptly at an altitude of about 9 mi (14 km). Instead of continuing to fall, temperature here remains constant with increasing altitude. This level of change is the *tropopause*; it marks a transition into the next higher temperature zone, known as the *stratosphere*. Upward into the stratosphere the temperature shows a gradual increase.

The troposphere and Man

The lowermost atmospheric layer, the troposphere, is of most direct importance to humans in their environment at the bottom of the atmosphere. Almost all phenomena of weather and climate that physically affect Man take place within the troposphere (Figure 2.5).

figure 2.5
The troposphere, seen in its proper perspective from an orbiting space vehicle, is a very shallow layer holding the earth's cloud cover and most of the free water vapor. (NASA Gemini V photo.)

In addition to pure dry air, the troposphere contains *water vapor,* a colorless, odorless gaseous form of water that mixes perfectly with the other gases of the air. The quantity of water vapor present in the atmosphere is of primary importance in weather phenomena. Water vapor can condense into clouds and fog. When condensation is rapid, rain, snow, hail, or sleet, collectively termed precipitation, are produced and fall to earth. Where water vapor is present only in small proportions, extremely dry deserts result. In addition, a most important function is performed by water vapor. Like carbon dioxide, it is a gas capable of absorbing heat in the radiant form coming from the sun and from the earth's surface. Water vapor gives to the troposphere the qualities of an insulating blanket, which inhibits the escape of heat from the earth's surface.

The troposphere contains myriads of tiny dust particles, so small and light that the slightest movements of the air keep them aloft. They have been swept into the air from dry desert plains, lake beds and beaches, or explosive volcanoes. Strong winds blowing over the ocean lift droplets of spray into the air. These may dry out, leaving as residues extremely minute crystals of salt that are carried high into the air. Forest and brush fires are another important source of atmospheric dust particles. Countless meteors, vaporizing from the heat of friction as they enter the upper layers of air, have contributed dust particles. Industrial processes involving combustion of fuels are also a major source of atmospheric dust.

Dust in the troposphere contributes to the occurrence of twilight and the red colors of sunrise and sunset, but the most important function of dust particles cannot be seen and is rarely appreciated. Certain types of dust particles serve as nuclei, or centers, around which water vapor condenses to produce cloud particles. In contrast, the stratosphere is almost entirely free of water vapor and dust. Clouds are rare in the stratosphere, but there are high-speed winds in narrow zones.

The ozone layer — an inner shield to life

Of vital concern to Man and all other life forms on earth is the presence of an *ozone layer* occurring within the stratosphere. This layer sets in an altitude of about 12 mi (20 km) and extends upward to about 35 mi (55 km). The ozone layer is a region of concentration of the form of oxygen molecule known as *ozone* (O_3) in which three oxygen atoms are combined instead of the usual two atoms (O_2). Ozone is produced by the action of ultraviolet rays on ordinary oxygen atoms.

The ozone layer serves as a shield, protecting the troposphere and earth's surface from most of the ultraviolet radiation found in the sun's rays. If these ultraviolet rays were to reach the earth's surface in full intensity, all exposed bacteria would be destroyed and animal tissues severely damaged. In this protective role the presence of the ozone layer is an essential factor in Man's environment.

This brings us to an interesting environmental issue, the debate over approval of government appropriations to develop a supersonic transport (SST) jet aircraft in the United States. The project was not approved, but a French SST aircraft, the Concorde, is presently in operation. The Concorde flies at an altitude of about 60,000 ft (18 km); the proposed American SST aircraft would have operated at about 70,000 ft (22 km), so that both levels are well within the stratosphere. Much concern has been expressed by atmospheric scientists that exhaust emissions from these aircraft could, in time, alter the composition of the stratosphere. Should the effect be to reduce the ozone content, increased penetration of ultraviolet radiation would result, with possible undesirable effects to life and with side effects in the form of changes of climate in the troposphere. Whether SST aircraft operation will, in fact, lead to reduction of stratospheric ozone remains undecided, but this effect has been recognized as a distinct possibility by an investigative panel of the National Academy of Sciences.

Another threat to the ozone layer is through the release of aerosol sprays into the atmosphere. Aerosol spray cans used in the household are usually charged with Freon, a compound containing carbon, fluorine, and chlorine atoms. Compounds of this class are often called fluorocarbons; they are also widely used as refrigerants. Molecules of fluorocarbons drift upward through the troposphere and eventually reach the stratosphere. As these compounds absorb ultraviolet radiation, they are decomposed, and chlorine is released. The chlorine in turn attacks molecules of ozone, converting them in large numbers by a chain reaction into ordinary oxygen molecules. In this way the ozone concentration within the stratospheric ozone layer can be reduced, and the intensity of ultraviolet radiation reaching the earth's surface can be increased. An increase in the incidence of skin cancer in humans is one of the predicted effects.

Man and the oceans

The importance of the oceans to Man is felt in a wide range of dimensions and scales. One environmental role played by the oceans is climatic. The huge water mass of the oceans stores a large quantity of heat. This heat is gained or lost very slowly. As we shall see, the oceans effectively moderate the seasonal extremes of temperature over much of the earth's surface. The oceans supply water vapor to the atmosphere and are the basic source of all rain that falls on the lands. This rainfall, the source of Man's vital freshwater supplies, originates from the ocean surface by a process of distillation of salt water.

The oceans sustain a vast and complex assemblage of marine life forms, both plant and animal. This organic production provides humans with a modest but important share of their food. Throughout history the oceans have served as trackless surfaces of transport of people and the commodities that sustain their civilizations. Winds, waves, currents, sea ice, and fog are environmental factors — sometimes favorable and sometimes hazardous — that the oceans impose on humans and their ships at sea.

The zone of contact between oceans and lands is a unique environment for Man. In Chapter 23 we shall investigate the processes by which ocean waves and currents shape coastal land features. Man uses and modifies the coastal zone in various ways, ranging from ports to recreational facilities.

The world ocean

We shall use the term *world ocean* to refer to the combined ocean bodies and seas of the globe. Let us consider some statistics that emphasize the enormous extent and bulk of this great saltwater layer. The world ocean covers about 71 percent of the global surface (Figure 2.6); its average depth is about 12,500 ft (3800 m), when shallow seas are included with the deep main ocean basins. For major portions of the Atlantic, Pacific, and Indian Oceans the average depth is about 13,000 ft (4000 m).

The total volume of the world ocean is about 317 million cu mi (1.4 billion cu km), a

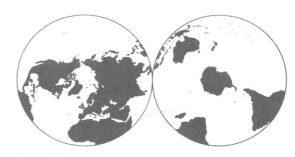

figure 2.6

Northern and southern hemispheres — a study in contrasts in the distribution of lands and oceans. One pole bears a deep ocean, the other a great continental landmass.

quantity just over 97 percent of the world's free water. Of the small remaining volume about 2 percent is locked up in the ice sheets of Antarctica and Greenland, and about 1 percent is represented by fresh water of the lands. These figures show the extent of the *hydrosphere,* a general term for the total free water of the earth, in all three of its states, as gas, liquid, or solid. The hydrosphere is largely represented by the world ocean. To place the masses of the atmosphere and oceans in their proper planetary perspective, compare the following figures: (the unit of mass used here is 10^{21} kg)

Entire earth	6000
World ocean	1.4
Atmosphere	0.005

What are the basic differences between the world ocean and the atmosphere in terms of properties and behavior? How do atmosphere and ocean interact in the region of their interface? The answers to these question are vital to an understanding of environmental processes, because marine life depends on the exchanges of matter and energy across the atmosphere-ocean interface. It is also significant that the earliest life forms originated and developed in the shallow layer of water immediately beneath the interface.

The atmosphere, being composed of a gas that is easily compressed, has no distinct upper boundary; it becomes progressively denser toward its base under the load of the overlying gas. The world ocean has a sharply defined upper surface in contact with the densest layer of the overlying atmosphere.

While the most active region of the atmosphere is the lowermost layer — the troposphere — the most active region of the ocean is its uppermost layer. At great ocean depths, water moves extremely slowly and maintains a uniformly low temperature. One reason for intense physical and biological activity in the uppermost ocean layer is that the input of energy from the overlying atmosphere drives water motions in the form of waves and currents. Another reason is that oxygen and carbon dioxide, gases vital to animal and plant growth, enter the ocean from the atmosphere. The atmosphere is also the source layer of heat and of condensed fresh water initially evaporated from the ocean. But the ocean surface also returns heat and water (in vapor form) to the lower atmosphere, and this phenomenon is of primary importance in driving atmospheric motions. Interaction between atmosphere and ocean surface is a topic we shall explore in later chapters.

The oceans are formed into compartments by intervening continental masses. The land barriers inhibit the free global interchange of ocean waters, whereas the atmosphere is free to move globally. Another difference in the two bodies is that the atmosphere has little ability to resist unbalanced forces. The air therefore moves easily and rapidly, changing its velocity very quickly from place to place. In contrast, ocean water can move only sluggishly and is very slow to respond to the changes in force applied by winds.

Composition of seawater

Seawater is a solution of salts — a brine — whose ingredients have maintained approximately fixed proportions over a considerable span of geologic time. The principal constituents of seawater are listed in Table 2.1.

Besides their importance in the chemical environment of marine life, these salts constitute a vast reservoir of mineral matter from which certain constituents may be extracted by Man for his use. One way to describe the composition of seawater is to state the principal ingredients that would be required to make an artificial brine approximately like seawater. These are listed in Table 2.1. Of the various elements combined in these salts, chlorine alone makes up 55 percent by weight of all the dissolved matter, and sodium makes up 31 percent. Magnesium, calcium, sulfur, and potassium are the other four major elements in these salts. Important, but less abundant than elements of the five salts in Table 2.1, are bromine, carbon, strontium, boron, silicon, and fluorine. At least some trace of half of the known elements can be found in seawater. Seawater also holds in solution small amounts of all of the gases of the atmosphere. Of these, oxygen and carbon dioxide are vital to organic processes in marine animals and plants.

table 2.1
Composition of seawater

Name of Salt	Grams of Salt per 1000 g of Water
Sodium chloride $NaCl$	23
Magnesium chloride $MgCl$	5
Sodium sulfate Na_3SO_4	4
Calcium chloride $CaCl_2$	1
Potassium chloride KCl	0.7
With other minor ingredients, To total	34.5

Layered structure of the oceans

As with the atmosphere, the ocean has a layered structure. Ocean layers are recognized in terms of both temperature and oxygen content. In the troposphere air temperatures are generally highest at ground level and diminish upward. In the oceans, temperatures are generally highest at the sea surface and decline with depth. This trend is to be expected, since the source of heat is from the sun's rays and from heat supplied by the overlying atmosphere.

With respect to temperature, the ocean presents a three-layered structure in cross section, as shown in the left-hand diagram of Figure 2.7. At low latitudes throughout the year and in midlatitudes in the summer, a warm surface layer develops. Here wave action mixes heated surface water with the water below it to give a warm layer that may be as thick as 1600 ft (500 m) with a temperature of 70 to 80° F (20 to 25° C) in oceans of the equatorial belt. Below the warm layer temperatures drop rapidly, making a second layer known as the _thermocline_. Below the thermocline is a third layer of very cold water extending to the deep ocean floor. Temperatures near the base of the deep layer range from 32 to 40° F (0 to 5° C). In arctic and antarctic regions, the three-layer system is replaced by a single layer of cold water, as Figure 2.8 shows. Temperature is a prime environmental factor controlling the abun-

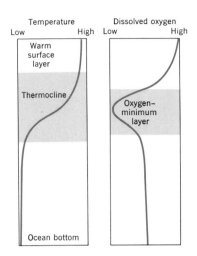

figure 2.7
This schematic diagram shows the changes of temperature and oxygen content with depth typical of broad expanses of the oceans in lower latitudes.

dance and variety of marine life, the bulk of which thrives in the shallow upper layer.

The content of free oxygen dissolved in seawater shows an oxygen-rich surface layer, accounted for by the availability of atmospheric oxygen and the activity of oxygen-releasing plant life in the sea. As shown in the right-hand diagram of Figure 2.7, oxygen content falls rapidly with depth; over large ocean areas there is a distinct minimum zone of low oxygen content. Here oxygen has been consumed by biological activity. In deep water the oxygen content holds to a uniform and moderate value down to the ocean floor.

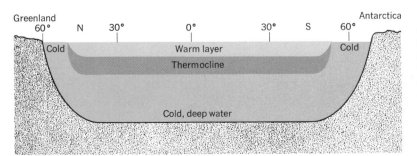

figure 2.8
A schematic north-south cross section of the world ocean shows that the warm surface water layer disappears in arctic latitudes, where very cold water lies at the surface.

Atmosphere
and oceans in review

The broad overview of atmosphere and oceans given in this chapter shows us the major elements of physical structure and chemical composition of those two great global layers so vital to life on earth. Most of the information has been about static conditions we would encounter when probing upward into the atmosphere and downward into the oceans.

In the next several chapters we turn to the great systems of flow of matter and energy that continually involve the atmosphere and oceans, making them dynamic rather than static bodies. First, we shall trace the course of radiant energy from the sun as it passes through the atmosphere, reaches the earth, and is returned to outer space. The solar radiation system controls the thermal environment of the life layer, or biosphere. Solar radiation supplies the life layer with the energy needed for biological processes. We shall then follow with accounts of the vast systems of transport of the atmosphere and oceans. These great flows of air and water redistribute heat over the globe to provide more moderate and more favorable life conditions than would otherwise exist on our planet. Accompanying these circulation systems are intense disturbances of the air and sea — storms that are often severe and hazardous environmental stresses.

Looking over the other planets of the solar system, the uniqueness of Planet Earth as a life environment is most striking. Only Earth has both a great world ocean and a comparatively dense oxygen-rich atmosphere combined with a favorable temperature range. Mars, our nearest planet, has practically no free water in any form and only a very rarefied atmosphere with little oxygen. Venus, matching us closely in size, has a much denser atmosphere than Earth. But water in any form is almost totally lacking on Venus, and surface temperatures there are very much higher than on Earth. Little Mercury, with no atmosphere and no water, roasts in the sun's rays. The great outer planets — Jupiter, Saturn, Uranus, and Neptune — probably have large quantities of water, but it is frozen solid. Their atmospheres, composed in part of ammonia and methane, would be lethal to life such as ours even if the surface temperatures were not impossibly cold. The Moon's surface has no free water or atmosphere to offer. So there is really no other place for humans to live in large numbers but on Planet Earth.

The Earth's Radiation Balance

Chapter 3

ALL LIFE PROCESSES on Planet Earth are supported by radiant energy coming from the sun. The planetary circulation systems of atmosphere and oceans are driven by solar energy. Exchanges of water vapor and liquid water from place to place over the globe depend on this single energy source. It is true that some heat flows upward through the earth's crust to reach the surface, but the amount is trivial in comparison with the amount of energy that the earth intercepts from the sun's rays.

The flow of energy from sun to earth and then out into space is a complex system. It involves not only energy transmission by a form of radiation, but also energy storage and transport. Both storage and transport of heat occur in the gaseous, liquid, and solid matter of the atmosphere, hydrosphere, and lithosphere. However, we can simplify the study of this total system by first examining each of its parts. We shall start with the radiation process itself and develop the concept of a *radiation balance,* in which energy absorbed by our planet is matched by the planetary output of energy into outer space.

Solar energy is intercepted by our spherical planet, and the level of heat energy tends to be raised. At the same time, our planet radiates energy into outer space, a process that tends to diminish the level of heat energy. Incoming

and outgoing radiation processes are simultaneously in action (Figure 3.1). In one place and time more energy is being gained than lost; in another place and time more energy is being lost than gained.

The equatorial region receives much more energy through solar radiation than is lost directly to space. In contrast, polar regions lose much more energy by radiation into space than is received. So mechanisms of energy transfer must be included in the energy system. These mechanisms must be adequate to export energy from the region of surplus and to carry that energy into the regions of deficiency. On our planet, motions of the atmosphere and oceans act as heat-transfer mechanisms. A study of the earth's energy balance will not be complete until the global patterns of air and water circulation are described and explained in Chapter 5.

To the inorganic phase, or physical energy cycle, there must be added an organic phase, the biochemical energy cycle, in which a part of the incoming solar energy is used and given up by plants and animals. Briefly, this cycle consists of the absorption of solar energy by plants in the manufacture of carbohydrate compounds, which provide the food for animals. Eventually, after much recycling, most of these compounds are oxidized in a process known as respiration, and the energy is

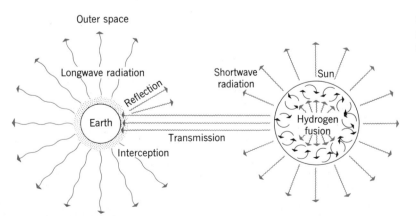

figure 3.1
A schematic diagram of the global energy balance.

returned to the atmosphere. However, some fraction of the carbohydrate compounds may be stored in soil layers, for example, as peat in bogs. In the geologic past the total accumulation of organic compounds (hydrocarbon compounds) in rock strata as coal and petroleum has been enormous and represents a great storage bank of solar energy. While the biochemical energy cycle involves only a tiny fraction (about one tenth of a percent) of the solar energy received by the earth, the process is of the first order of importance in the biosphere, since it represents the energy cycle of the whole organic world. We shall examine this energy cycle in more detail in Chapter 11.

Thinking further, we realize that the movement of water through atmosphere and oceans and on the lands comprises a global system of transport of matter. This system is equal in environmental importance to the flow of energy. We also realize that the activities of these two systems are closely intermeshed. The

concept of a water balance will be developed in Chapter 9 and will take its place beside the energy balance. Together, these two great flow systems of energy and matter form a single, grand planetary system and permit us to relate and explain many of the environmental phenomena of our earth within a single unified framework.

A systematic approach to the earth's energy balance begins with an examination of the input, or source, of energy from solar radiation. We shall trace this radiation as it penetrates the earth's atmosphere and is absorbed or transformed. We then turn to the mechanism of output of energy by the earth as a secondary radiator.

Solar radiation

Our sun, a star of about average size and temperature as compared with the overall range of stars, has a surface temperature of about 11,000° F (6000° K). The highly heated, incandescent gas that comprises the sun's surface emits a form of energy known as *electromagnetic radiation*. This form of energy transfer can be thought of as a collection, or spectrum, of waves of a wide range of lengths traveling at the uniform velocity of 186,000 mi (300,000 km) per second. The energy travels in straight lines radially outward from sun and requires about 8⅓ minutes to travel the 93 million mi (150 million km) from sun to earth. Although the solar radiation travels through space without energy loss, the rays are diverging as they move away from the sun. Consequently, the intensity of radiation within a beam of given cross section (such as 1 sq in.) decreases inversely as the square of the distance from the sun. The earth intercepts only about one two-billionth of the sun's total energy output.

The sun's electromagnetic spectrum can be divided into three major portions, based on

figure 3.2

Wavelength, L, is the crest-to-crest distance between successive wave crests.

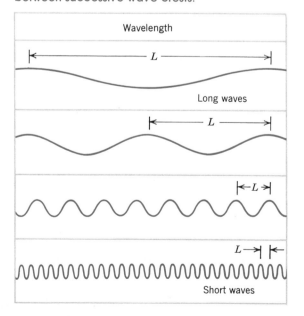

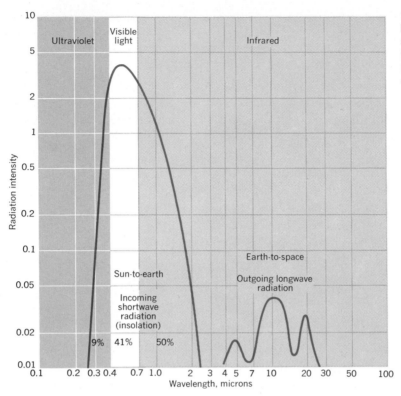

figure 3.3
**Intensity of incoming
shortwave radiation at the
top of the atmosphere (left);
longwave radiation from
earth to outer space (right).**

wavelengths. _Wavelength_ describes the distance separating one wave crest from the next (Figure 3.2). Long waves have a comparatively large separating distance; short waves are closely spaced. Figure 3.3 shows these portions. The unit of wavelength is the micron. (A micron is one ten-thousandth of a centimeter; 0.0001 cm.) At the left lies the _ultraviolet radiation_ portion of the spectrum, with wavelengths ranging from 0.2 to 0.4 microns. In the center is the _visible light_ portion in the range of 0.4 to 0.7 microns. At the right lies the _infrared radiation_ region, ranging upward of 0.7 microns. The scales used on this graph increase by powers of ten, meaning that the scale becomes tightly compressed, both toward the right and the top of the graph.

We are primarily concerned with the energy of the solar spectrum and how it is distributed. The high curve at the left on the graph shows the relative energy intensity from one end of the sun's spectrum to the other, if it were to be measured at the top of the atmosphere. Energy intensity is shown on the graph (Figure 3.3) on the vertical scale. The units used here do not have to be explained, because we are only interested in making rough comparisons of how intensities are related to the bands of the spectrum. Energy intensity rises rapidly through the ultraviolet portion of the spectrum. This portion accounts for about 9 percent of the total incoming solar energy. Also included in the ultraviolet division are X rays and gamma rays. These are forms of radiation with shorter wavelengths than the ultraviolet and are not shown on the graph. The energy curve peaks in the visible light range, which accounts for about 41 percent of the total energy. The infrared portion of the spectrum accounts for 50 percent of the total energy. Very little energy arrives in wavelengths longer than 2 microns. Scientists refer to the entire solar spectrum as _shortwave radiation_, because the peak of intensity lies in the shorter wavelengths. In contrast, radiation within the infrared region is called _longwave radiation_.

The source of solar energy is in the sun's interior. Here, under enormous confining pressure and high temperature, hydrogen is converted to helium. In this nuclear fusion process, a vast quantity of heat is generated and finds its way to the sun's surface. Because the rate of production of nuclear energy is constant, the output of solar radiation is almost unvarying. So, at the average distance of earth from the sun, the amount of solar energy received on a unit area of surface held at right angles to the sun's rays is almost unvarying. We are assuming, of course, that the radiation intensity is measured beyond the limits of the earth's atmosphere so that none has been lost. Known as the _solar constant_, this radiation rate has a value of 2 gram calories per square centimeter per minute. The gram calorie is that quantity of heat required to raise by 1C° the temperature of 1 gram of pure water. One gram calorie per square centimeter constitutes a unit measure of radiation intensity known as the _langley_. Therefore, we can say that the solar constant is equal to 2 langleys per minute.

Insolation over the globe

Referring back to Chapter 1 and Figure 1.13, showing equinox conditions, recall that at only one point, the subsolar point, does the earth's spherical surface present itself at right angles to the sun's rays. In all directions away from the subsolar point, the earth's curved surface becomes turned at a decreasing angle with respect to the rays until the circle of illumination is reached. Along that circle the rays are parallel with the surface.

Let us now assume that the earth is a perfectly uniform sphere with no atmosphere. Only at

the subsolar point will solar energy be intercepted at the full value of the solar constant, 2 langleys per minute. We will now use the term _insolation_ to mean the interception of solar shortwave energy by an exposed surface. At any particular place on the earth, insolation received in one day will depend on two factors: (1) the angle at which the sun's rays strike the earth; and (2) the length of time of exposure to the rays. These factors are varied by latitude and by the seasonal changes in the path of the sun in the sky.

Figure 3.4 shows that intensity of insolation is greatest where the sun's rays strike vertically, as they do at noon at some parallel within the belt lying between the tropic of cancer and the tropic of capricorn. With diminishing angle, the same amount of solar energy spreads over a greater area of ground surface. So, on the average, the polar regions receive the least insolation over a year's time.

In Chapter 1 we considered a hypothetical situation in which the earth's axis is perpendicular to the plane of the eclipitic as the earth revolves around the sun. Under such

figure 3.4

The angle of the sun's rays determines the intensity of insolation on the ground. The energy of vertical rays <u>A</u> is concentrated in square <u>a</u>, but the same energy in the slanting rays <u>B</u> is spread over a rectangle, <u>b</u>.

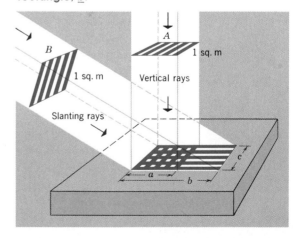

conditions, the poles would not receive any insolation, regardless of time of year, whereas the equator would receive an unvarying maximum. In other words, equinox conditions would prevail year-around. But we know that the earth's axis is not perpendicular to the plane of the orbit. The axis is tilted with respect to the orbital plane by an angle that measures 23½° from the perpendicular (Figure 1.11). Moreover, the axis at all times holds its orientation in space. Consequently, as the earth travels in its orbit about the sun, the tilted globe assumes different positions with respect to the sun's rays (Figure 1.12). At this point, a review of the conditions at solstices and equinoxes may be helpful (Chapter 1).

The subsolar point, representing the sun's noon rays, shifts through a total latitude range of 47° from one solstice to the next. This cycle does not make the yearly total of insolation for the entire globe different from an imaginary situation in which the earth's axis were not inclined, but it does cause a great difference in the quantities received at various latitudes.

The total annual insolation from equator to poles in thousands of langleys (kilolangleys) per year is shown in Figure 3.5 by a solid line. A dashed line shows the insolation that would result if the earth's axis had no tilt. Notice how much insolation the polar regions actually receive — over 40 percent of the equatorial value. The added insolation at high latitudes is of major environmental importance because it brings heat in summer to allow forests to flourish and crops to be grown over a much larger portion of the earth than would otherwise be the case.

A second effect of the axial tilt is to produce seasonal differences in insolation at any given latitude. These differences increase toward the poles, where the ultimate in opposites (six months of day, six of night) is reached. Along with the variation in angle of the sun's rays, another factor, the duration of daylight,

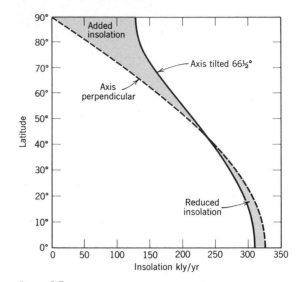

figure 3.5
Total annual insolation from equator to pole (solid line) compared with insolation on a globe with axis perpendicular to the ecliptic plane (dashed line).

operates. At the season when the sun's path is highest in the sky, the length of time it is above the horizon is correspondingly greater. The two factors thus work hand in hand to intensify the contrast between amounts of insolation at opposite solstices. We reinforce this concept with an example.

Figure 3.6 is a three-dimensional diagram showing how the sun's path in the sky changes from season to season. The diagram is drawn for lat. 40° N, and is typical of conditions found in midlatitudes in the northern hemisphere. The diagram shows a small area of the earth's surface bounded by a circular horizon. This is the way things look to human beings standing on a wide plain. The earth's surface appears to be flat, and the celestial dome on which the sun, moon, and stars move seems to be a hemispherical dome. Actually, this situation resembles a planetarium with its domed ceiling.

At equinox, the sun rises in the east and sets in the west; while at noon, the sun is at a point in

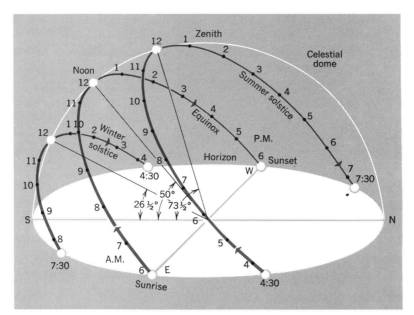

figure 3.6
The sun's path in the sky changes greatly in position and height above horizon from summer to winter. This diagram is for a place located at lat. 40° N.

the southern sky at an angle of 50° above the horizon. At equinox the sun is above the horizon for just 12 hours, as shown by the hour numbers on its path. At summer solstice the sun is above the horizon for about 15 hours, and its path rides much higher in the sky than at equinox. At winter solstice the path is low, and the sun is above the horizon for only about 9 hours. Obviously the insolation reaching a square centimeter of horizontal surface is much greater during a day at summer solstice than during a day at winter solstice. When the daily values of insolation are plotted for a full year, they form a wavelike curve on a graph. Figure 3.7 is just such a graph, and the curve for lat. 40° N is the one to which we are referring.

A three-dimensional diagram (Figure 3.8) shows how insolation varies with latitude and with season of year. These diagrams show insolation at the outer limits of the atmosphere and would apply at the ground surface only for an earth imagined to have no atmosphere to absorb or reflect radiation. Notice that the equator receives two maximum periods (corresponding with the equinoxes, when the

sun is overhead at the equator). There are also two minimum periods (corresponding to the solstices, when the subsolar point shifts farthest

figure 3.7

Insolation curves at various latitudes in the northern hemisphere. (From A. N. Strahler, 1971, The Earth Sciences, 2nd ed., Harper & Row, New York.)

Langleys per day:

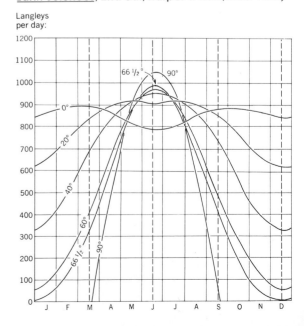

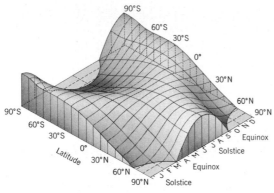

figure 3.8
The effect of both latitude and season of year on the intensity of insolation show on this diagram for the whole globe. At any given latitude and date the relative amount of insolation is proportional to the height of the surface point above the flat base of the block. (After W. M. Davis.)

north and south from the equator). At the arctic circle (66½° N), insolation is reduced to nothing on the day of the winter solstice and, with increasing latitude poleward, this period

of no insolation becomes longer. All latitudes between the tropic of cancer (23½° N) and the tropic of capricorn (23½° S) have two maxima and two minima, but one maximum becomes dominant as the tropic is approached. Poleward of the two tropics there is a single continuous insolation cycle with a maximum at one solstice and a minimum at the other.

World latitude zones

The angle of attack of the sun's rays determines the flow of solar energy reaching a given unit area of the earth's surface and so governs the thermal environment of the life layer. This concept provides a basis for dividing the globe into latitude zones (Figure 3.9). We don't intend that the specified zone limits be taken as absolute and binding but, instead, that the system be considered as a convenient terminology for identifying world geographical belts throughout this book.

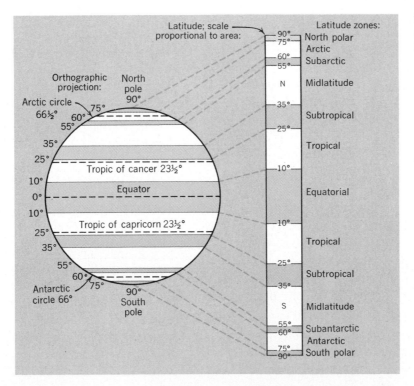

figure 3.9
A geographical system of latitude zones.

The _equatorial zone_ lies astride the equator and covers the latitude belt roughly 10° north to 10° south. Within this zone, the sun throughout the year provides intense insolation, while day and night are of roughly equal duration. Astride the tropics of cancer and capricorn are the _tropical zones,_ spanning the latitude belts 10° to 25° north and south. In these zones, the sun takes a path close to the zenith at one solstice and is appreciably lower at the opposite solstice. Thus a marked seasonal cycle exists, but it is combined with a large total annual insolation.

Literary usage and that of many scientific works differ from what is described here; the terms "tropical zone" and "tropics" have been widely used to denote the entire belt of 47 degrees of latitude between the tropics of cancer and capricorn. That is the definition of "tropics" you will find in most dictionaries. Even though correct, that definition is not well suited to a study of Man's physical environment, because it combines belts of very unlike climatic properties; it lumps the greatest of the world's deserts with the wettest of climates.

Immediately poleward of the tropical zones are transitional regions that have become widely accepted in geographers' usage as the _subtropical zones_. For convenience, we have assigned these zones the latitude belts 25° to 35° north and south, but it is understood that "subtropical" may be extended a few degrees farther poleward or equatorward of these parallels.

The _midlatitude zones,_ lying between 35° and 55° north and south latitude, are belts in which the sun's angle of attack shifts through a relatively large range, so that seasonal contrasts in insolation are strong. Strong seasonal differences in lengths of day and night exist as compared with the tropical zones.

Bordering the midlatitude zones on the poleward side are the _subarctic_ and _subantarctic zones,_ 55° to 60° north and south latitudes. Astride the arctic and antarctic circles, 66½° north and south latitudes, lie the _arctic_ and _antarctic zones_. The arctic zones have an extremely large yearly variation in lengths of day and night, yielding enormous contrasts in insolation from solstice to solstice. Notice that dictionaries define "arctic" as the entire area from arctic circle to north pole, and "antarctic" in a corresponding sense for the southern hemisphere.

The _polar zones,_ north and south, are circular areas between about 75° latitude and the poles. Here the polar regime of a six-month day and six-month night is predominant. These zones experience the ultimate in seasonal contrasts of insolation.

Insolation losses in the atmosphere

As the sun's radiation penetrates the earth's atmosphere, its energy is absorbed or diverted in various ways. At an altitude of 95 mi (150 km), the radiation spectrum possesses almost 100 percent of its original energy but, by the time rays have penetrated to an altitude of 55 mi (88 km), absorption of X rays is almost complete, and some of the ultraviolet radiation has been absorbed as well.

As solar rays penetrate more deeply into the atmosphere, they reach the stratosphere. Here, as we explained in Chapter 2, the ozone layer is formed by action of ultraviolet rays on molecules of oxygen. The environmental shielding provided by this layer has already been stressed.

As radiation penetrates into deeper and denser atmospheric layers, gas molecules cause the visible light rays to be turned aside in all possible directions, a process known as _scattering_. Dust and cloud particles in the troposphere cause further scattering, described as _diffuse reflection_. Scattering and diffuse reflection send some energy into outer space and some down to the earth's surface.

As a result of all forms of shortwave scattering, about 6 percent of the total insolation is returned to space and forever lost, as shown in Figure 3.10. Scattered shortwave energy directed earthward is referred to as *down-scatter*.

Another form of energy loss, *absorption*, takes place as the sun's rays penetrate the atmosphere. Both carbon dioxide and water vapor are capable of directly absorbing infrared radiation. Absorption results in a rise of temperature of the air. In this way some direct heating of the lower atmosphere takes place during incoming solar radiation. Although carbon dioxide is a constant quantity in the air, the water vapor content varies greatly from place to place. Absorption correspondingly varies from one global environment to another.

All forms of direct energy absorption — by air molecules, including molecules of water vapor and carbon dioxide, and by dust — are estimated to total as little as 10 percent for conditions of clear, dry air to as high as 30 percent when a cloud cover exists. A global average of 14 percent for absorption generally is shown in Figure 3.10. When skies are clear, reflection and absorption combined may total about 20 percent, leaving as much as 80 percent to reach the ground.

Yet another form of energy loss must be brought into the picture. The upper surfaces of clouds are extremely good reflectors of shortwave radiation. Air travelers are well aware of how painfully brilliant the sunlit upper surface of a cloud deck can be when seen from above. *Cloud reflection* can account for a direct turning back into space of from 30 to 60 percent of total incoming radiation (Figure 3.10). Clouds also absorb radiation. So we see that, under conditions of a heavy

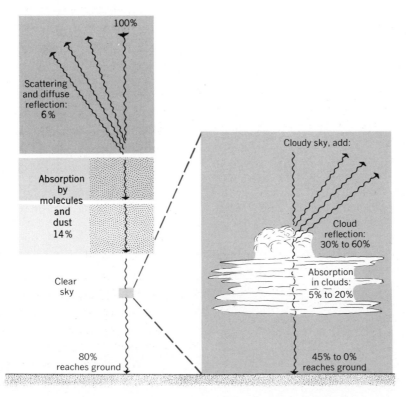

figure 3.10
Losses of incoming solar energy are much less with clear skies (left) than with cloud cover (right). (From A. N. Strahler, 1971, The Earth Sciences, 2nd ed., Harper and Row, New York.)

100%

Scattering and diffuse reflection: 6%

Absorption by molecules and dust 14%

Clear sky

80% reaches ground

Cloudy sky, add:

Cloud reflection: 30% to 60%

Absorption in clouds: 5% to 20%

45% to 0% reaches ground

cloud layer, the combined reflection and absorption from clouds alone can account for a loss of 35 to 80 percent of the incoming radiation and allow from 45 to 0 percent to reach the ground. A world average value for reflection from clouds to space is about 21 percent of the total insolation. Average absorption by clouds is much less — about 3 percent.

The surfaces of the land and ocean reflect some shortwave radiation directly back into the atmosphere. This small quantity, about 6 percent as a world average, may be combined with cloud reflection in evaluating total reflection losses. Figure 3.11 lists the percentages given so far for the energy losses by reflection and absorption. Altogether the losses to space by reflection total 32 percent of the total insolation.

The percentage of radiant energy reflected back by a surface is the *albedo*. This is an important property of the earth's surface, because it determines how fast a surface heats up when exposed to insolation. Albedo of a water surface is very low (2 percent) for nearly vertical rays but high for low-angle rays. It is also extremely high for snow or ice (45 to 85 percent). For fields, forests, and bare ground the albedos are of intermediate value, ranging from as low as 3 percent to as high as 25 percent.

Orbiting satellites, suitably equipped with instruments to measure the energy levels of shortwave and infrared radiation, both incoming from the sun and outgoing from the atmosphere and earth's surface below, have provided data for estimating the earth's average albedo. Values between 29 and 34 percent have been obtained. The value of 32 percent given in Figure 3.11 lies between these limits.

The earth's albedo lies in a range intermediate between low values of the moon and inner

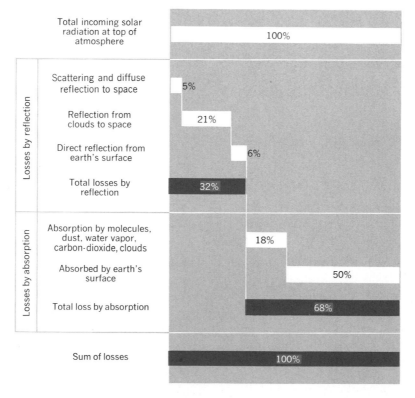

figure 3.11
A summary of the energy losses by reflection and absorption as solar energy penetrates the atmosphere to reach the earth's surface.

Total incoming solar radiation at top of atmosphere — 100%

Losses by reflection

Scattering and diffuse reflection to space — 5%

Reflection from clouds to space — 21%

Direct reflection from earth's surface — 6%

Total losses by reflection — 32%

Losses by absorption

Absorption by molecules, dust, water vapor, carbon-dioxide, clouds — 18%

Absorbed by earth's surface — 50%

Total loss by absorption — 68%

Sum of losses — 100%

planets (Mercury, 6 percent; Mars, 16 percent; Moon, 7 percent) and high values of Venus (76 percent) and the great outer planets (73 to 94 percent). Here, again, we find another unique element of the environment of Planet Earth as compared with that of the other planets and the moon.

Continuing down the bar chart of Figure 3.11, we come to the summary of energy losses by absorption. A global figure of 18 percent is given for the combined losses through absorption by molecules, dust, water vapor, carbon dioxide, and clouds. There remains 50 percent of the original solar energy, and this amount is absorbed by the earth's land and water surfaces. We have now accounted for all of the energy of insolation. Our next step is to deal with outgoing energy so as to balance the global energy budget.

Longwave radiation

Any warm substance sends out energy of electromagnetic radiation from its surface. The amount of energy so radiated increases rapidly as temperature goes up. Another important principle is that at low temperatures the radiating substance emits mostly long wavelengths in the infrared region of the spectrum. As temperature rises, the emitted radiation is shifted toward the shorter wavelengths.

The ground and ocean surfaces possess heat derived originally from absorption of the sun's rays. These surfaces continually radiate longwave energy back into the atmosphere. The process is known as *ground radiation*. As Figure 3.3 shows, longwave radiation consists of infrared wavelengths longer than 3 microns. The atmosphere also radiates longwave energy both downward to the ground and outward into space, where it is lost. Be sure to understand that longwave radiation is quite different from reflection, in which the rays are turned back directly without being absorbed.

Longwave radiation from both ground and atmosphere continues during the night, when no solar radiation is being received.

Because longwave radiation goes on through the night, it is possible to obtain *infrared imagery* of ground features using special cameras mounted on aircraft. Figure 3.12 is an example. This image was obtained in the early morning hours in total darkness. Pavements and streams appear bright on the print because they are warm and radiate more intensely. In contrast, moist soil surfaces of fields are cooler and appear dark. Infrared imagery is one of the kinds of information used in *remote sensing*, a new field of major importance in geography. Remote sensing makes use of electromagnetic radiation in various bands over a wide range of the total spectrum.

Longwave radiation from the earth's land and ocean surfaces spans a wavelength range of 4 to 30 microns, as shown in Figure 3.3. The emission of longwave radiation into outer space takes place largely in a wavelength band between 8 and 13 microns, often referred to as a *window*. There are also lesser windows in the range of 4 to 6 microns and 17 to 21 microns. These windows show as peaks on the longwave radiation curve in Figure 3.3. Between the windows most of the longwave radiation leaving the ground is absorbed by water vapor and carbon dioxide, and this absorbed energy is converted into sensible heat.

As Figure 3.3 shows, the intensity of longwave energy leaving our planet is only a small fraction of the shortwave solar energy. Keep in mind that longwave radiation is constantly emitted from the entire spherical surface of the earth, whereas insolation falls on only one hemisphere. A single hemisphere presents the equivalent of only a cross section of the sphere to full insolation. Also, do not forget that about 32 percent of the incoming solar radiation is reflected directly back into space. Only the

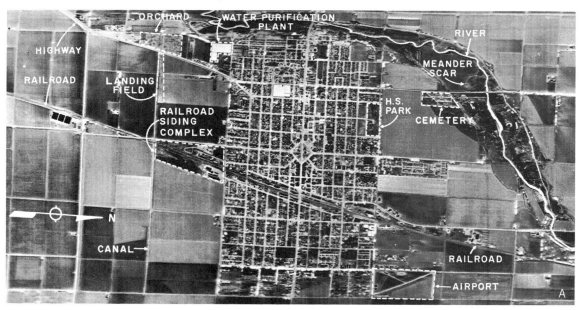

figure 3.12
This infrared imagery of Brawley, a small town in the Imperial Valley of California, looks like an air photograph. It was taken in darkness between 2 and 4 a.m. Special lenses and films can record the infrared rays coming up from the surface. (Environmental Analysis Department, HRB — Singer, Inc.)

remaining 68 percent that is absorbed must be disposed of by longwave radiation. On the average, year in and year out, for the planet as a whole, the quantity of absorbed solar energy is balanced by an equal longwave emission to space.

Part of the ground radiation absorbed by the atmosphere is radiated back toward the earth surface, a process called _counterradiation_. For this reason, the lower atmosphere with its water vapor and carbon dioxide acts as a blanket that returns heat to the earth. This mechanism helps to keep surface temperatures from dropping excessively during the night or in winter at middle and high latitudes.

Somewhat the same principle is used in greenhouses and in homes using solar heating. Here, the glass windows permit entry of shortwave energy. Accumulated heat cannot escape by mixing with cooler air outside. Meteorologists use the expression _greenhouse effect_ to describe this atmospheric heating principle. Cloud layers are even more important in producing a blanketing effect to retain heat in the lower atmosphere, since they are excellent absorbers and emitters of longwave radiation.

Figure 3.13 shows the components of outgoing longwave radiation for the planet as a whole, using the same percentage units as for the incoming radiation. A small percentage (8 percent) passes directly out into space, while the remainder is absorbed by the atmosphere. In turn, the atmosphere emits longwave radiation, both to outer space and back to the earth's surface as counterradiation.

Now a new concept enters the radiation balance picture. Much of the heat energy passing from the ground to the atmosphere is carried upward by two mechanisms other than longwave radiation. First, sensible heat is conducted upward by turbulent motions of the air. As Figure 3.13 shows, this heat transport amounts to about 9 percentage units. Second, latent heat is conducted upward in water vapor that has evaporated from land and ocean surfaces. Latent heat transport accounts for about 20 percentage units. (Latent heat is explained in Chapter 6.)

Summarizing the information of the previous paragraphs, energy leaving the ground (both land and water surfaces) is divided up as follows.

	Percent
Longwave radiation (net value)	21
Mechanical transport as sensible heat	9
Transport as latent heat	20
Total	50

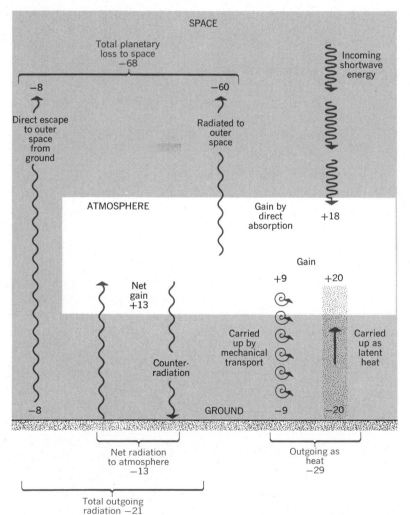

figure 3.13
Energy passes from ground to atmosphere by longwave radiation, by mechanical transport, and as latent heat. From the atmosphere energy passes out into space by longwave emission.

Recall from Figure 3.11 that 50 percent is the amount of energy absorbed by the ground. We have now balanced the annual energy budget so far as the ground surface is concerned. This is the interface of vital importance to Man and all life forms; it is the life layer we have emphasized before.

The atmosphere in turn disposes of its heat by longwave radiation into outer space. Together with direct longwave emission from the ground, the total emission to space is 68 percent. Add this figure to 32 percent losses by reflection (the earth's albedo), and the total is 100 percent. The global radiation budget stands balanced.

Latitude and the radiation balance

Earlier in this chapter, in relating latitude to insolation, we showed that the tilt of the earth's axis causes a poleward redistribution of insolation, as compared with conditions that would apply if the axis were perpendicular to the orbital plane (Figure 3.5). Let us now look deeper into the wide range in rates of incoming and outgoing energy in terms of a profile spanning the entire latitude range 90° N to 90° S. In this analysis yearly averages are used so that the effect of seasons is concealed.

We are going to examine the _net radiation,_ which is the difference between all incoming energy and all outgoing energy carried by both shortwave and longwave radiation. Our analysis of the radiation balance has already shown that for the entire globe as a unit, the net radiation is zero on an annual basis. However, in some places, energy is coming in faster than it is going out, and the energy balance is a positive quantity, or _energy surplus_. In other places energy is going out faster than it is coming in, and the energy balance is a negative quantity, or _energy deficit._

Figure 3.14 is a global profile of the net radiation from pole to pole. Between about lat. 40° N and lat. 40° S there is an energy surplus, as you would expect, because at low latitudes insolation is intense throughout the year. Poleward of lat. 40° N and S a deficit sets in and becomes more and more severe as the poles are reached. The two areas labeled "deficit" have a combined area on the graph equal to the area labeled "surplus." In other words, the net radiation for the entire globe is zero.

It is obvious from Figure 3.14 that the earth's radiation balance can be maintained only if heat is transported from the low-latitude belt of surplus to the two high-latitude regions of

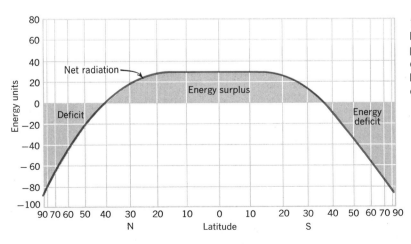

figure 3.14
Net radiation from pole to pole shows two polar regions of energy deficit, matching a large low-latitude region of energy surplus.

deficit. This poleward movement of heat is described as *meridional transport.* "Meridional" means moving north or south along the meridians of longitude. The rate of meridional heat transport is greatest in midlatitudes between the 30th and 50th parallels.

The meridional flow of heat is carried out by circulation of the atmosphere and oceans. In the atmosphere, heat is transported both as sensible heat and as latent heat. Meridional energy transport will be a topic in Chapters 6 and 7.

Man's impact on the earth's radiation balance

Although our analysis of the earth's radiation balance is far from complete, it should be obvious by now that the balance is a sensitive one, involving as it does a number of variable factors that determine how energy is transmitted and absorbed. Has Man's industrial activity already altered the components of the planetary radiation balance? An increase in carbon dioxide may be expected to increase the absorption of longwave radiation by the atmosphere. Will this change cause a rise in atmospheric temperature? Has such a change already occurred? If continued, where will it lead? An increase in stratospheric water vapor might cause that layer to become warmer. Increase in atmospheric dusts will increase scatter of incoming shortwave radiation and perhaps reduce the atmospheric temperature; but, on the other hand, increased dust content will act to absorb more longwave radiation and so raise the atmospheric temperature. Has either change occurred?

Man has profoundly altered the earth's land surfaces by cultivation and urbanization. Have these changes in surface albedo and in the capacity of the ground to absorb and to emit longwave radiation resulted in a changed energy balance? Answers to such questions require further study of the processes of heating and cooling of the earth's atmosphere, lands, and oceans; these will be the subject of the next chapter.

Heat and Cold in the Life Layer

Chapter 4

TEMPERATURE OF THE air and the soil layer is of major interest to the physical geographer. Temperature is a measure of heat energy available in the air and the soil. Organisms respond directly to heating and cooling of the environmental substance that surrounds them. We can refer to this influence as the *thermal environment*.

Our study of the radiation balance has shown that temperature changes result from the gain or loss of energy by the absorption or emission of radiant energy. When a substance absorbs radiant energy, the surface temperature of that substance is raised. This process represents a transformation of radiant energy into the energy of sensible heat, which is the physical property measured by the thermometer. Heat can also enter or leave a substance by conduction or can be lost as latent heat during evaporation.

Many of the biochemical processes taking place within organisms as well as many common inorganic chemical reactions are intensified by an increase in temperature of the solutions in which these reactions are occurring. Severe cold, which is simply the lack of heat energy within matter, may greatly reduce or even completely stop biochemical and inorganic reactions. This is why the vital environmental ingredient of heat — heat in the air, water, and soil — needs to be thoroughly understood. The thermal environment is surely a dominant part of Man's total physical environment.

We are all familiar with natural cycles of temperature change. There is a daily rhythm of rise and fall of air temperature as well as a seasonal rhythm. There are also systematic average changes in air temperatures from equatorial to polar latitudes and from oceanic to continental surfaces. These temperature changes require that the lower atmosphere and the surfaces of the lands and oceans receive and give up heat in daily and seasonal cycles. There must also be great differences in the quantities of heat received and given up annually in low latitudes as compared to high latitudes. Temperature cycles — daily and seasonal — and the influence of latitude on temperature will be dominant themes of this chapter.

Measurement of air temperatures

Air temperature is one of the most familiar bits of weather information we hear and read daily through the news media. This information comes from standard National Weather Service observing stations and is taken following a carefully standardized procedure. Thermometers are mounted in a standard instrument shelter, shown in Figure 4.1. The shelter shades the instruments from sunlight, but louvers allow air to circulate freely past the thermometers. The instruments are mounted from 4 to 6 ft (1.2 to 1.8 m) above ground level, at a height easy to read.

figure 4.1
This thermometer shelter stands in a grass-covered area, cleared of trees. A rain gauge is at the left. (National Weather Service.)

At most stations, only the highest and lowest temperatures of the day are recorded. To save the observer's time, the *maximum-minimum thermometer* is used. This instrument uses two thermometers; one to show the highest temperature, since it was last reset; the other to show the lowest. Another labor-saving device is the recording thermometer, which draws a continuous record on graph paper attached to a slowly moving drum. Figure 4.2 shows typical traces for several days of record.

When the maximum and minimum temperatures of a given day are added together and divided by two, we obtain the *mean daily temperature*. The mean daily temperatures of an entire month can be averaged to give the *mean monthly temperature*. Averaging the daily means (or monthly means) for a whole year gives the *mean annual temperature*. Usually such averages are compiled for records of many years' duration at a given observing station. These averages are used in describing the climate of the station and its surrounding area.

The daily cycle of air temperature

Because the earth turns on its axis, the net radiation undergoes a daily cycle of change. This cycle determines the daily cycle of rising and falling air temperatures with which we are all familiar. Let us see how net radiation and air temperature are linked in this cycle.

The upper graph of Figure 4.3 shows insolation on a clear day at a city in the United States located about lat. 40° N, such as Chicago or New York. At equinox insolation begins at sunrise, 6 a.m. local time, and reaches a peak at noon, when the sun is highest in the sky. Insolation then falls off and ceases at sunset, 6 p.m.

The middle graph of Figure 4.3 shows the net radiation close to the ground. Throughout the

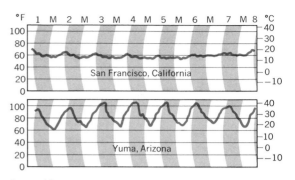

figure 4.2
The recording thermometer made these continuous records of the rise and fall of air temperature over a period of one week. The daily cycle is strongly developed at Yuma, a desert station. In contrast, the record for San Francisco shows a very weak daily cycle.

figure 4.3
Daily cycles of insolation and net radiation control the daily rhythm of air temperature.

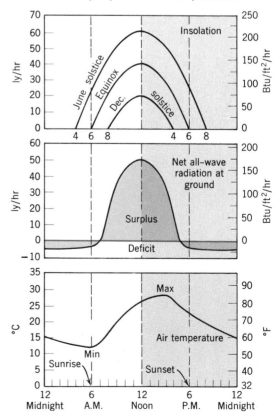

night there has been a steady energy deficit, because outgoing longwave radiation continues during darkness when there is no incoming solar radiation. Just after sunrise, the situation is reversed and an energy surplus sets in. This surplus builds as the sun rises higher, reaching a maximum at noon. More energy is entering the lower air layer than is leaving, so the air temperature rises rapidly, as the lower graph shows.

In the afternoon insolation declines and, just before sunset, the energy surplus gives way to a deficit. The air temperature does not decline after noon but, instead, keeps rising into the middle afternoon. As long as there is an energy surplus, the air temperature should continue its rise. However, by 3 or 4 p.m., another activity begins to take effect. Mixing of the lower air in eddies distributes heat upward. As a result, the temperature begins to fall in late afternoon. The temperature decline continues through sunset and into the night. All night long air temperature slowly falls, reaching its minimum just before sunrise.

Next, we must take into account changes in insolation from summer to winter. As the upper diagram in Figure 4.3 shows, insolation at June solstice is not only more intense than at equinox, but it also begins earlier in the

morning and lasts later into the evening. The result is to make the energy surplus larger and cover a longer period of hours. The effect on the air temperature curve is shown clearly in Figure 4.4. At El Paso, Texas, air temperatures in July are much higher at all hours than in spring and fall (April and October). Moreover, in summer the minimum temperature occurs an hour or so earlier in the morning. At the time of winter solstice insolation is greatly reduced and covers a much shorter period than at equinox. As a result, air temperatures are lowered, as the January curve for El Paso shows. The time of minimum temperature is now advanced by an hour. Notice that the time of occurrence of the maximum daily temperature is little affected by the season. The graphs for North Head, Washington, illustrate the same seasonal effects, but the cycles are much weaker for reasons that we will explain in a later paragraph.

The daily cycle of air temperature can be strongly intensified close to the ground surface, as Figure 4.5 shows. Under the direct sun's rays, a bare soil surface or pavement heats to much higher temperatures than one reads in the thermometer shelter. At night, this surface layer cools to temperatures much lower than in the shelter. These effects are quite weak under

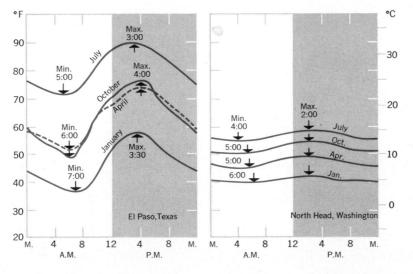

figure 4.4
The average daily cycle of air temperature for four different months shows the effect of the seasons. Seasonal changes are great at El Paso, a station in the continental interior, but only weakly developed at North Head, Washington, close to the Pacific Ocean.

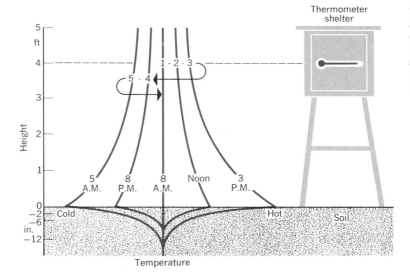

figure 4.5
A highly simplified schematic diagram of the temperature profile close to the ground surface. The soil surface becomes very hot at midday but very cold at night.

a forest cover where the ground is shaded and moist. On the dry desert floor and on the pavements of city streets and parking lots the surface temperature extremes are large.

Urbanization and the heat island

Applying principles of insolation and temperature, we can anticipate the impact of Man as cities spread, replacing a richly vegetated countryside with blacktop and concrete. In the urban environment the absorption of solar radiation causes higher ground temperatures for two reasons. First, foliage of plants is absent, so that the full quantity of solar energy falls on the bare ground. Absence of foliage also means absence of transpiration (evaporation from leaves), which elsewhere produces a cooling of the lower air layer. A second factor is that roofs and pavements of concrete and asphalt hold no moisture, so that evaporative cooling cannot occur as it would from a moist soil.

The thermal effect is that of converting the city into a hot desert. The summer temperature cycle close to the pavement of a city may be

almost as extreme as that of the desert floor. This surface heat is conducted into the ground and stored there. The thermal effects within a city are actually more intense than on a sandy desert floor because the capacity of solid concrete, stone, or asphalt to conduct and hold

figure 4.6
A heat island over Washington, D.C. Air temperatures were taken at 10 p.m. on a day in early August. (Data of H. E. Landsberg.)

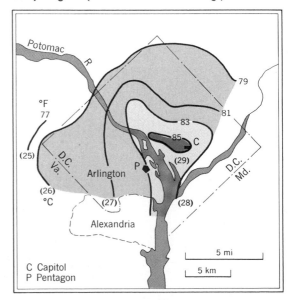

heat is greater than that of loose, sandy soil. An additional thermal factor is that vertical masonry surfaces absorb insolation or reflect it to the ground and to other vertical surfaces. The absorbed heat is then radiated back into the air between buildings.

As a result of these changes in the energy balance, the central region of a city typically shows summer air temperatures several degrees higher than for the surrounding suburbs and countryside. Figure 4.6 is a map of the Washington, D.C. area, showing air temperatures for a typical August afternoon. The lines of equal air temperature delineate a _heat island_. The heat island persists through the night because of the availability of a large quantity of heat stored in the ground during the daytime hours. In winter additional heat is radiated by walls and roofs of buildings, which conduct heat from the inside. Even in summer, Man adds to the city heat output through use of air conditioners, which expend enormous amounts of energy at a time when the outside air is at its warmest.

In a later chapter we shall make further observations about the impact of urbanization on climate of the cities. Other weather phenomena, such as wind speeds, fog, clouds, and precipitation, show measurable changes as a result of heat output and the injection of pollutants into the air.

Temperature inversion and frost

During the night, when the sky is clear and the air calm, the ground surface rapidly radiates longwave energy into the atmosphere above it. As we have just explained, soil temperatures drop rapidly, and the overlying air layer becomes colder. When we plot temperature against altitude, as in Figure 4.7, the straight, slanting line of the normal environmental lapse rate becomes bent to the left in a "J" hook. In the case shown, the air temperature at the surface, point A, has dropped to 30° F (−1° C). This value is the same as at point B, some 2500 ft (750 m) aloft. As we move up from ground level, temperatures become warmer up to about 1000 ft (300 m). Here the curve reverses itself and the normal lapse rate takes over.

The lower, reversed portion of the lapse rate curve is called a _low-level temperature_

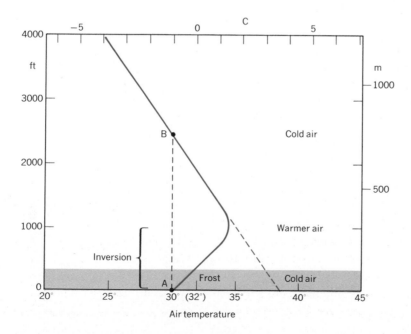

figure 4.7
A low-level temperature inversion. In this case a layer of air close to the ground has dropped below the freezing temperature, while the air above remains warmer.

inversion. In the case we have shown you temperatue of the lowermost air has fallen below the freezing point (32° F, 0° C). For sensitive plants this condition is a _killing frost_ when it occurs during the growing season. Killing frost can be prevented in citrus groves by setting up an air circulation that mixes the cold basal air with warmer air above. One method is the use of oil-burning heaters; another is to operate powerful motor-driven propellors to circulate the air.

Low-level temperature inversion is also prevalent over snow-covered surfaces in winter. Inversions of this type are very intense and often extend several thousand feet into the air.

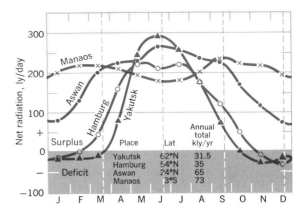

figure 4.8
Net radiation for four stations ranging from the equatorial zone to the arctic zone. (Data by courtesy of David H. Miller.)

Place	Lat	Annual total kly/yr
Yakutsk	62°N	31.5
Hamburg	54°N	35
Aswan	24°N	65
Manaos	3°S	73

The annual cycle of air temperature

As the earth revolves about the sun, the tilt of the earth's axis causes an annual cycle of incoming solar radiation, as we have already learned. This cycle is also felt in an annual cycle of the net radiation which, in turn, causes an annual cycle in the mean daily and monthly air temperatures. In this way the climatic seasons are generated.

Figure 4.8 shows the yearly cycle of net radiation for four stations, ranging in latitude from the equator almost to the arctic circle. Figure 4.9 shows mean monthly air temperatures for these same stations. Starting with Manaus, a city on the Amazon River in Brazil, let us compare the net radiation graph with the air temperature graph. At Manaus, almost on the equator, the net radiation shows a large surplus in every month. The average surplus is about 200 langleys per day (200 ly/day), but there are two minor maxima, approximately coinciding with the equinox, when the sun is nearly straight overhead. A look at the temperature graph of Manaus shows monotonously uniform air temperatures,

averaging about 81° F (27° C) for the year. The _annual temperature range,_ or difference between the highest mean monthly temperature and lowest mean monthly temperature, is only 3 F° (1.7 C°). In other words, near the equator one month is about like the next, thermally speaking. There are no temperature seasons.

We go next to Aswan, United Arab Republic (Egypt), on the Nile River at lat. 24° N. Here we are in a very dry desert. The net radiation curve has a strong annual cycle, and the surplus is large in every month. The surplus rises to more than 250 ly/day in June and July, but falls to less than 100 ly/day in December and January. The temperature graph shows a corresponding annual cycle, with an annual range of about 30 F° (17 C°). June, July, and August are terribly hot, averaging over 90° F (32° C).

Moving further north, we come to Hamburg, Germany, lat. 54° N. The net radiation cycle is strongly developed. The surplus lasts for 9 months, and there is a deficit for 3 winter months. The temperature cycle reflects the reduced total insolation at this latitude. Summer months reach a maximum of just over

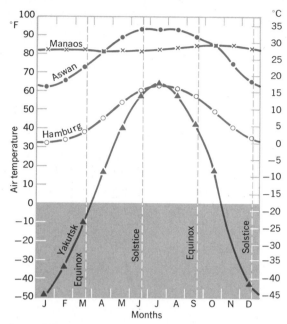

figure 4.9
The annual cycles of monthly mean temperatures for the same stations shown in Figure 4.8.

Land and water temperature contrasts

What gives our planet its great variety of climates is the odd distribution of continents and ocean basins. Look back at the two global hemispheres — northern and southern — outlined in Figure 2.6. The northern hemisphere displays a polar sea surrounded by massive continents; the southern hemisphere shows the very opposite — a pole-centered continent surrounded by a vast ocean. The Americas form a north-south barrier between two oceans — Atlantic and Pacific. The continents of Eurasia and Africa together form another great north-south barrier. Oceans and continents have quite different properties when it comes to absorbing and radiating energy.

Land surfaces behave quite differently from water surfaces. The important principle is this: the surface of any extensive deep body of water heats more slowly and cools more slowly than the surface of a large body of land when both are subject to the same intensity of insolation.

The slower rise of water-surface temperature can be attributed to four causes (Figure 4.10): (1) solar radiation penetrates water, distributing the absorbed heat throughout a substantial water layer; (2) the specific heat of water is large (a gram of water heats up much more

60° F (16° C); winter months reach a minimum of just about freezing (32° F; 0° C). The annual range is 30 F° (17 C°).

Finally, we travel to Yakutsk, Siberia, lat. 62° N. During the long, dark winters, there is an energy deficit; it lasts about 6 months. During this time air temperatures drop to extremely low levels. For three of the winter months monthly mean temperatures are between −30 and −50° F (−35 and −45° C). Actually, this is one of the coldest places on earth. In summer, when daylight lasts most of the 24 hours, the energy surplus rises to a strong peak, reaching 300 ly/day. This is a value higher than any of the other three stations. As a result, air temperatures show a phenomenal spring rise to summer-month values of over 55° F (13° C). In July the temperature is about the same as for Hamburg. The annual range at Yakutsk is enormous — over 110 F° (61 C°). No other region on earth, even the south pole, has so great an annual range.

figure 4.10
Four reasons why a land surface heats more rapidly and more intensely than the surface of a deep ocean body.

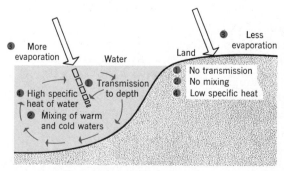

slowly than a gram of rock); (3) water is mixed through eddy motions, which carry the heat to lower depths; and (4) evaporation cools the water surface.

In contrast, the more rapid rise of land surface temperature can be attributed to these causes: (1) soil or rock is opaque, concentrating the heat in a shallow layer; there is little transmission of heat downward; (2) specific heat of mineral matter is much lower than water; (3) soil, if it is dry, is a poor conductor of heat; and (4) no mixing occurs in soil and rock.

The effect of land and water contrasts is seen in the two sets of daily air temperature curves shown in Figure 4.4. El Paso, Texas, exemplifies the temperature environment of an interior desert in midlatitudes. Soil moisture content is low, vegetation sparse, and cloud cover generally light. Responding to intense heating and cooling of the ground surface, air temperatures show an average daily range of 20 to 25 F° (11 to 14 C°). North Head, Washington, is a coastal station strongly influenced by air brought from the adjacent Pacific Ocean by prevailing westerly winds. Consequently, North Head exemplifies a

maritime temperature environment. The average daily range at North Head is a mere 5 F° (3 C°) or less. Persistent fogs and cloud cover also contribute to the small daily range. Refer also to Figure 4.2, which shows the same environmental contrast when the record of Yuma is compared with that of San Francisco.

The principle of contrasts in heating and cooling of water and land surfaces also explains the differences in the annual temperature cycles of coastal and interior stations. Let us examine the seasonal temperature curves for two places at approximately the same latitude, where insolation is about the same for both. Two such places are Winnipeg, Manitoba, located in the heart of the continent, and the Scilly Islands, England, exposed to the Atlantic Ocean. Figure 4.11 shows the average daily temperatures through an entire year as well as the insolation curve for lat. 50° N.

Although insolation reaches a maximum at summer solstice, the hottest part of the year for inland regions is about a month later, since heat energy continues to flow into the ground well into August. The air temperature maximum, closely corresponding with

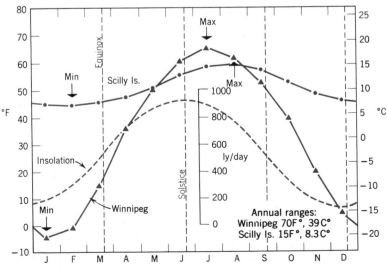

figure 4.11
Annual cycles of monthly mean air temperature for two stations at lat. 50° N: Winnipeg, Manitoba, Canada, and Scilly Islands, England.

maximum ground output of longwave radiation, is correspondingly delayed. (Bear in mind that this cycle applies to middle and high latitudes, but not to the region between the tropics of cancer and capricorn.) Similarly, the coldest time of year for large land areas is January, about a month after winter solstice, because the ground continues to lose heat even after insolation begins to rise.

Over the oceans there are two differences: (1) maximum and minimum temperatures are reached about a month later than on land — in August and February, respectively — because water bodies heat or cool much more slowly than land areas; and (2) the yearly range is less than over land, following the law of temperature differences between land and water surfaces. Coastal regions are usually influenced by the oceans to the extent that maximum and minimum temperatures occur later than in the interior. This principle shows nicely for monthly temperatures of the Scilly Islands. February is slightly colder than January (Figure 4.10).

Figure 4.4 reinforces the evidence of Figure 4.11. The annual temperature range at El Paso is about 35 F° (20 C°), while that at North Head is only about 15 F° (8 C°).

Air temperature maps

The distribution of air temperatures over large areas can best be shown by a map composed of *isotherms*, lines drawn to connect all points having the same temperature. Figure 4.12 shows a map on which the observed air temperatures have been recorded in the correct places. These may represent single readings taken at the same time everywhere, or they may represent the averages of many years of records for a particular day or month of a year, depending on the purposes of the map.

Usually, isotherms representing 5 or 10° differences are chosen, but they can be drawn

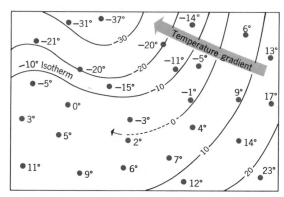

figure 4.12
Isotherms are used to make temperature maps. Each line connects those points having the same temperature.

for any selected temperatures. The value of isothermal maps is that they make clearly visible the important features of the prevailing temperatures. Centers of high or low temperature are clearly outlined.

World patterns of air temperature

World maps of air temperature (Figure 4.13) allow us to compare thermal conditions in the two months of greatest extremes — January and July. Patterns taken by the isotherms are largely explained by three controls: (1) latitude; (2) continent-ocean contrasts; and (3) altitude. (Ocean currents play an important, but secondary role.) Here are the important features you should identify when interpreting these maps.

1. The trend of isotherms is east-west, with temperatures decreasing from equatorial zone to polar zones. This feature shows best in the southern hemisphere, because the great Southern Ocean girdles the globe as a uniform expanse of water, while the continent of Antarctica is squarely centered on the south pole. The parallel pattern of isotherms is

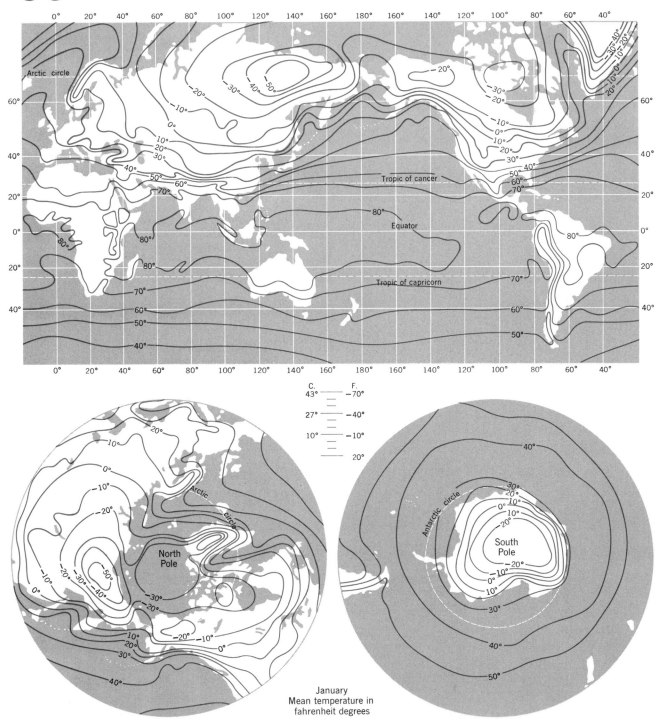

January
Mean temperature in
fahrenheit degrees

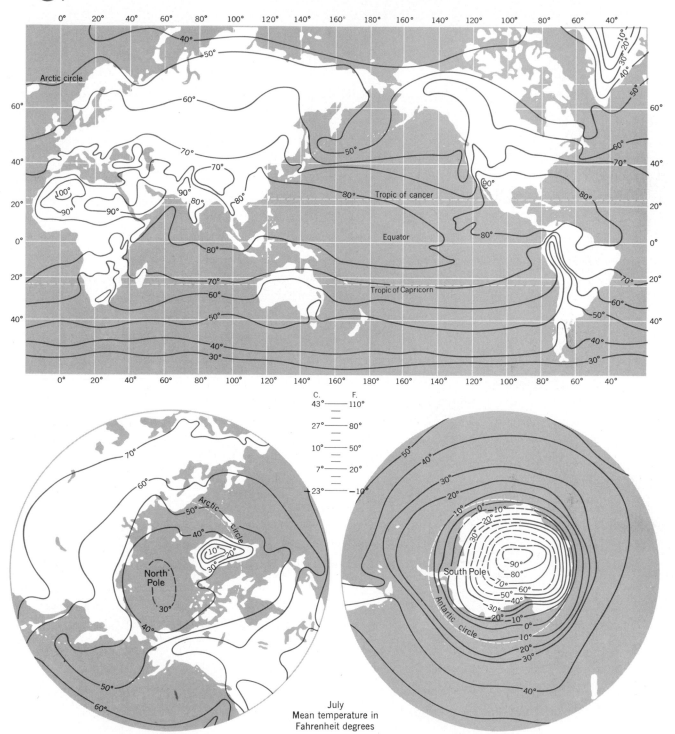

July
Mean temperature in
Fahrenheit degrees

figure 4.13
(<u>above and left</u>) January and July air temperatures in degrees
Fahrenheit. (Compiled by John E. Oliver.)

explained, of course, by the general decrease in intensity of insolation from equator to poles.

2. Large landmasses located in the subarctic and arctic zones develop centers of extremely low temperatures in winter. The two large landmasses we have in mind are North America and Eurasia. Find these centers on the January map.

3. Isotherms change very little in position from January to July over the equatorial zone, particularly over the oceans. This feature reflects the uniformity of insolation throughout the year near the equator.

4. Isotherms make a large north-south shift from January to July over continents in the midlatitude and subarctic zones. Figure 4.14 shows this principle. Check this principle for North America. In January the 60° F isotherm lies over central Florida; in July this same isotherm has moved far north, cutting the southern shore of Hudson Bay and then looping far up into northwestern Canada. Check the same 60° F isotherm in the Eurasian continent for this same effect. We can explain the large north-shift in position over these continents by the law of land and water contrasts.

5. Highlands are always colder than surrounding lowlands. Look for the Andes range, running along the western side of South America. The isotherms loop equatorward in long fingers over this lofty mountain chain.

6. Areas of perpetual ice and snow are always intensely cold. Greenland and Antarctic are the two great ice sheets. Notice how they stand out as cold centers both in January and July. Not only do these ice sheets have high surfaces, rising to over 10,000 ft (3000 m) in their centers, but the white snow surfaces have a high albedo and reflect away much of the insolation. The Arctic Ocean, bearing a cover of floating ice, also maintains its cold throughout the year, but the cold is much less intense in July than over the Greenland Ice Sheet.

The annual range of air temperatures

Figure 4.15 is a world map showing the annual range of air temperatures. The lines resembling isotherms tell the difference between the January and July monthly means. This map serves as a summary of concepts we have previously emphasized. Note these features.

1. The annual range is extremely great in the subarctic and arctic zones of both Asia and North America. This effect is shown well in the annual temperature graph for Yakutsk (Figure 4.9).

2. The annual range is moderately large on land areas in the tropical zone, near the tropics of cancer and capricorn. North Africa, southern Africa, and Australia are examples. Here the annual ranges are substantially greater than over adjacent oceans.

3. The annual range is very small over oceans in the equatorial zone. At all latitudes the annual range over all oceans is generally less than over lands.

This review of air temperatures over the globe will form the groundwork for your understanding of climates, developed in later chapters.

figure 4.14
The seasonal migration of isotherms is much greater over the continents than over the oceans.

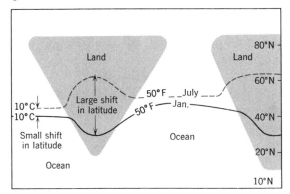

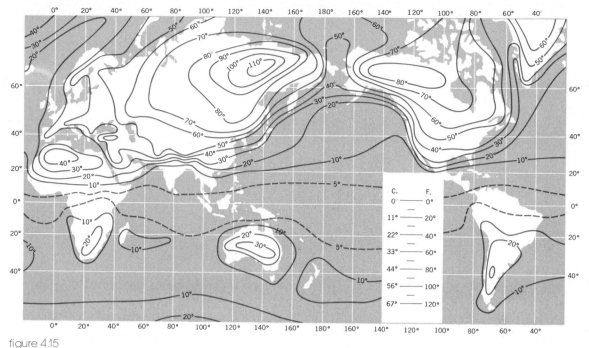

figure 4.15
Annual range of air temperatures. Lines show the difference between January and July means.

Carbon dioxide, dust, and global climate change

Atmospheric changes induced by Man fall into four categories with respect to basic causes: (1) changes in concentrations of the natural component gases of the lower atmosphere; (2) changes in the water vapor content of the troposphere and stratosphere; (3) alteration of surface characteristics of the lands and oceans in such a way as to change the interaction between the atmosphere and those surfaces; (4) introduction of finely divided solid substances into the lower atmosphere, along with gases not normally found in substantial amounts in the unpolluted atmosphere.

Under preindustrial conditions of recent centuries, the atmospheric content of carbon dioxide (CO_2) had been fixed at a level of roughly 0.030 percent by volume. The problem is, of course, that Man has recently begun to extract and burn hydrocarbon fuels at a rapid rate, releasing into the atmosphere the combustion products (principally water and CO_2) and also a great deal of heat. How does this activity affect our environment?

During the past 110 years, there has been an increase in atmospheric CO_2 of about 10 percent. Moreover, the rate of increase in this period, while slow at first, has become much more rapid toward the end of the period. Projection of the present curve of increase into the future shows a CO_2 value of 0.037 percent in the year 2000. At that time, if the predictions are correct, the content of CO_2 will have increased by about 25 percent over the 1860 value.

Consider now the environmental effects to be anticipated from an increase of atmospheric CO_2. Because CO_2 is an absorber and emitter of longwave radiation, its presence in larger proportions would raise the level of absorption of both incoming and outgoing radiation, changing the energy balance. The result would be to raise the average level of sensible heat in the atmosphere. In other words, a general air temperature rise is the anticipated result. We shall look next at available temperature data to see if such an increase has occurred.

Figure 4.16 shows the observed change in mean air temperature for approximately the past century. From 1880 to 1940, when fuel consumption was rising rapidly, the average temperature increased by about 0.6 F° (0.4 C°). This relationship follows the predicted pattern. However, since 1940, the graph shows a drop in temperature, despite the rising rate of fuel combustion. Assuming that the atmospheric warming because of increased CO_2 is a valid effect in principle, some other factor, working in the opposite direction, has entered the picture, and its cooling effect has outweighed that of warming by CO_2.

Downturn in average air temperatures since about 1940 may be the result of greatly increased input of dust into the upper atmosphere by a number of major volcanic eruptions, the first of which was the 1947 eruption of the Icelandic volcano, Hekla. It is also suspected by some observers that man-made dusts from industrial sources have increased sufficiently to cause, or at least

contribute to, the temperature downturn. Increased fallout of dust from the lower troposphere is well documented and is related to industrial activities.

An increase in volcanic dust at high elevations increases the proportion of solar energy absorbed in the stratosphere. The result is that less energy arrives at lower atmosphere levels. A reduction in level of sensible heat of the lower atmosphere follows. At this time, it is not possible to decide the extent of Man's activity in increasing atmospheric dust content at high levels, but the climatic impact may prove to be important eventually.

The thermal environment in review

In this chapter we have covered a vital environmental factor in the life layer. Heat, as recorded by the thermometer, is an essential ingredient of climate near the ground. All organisms respond to changes in temperatures of the medium that surrounds them, whether it be air, soil, or water.

Cycles of air temperature change are particularly important in the thermal environment. Both daily and seasonal changes in air temperature can be explained by daily and seasonal cycles of insolation and the energy balance. Superimposed on these astronomical rhythms is the powerful effect of latitude. A zoning of thermal environments from the equatorial zone to the polar zones is one of the most striking features of the global climate. But equally important is the way in which large land areas, especially the continents of North America and Eurasia, subvert the simple latitude zones to cause great seasonal extremes of temperature. In contrast, the southern hemisphere is strongly dominated by the simple effects of latitude.

From the long-range standpoint, Man's combustion of fuels has some serious implications, and these must be studied with care.

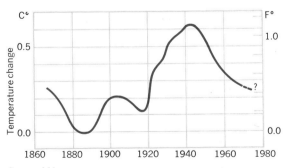

figure 4.16

Changes in the mean annual air temperature in the zone between the equator and lat. 80° N. (Generalized from data of J. M. Mitchell, Jr. and NOAA.)

Global Circulation of the Atmosphere and Oceans

Chapter 5

MAN'S PHYSICAL ENVIRONMENT at the earth's surface depends for its quality as much on atmospheric motions as it does on the flow of heat energy by radiation. In the form of very strong winds — those in hurricanes and tornadoes — air in motion is a severe environmental hazard. Winds also transfer energy to the surface of the sea, as wind-driven waves. Wave energy, in turn, travels to the shores of continents, where it is transformed into vigorous surf and coastal currents capable of reshaping the coastline.

But air in motion has another, more basic role to play in the planetary environment. Large-scale air circulation transports heat, both as sensible heat and as latent heat present in water vapor. Because of the global radiation imbalance — a surplus in low latitudes and a deficit in high latitudes — atmospheric circulation must transport heat across the parallels of latitude from the region of surplus to the regions of deficit. Figure 5.1 shows this meridional transport in schematic form. Notice that circulation of ocean waters is also involved in transport of sensible heat. But this oceanic circulation is a secondary mechanism, driven largely by surface winds.

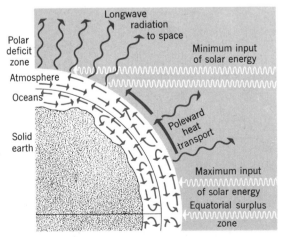

figure 5.1
The global energy balance is maintained by transport of heat from low to high latitudes.

Equally important to Man's environment are the rising and sinking motions of air. We shall find in Chapter 6 that precipitation, the source of all fresh water on the lands, requires massive lifting of large bodies of air heavily charged with water vapor. In reverse, large-scale sinking motions in the atmosphere lead to aridity and the occurrence of deserts. In this way the earth's surface becomes differentiated into regions of ample fresh water and regions of water scarcity.

With these broad concepts in mind, we examine the forces that set the atmosphere in motion and drive the ceaseless global circuits of air circulation.

Winds and the pressure gradient force

Wind is air motion with respect to the earth's surface, and it is dominantly horizontal. (Dominantly vertical air motions are referred to by other terms, such as updrafts or downdrafts.) To explain winds, we must first consider barometric pressure and its variations from place to place.

In Chapter 2 we found that barometric pressure falls with increasing altitude above the earth's surface. For an atmosphere at rest the barometric pressure will be the same within a given horizontal surface at any chosen altitude above sea level. In that case the surfaces of equal barometric pressure, called *isobaric surfaces,* are horizontal. In a cross section of a small portion of the atmosphere at rest isobaric surfaces appear as horizontal lines, as shown in Figure 5.2A.

Suppose, now, that the rate of upward pressure decrease is more rapid in one place than another, as shown in Figure 5.2B. As we proceed from left to right across the diagram, the upward rate of pressure decrease is more rapid. The isobaric surfaces now slope down toward the right. At a selected altitude, say

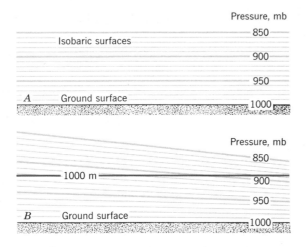

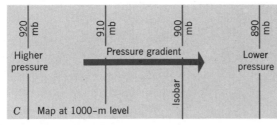

figure 5.2

Sloping isobaric surfaces mean that a pressure gradient exists.

Where a pressure gradient exists, air molecules tend to drift in the same direction as that gradient. This tendency for mass movement of the air is referred to as the *pressure gradient force*. The magnitude of the force is directly proportional to the steepness of the gradient. That is, a steep gradient is associated with a strong force. Wind is the horizontal motion of air in response to the pressure gradient force.

Sea and land breezes

Perhaps the simplest example of the relationship of wind to the pressure gradient force is a common phenomenon of coasts. It is the sea breeze and land breeze effect shown in Figure 5.3.

On a clear day, the sun heats the land surface rapidly and a shallow air layer near the ground is strongly warmed (Figure 5.3A). The warm air expands. Because it has become less dense than the cool air offshore, low pressure is formed over the coastal belt, while pressure remains higher over the water. Now there is a pressure gradient from ocean to land, and a *sea breeze* is set in motion. The sea breeze can become very strong in a shallow layer close to shore and is pleasantly cooling on a hot summer day.

At night, the land surface cools more rapidly than the ocean surface and a cool air layer of higher pressure develops (Figure 5.3B). Now the pressure gradient is from land to ocean, setting up a *land breeze*. These circulation systems are completed by rising and sinking motions of the air and by a weak return flow at higher levels, as indicated by the small arrows in the diagrams.

This illustration shows that a pressure gradient can be developed through unequal heating or cooling of a layer of the atmosphere. Air that is warmed expands and becomes less dense. Air that is cooled contracts and becomes denser. Heat energy drives this kind of circulation

1000 m (color line), barometric pressure declines from left to right. Figure 5.2C is a type of map; it shows that the 1000-m horizontal surface cuts across successive pressure surfaces. The trace of each pressure surface is a line on the map; the line is known as an *isobar*. The isobar is thus a line showing the location on a map of all points having the same barometric pressure.

The change in barometric pressure across the horizontal surface of a map constitutes a *pressure gradient*; its direction is indicated by the broad arrow in Figure 5.2C. The gradient is in the direction from higher pressure (at the left) to lower pressure (at the right). You can think of the sloping pressure surface as a sloping hillside; the downward slope of the ground surface is analogous to the pressure gradient.

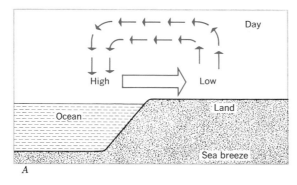

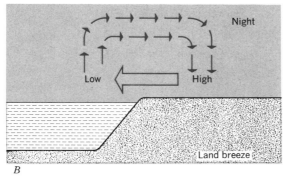

figure 5.3
Sea breeze and land breeze alternate in direction from day to night.

system by changing the air densities and setting up barometric pressure gradients. The entire mechanism is often described as a *heat engine*, since energy of air in motion is derived from the input of heat. The heat engine system of air circulation is also important on a large scale in the global atmosphere.

Measurement of surface winds

A description of winds requires measurement of two quantities: direction and speed. Direction is easily determined by a *wind vane*, commonest of the weather instruments. Wind direction is stated in terms of the direction from which the wind is coming (Figure 5.4). Thus an east wind comes from the east, but the direction of air movement is toward the west. The direction of movement of low clouds is an

excellent indicator of wind direction and can be observed without the aid of instruments.

Speed of wind is measured by an *anemometer*. There are several types. The commonest one seen at weather stations is the cup anemometer. It consists of three hemispherical cups mounted as if at the ends of spokes of a horizontal wheel (Figure 5.5). The cups travel with a speed proportional to that of the wind. One type of anemometer turns a small electric generator, and the current it produces can be transmitted to a meter calibrated in units of wind speed. Units are miles per hour or meters per second.

For wind speeds at higher levels a small hydrogen-filled balloon is released into the air and observed through a telescope. The rate of climb of the balloon is known in advance. Knowing the balloon's vertical position by measuring the elapsed time, an observer can calculate the horizontal drift of the balloon downwind. For upper-air measurements of wind speed and direction the balloon carries a target that reflects radar waves and can be followed when the sky is overcast.

figure 5.4
Winds are designated according to the compass point from which the wind comes. An east wind comes from the east, but the air is moving westward.

The Coriolis force and its effect on winds

If the earth did not rotate on its axis, winds would follow the direction of the pressure gradient. As mentioned in Chapter 1, earth rotation produces another force, the *Coriolis force*, which tends to turn the flow of air. The direction of action of this turning force can be stated thus: any object or fluid moving horizontally in the northern hemisphere tends to be deflected to the right of its path of motion, regardless of the compass direction of the path. In the southern hemisphere a similar deflection is toward the left of the path of motion. The Coriolis force is absent at the equator but increases in strength toward the poles.

On Figure 5.6 the small arrows show how an initial straight line of motion is modified by the Coriolis force. Note especially that the compass direction is not of any consequence. If we face down the direction of motion, turning will always be toward the right hand in

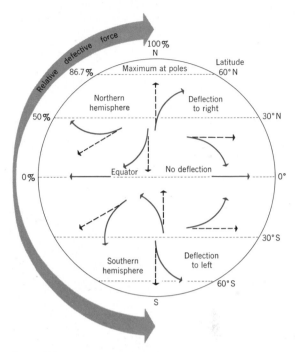

figure 5.6
The Coriolis force acts to turn winds or ocean currents to the right in the northern hemisphere and to the left in the southern hemisphere.

figure 5.5
This three-cup anemometer has a counter at the base to record the number of turns made by its shaft in a given interval of time. (National Weather Service.)

the northern hemisphere. Because the turning force is very weak, its action is conspicuous only in freely moving fluids such as air or water. Consequently, both winds and ocean current patterns are greatly affected by the Coriolis force.

Our next step is to apply the Coriolis principle to winds close to the ground surface. Figure 5.7 shows isobars running east-west, forming a ridge of high pressure in each hemisphere. From each ridge pressure decreases both to the north and south toward belts of low pressure. Broad arrows show the pressure gradient. The Coriolis force turns the wind so that it crosses the isobars at an angle. For surface winds, the angle of turning is limited by the force of friction of the air with the ground. The diagram shows the wind making an angle of 45° with the isobars. In nature the angle is subject to some variation, depending on the character of the ground surface.

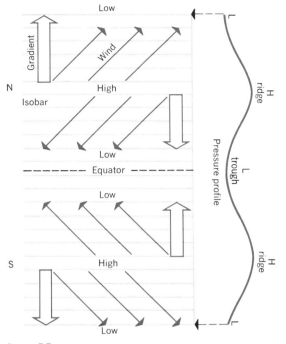

figure 5.7
Surface winds cross isobars at an angle as the air moves from higher to lower pressure. Turn the figure sideways to view the pressure profile.

Looking first at the northern hemisphere case, the deflection is to the right. The northward pressure gradient gives a southwest wind. The southward gradient gives a northeast wind. In the southern hemisphere winds are deflected to the left and the pattern is the mirror image of that in the northern hemisphere. We turn next to the case of concentric isobars, using the same rules.

Cyclones and anticyclones

In the language of meteorology, a center of low pressure is called a *cyclone*; a center of high pressure is an *anticyclone*. Cyclones and anticyclones may be of the stationary type, or they may be rapidly moving pressure centers such as those that create the weather disturbances described in Chapter 7.

For surface winds, which move obliquely across the isobars, the systems for cyclones and anticyclones in both hemispheres are shown in Figure 5.8. Winds in a cyclone in the northern hemisphere show an anticlockwise inspiral. In an anticyclone there is a clockwise outspiral. Note the reversal between the labels "anti-clockwise" and "clockwise" in the southern hemisphere.

In both hemispheres the surface winds spiral inward on the center of the cyclone, so that the air is converging on the center and must also rise to be disposed of at higher levels. For the anticyclone, by contrast, surface winds spiral out from the center. This motion represents a diverging of air flow and must be accompanied by a sinking (subsidence) of air in the center of the anticyclone to replace the outmoving air.

Global distribution of surface pressure systems

To understand the earth's surface wind systems, we must first study the global system of barometric pressure distribution. Once we grasp the patterns of isobars and pressure gradients, the prevailing or average winds can be predicted.

World isobaric maps are usually constructed to show average pressures for the two months of seasonal temperature extremes over large landmasses — January and July (Figure 5.9). Because observing stations lie at various altitudes above sea level, their barometric readings must be reduced to sea-level equiv-alents, using the standard rate of pressure change with altitude, explained in Chapter 2. When this has been done, and the daily readings are averaged over long periods of time, small but distinct pressure differences remain.

A reading of 1013 mb is taken as standard sea-level pressure. Readings higher than this will frequently be observed in midlatitudes, occasionally up to 1040 mb or higher. These pressures are designated as "high." Pressures ranging down to 982 mb or below are "low."

Over the equatorial zone is a belt of somewhat lower than normal pressure, between 1011 and 1008 mb, which is known as the *equatorial trough*. Lower pressure is conspicuous by contrast with belts of higher pressure lying to the north and south and centered at about lat. 30° N and S. These are the *subtropical belts of high pressure*. Pressures exceed 1020 mb. In the southern hemisphere this belt is clearly defined but contains centers of high pressure, called *pressure cells*.

In the southern hemisphere, south of the subtropical high-pressure belt, is a broad belt of low pressure, extending roughly from the midlatitude zone to the arctic zone. The axis of low pressure is centered at about lat. 65° S. This pressure trough is called the *subantarctic low-pressure belt*. Lying over the continuous

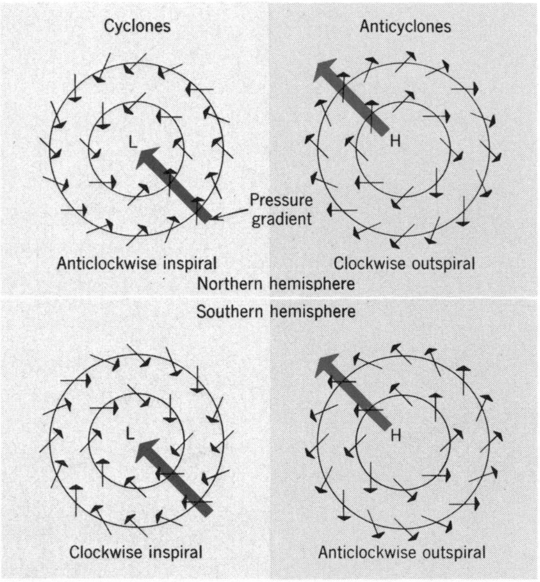

figure 5.8
Surface winds spiral inward toward the center of a cyclone, but outward from the center of an anticyclone.

expanse of Southern Ocean, this trough has average pressures as low as 984 mb. Over the continent of Antarctica is a permanent center of high pressure known as the *polar high*. It contrasts strongly with the encircling subantarctic low.

The pressure belts shift annually through several degrees of latitude, along with the isotherm belts. These annual pressure belt migrations are important in causing seasonal climate changes. We shall have several occasions to refer to these effects in analyzing world climates.

Northern hemisphere pressure centers

The vast continents of North America and Eurasia and the intervening North Atlantic and North Pacific oceans exert a powerful control over pressure conditions in the northern hemisphere. As a result, the belted arrangement typical of the southern hemisphere is absent.

In winter the large, very cold land areas develop high-pressure centers. At the same time, intense low-pressure centers form over the warmer oceans. Over north central Asia in winter we find the Siberian high, with pressure exceeding 1030 mb. Over central North America there is a clearly defined, but much less intense, ridge of high pressure, called the Canadian high. Over the oceans are the Aleutian low and the Icelandic low, named after the localities over which they are centered. These two low-pressure areas have much cloudy, stormy weather in winter. Figure 5.10 shows these pressure centers as they appear grouped around the north pole. Highs and lows occupy opposite quadrants.

In summer pressure conditions are exactly the opposite of winter conditions. In summer the land areas develop low-pressure centers because at this season land-surface temper-

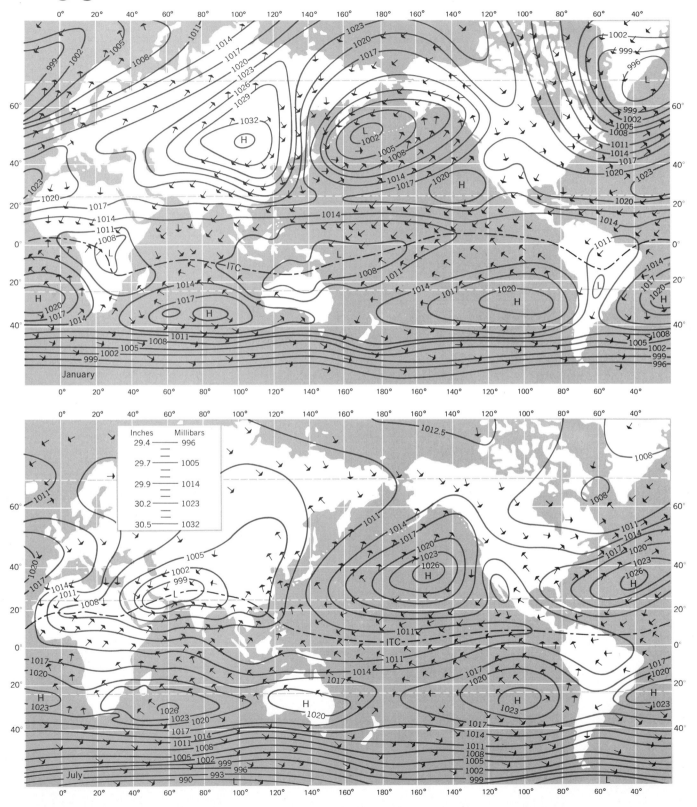

69

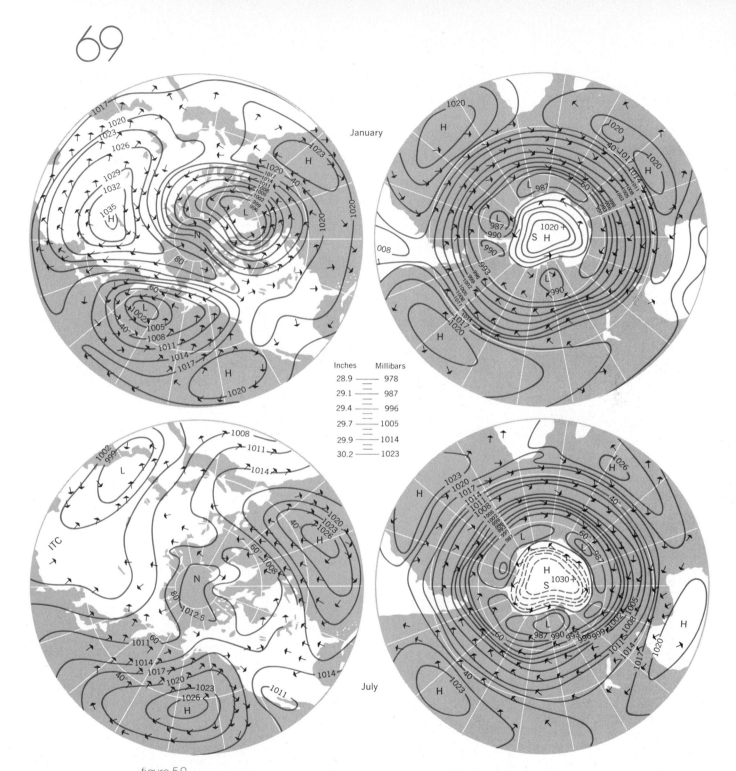

figure 5.9
World maps of average barometric pressure for the months of January
and July. The average direction of prevailing surface winds is shown
by small arrows. (Data compiled by John E. Oliver.)

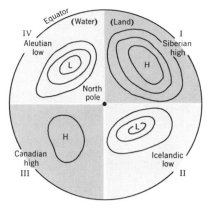

figure 5.10
This schematic map of the northern hemisphere in winter shows intense centers of high and low pressure occupying opposing quadrants.

atures rise sharply above temperatures over the adjoining oceans. At the same time, the ocean areas develop strong centers of high pressure. This system of pressure opposites is striking on both the January and July isobaric maps (Figure 5.9). The low in Asia is intense: it is centered in southern Asia. Over the Atlantic and Pacific oceans are two large, strong cells of the subtropical belt of high pressure. They have shifted northward of their winter position and are considerably expanded. They are called the Azores high (or Bermuda high) and the Hawaiian high.

The global pattern of surface winds

Prevailing surface winds during January and July are shown by arrows on the pressure maps of Figure 5.9. The basic wind patterns are stressed in Figure 5.11, which is a highly diagrammatic representation showing the earth as if no land areas existed to modify the belted arrangement of pressure zones.

Let us begin with winds of the tropical zones. From the two subtropical high-pressure belts

the pressure gradient is equatorward, leading down to the equatorial trough of low pressure. Following the simple model shown in Figure 5.7, air moving from high to low pressure is deflected by the Coriolis force. As a result, two belts of *trade winds,* or *trades,* are produced. These are labeled in Figure 5.11 as the northeast trades and southeast trades. The trades are very persistent winds, deviating very little from a single compass direction. Sailing vessels traveling westward made good use of the trades.

The pattern of the trades suggests that they must converge somewhere near the equator. Meeting of the trades takes place within a narrow zone called the *intertropical convergence zone.* This long term is usually abbreviated to *ITC.* The position of the ITC is marked on the January and July maps (Figure 5.9). Converging winds require a rise of air to dispose of the incoming volume of air. This rise takes the form of stalklike flow columns carrying the air toward the top of the troposphere.

Along parts of the equatorial trough of low pressure at certain times of year the trades do not come together in convergence. Instead, there forms a belt of calms and variable winds, called the *doldrums.* Mariners knew that crossing the doldrums was hazardous because of the likelihood of lying becalmed for long periods of time.

The trades, doldrums, and ITC all shift seasonally north and south, along with the shifting of pressure belts and isotherms. The ITC migrates north and south only a few degrees of latitude over the Pacific and Atlantic oceans, but covers as much as 20 to 30° of latitude over South America, Africa, and the large region of Southeast Asia and the Indian Ocean. Important seasonal changes in winds, cloudiness, and rainfall accompany these migrations of the ITC and the trades.

We now return to the subtropical high pressure belt, which ranges between lat. 25° and 40° N

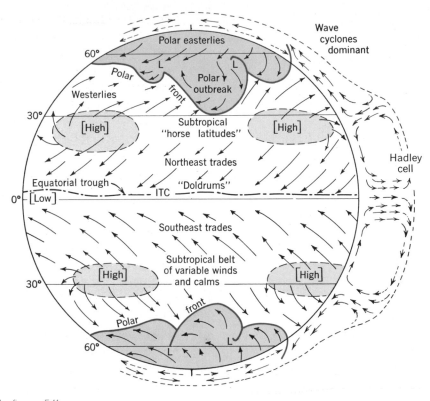

✳ figure 5.11

This schematic diagram of global surface winds disregards the disrupting effect of large continents in the northern hemisphere.

and S. Here we encounter large, stagnant, high-pressure cells or anticyclones. In the centers of the cells winds are weak and are distributed around a wide range of compass directions; calms prevail as much as one quarter of the time. Because of the high frequency of calms, mariners named this belt the *horse latitudes*. It is said that the name originated in colonial times from the experiences of New England traders, carrying cargoes of horses to the West Indies. When their ships were becalmed for long periods of time, the fresh water supplies ran low and it was necessary to throw the horses overboard. Figure 5.12 is a schematic map of large anticyclonic cells centered over the oceans in two hemispheres. Winds make an outspiralling pattern that feeds into the converging trades. On the western sides of the cells air flows

poleward; on the eastern sides the flow is equatorward. These flows have a strong influence on the climates of adjacent continental margins. Dryness of climate is a dominant general characteristic of the subtropical high-pressure belt and its cells, and we shall emphasize this feature again.

Between lat. 35° and 60° N and S is the belt of *prevailing westerly winds,* or *westerlies*. These surface winds are shown on Figure 5.11 as blowing from a southwesterly quarter in the northern hemisphere and from a northwesterly quarter in the southern hemisphere. This generalization is somewhat misleading, however, because winds from polar directions are frequent and strong. It is more nearly accurate to say that within the westerlies, winds blow from all directions of the compass, but that the westerly components are definitely

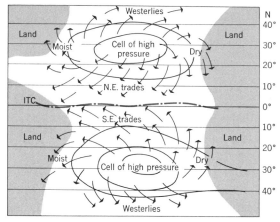

figure 5.12

Over the oceans surface winds spiral outward from the dominant cells of high pressure, feeding the trades and the westerlies.

predominant. Rapidly moving cyclonic storms are common in this belt.

In the northern hemisphere, landmasses disrupt the westerlies but, in the southern hemisphere, between lat. 40° and 60° S, there is an almost unbroken belt of ocean. Here the westerlies gather great strength and persistence. Sailors on the great clipper ships called these latitudes "roaring forties," "furious fifties," and "screaming sixties." This belt was extensively used by sailing vessels traveling eastward from the South Atlantic Ocean to Australia, Tasmania, New Zealand, and the southern Pacific Islands. From these places it was then easier to continue eastward around the world to return to European ports. Rounding Cape Horn was relatively easy on an eastward voyage but, in the opposite direction, in the face of prevailing stormy westerly winds, it was a very dangerous operation.

A wind system called the *polar easterlies* is often said to be characteristic of the arctic and polar zones (Figure 5.11). The concept is at best greatly oversimplified and is certainly misleading when applied to the northern hemisphere. Winds in these regions take a variety of directions, dictated by local weather disturbances. On the other hand, Antarctica is an ice-capped landmass resting squarely on the pole and surrounded by a vast oceanic expanse. Here the outward spiralling flow of polar easterlies seems to be a dominant feature of the circulation. The polar maps in Figure 5.9 show these easterly winds.

Monsoon winds of Southeast Asia

The powerful control exerted by the great landmass of Asia on air temperatures and pressures extends to the surface wind systems as well. In summer southern Asia develops a cyclone into which there is a strong flow of air (July map, Figure 5.13). From the Indian Ocean and the southwestern Pacific warm, humid air moves northward and northwestward into Asia,

figure 5.13

The Asiatic monsoon winds alternate in direction from January to July in response to reversals of barometric pressure over the large continent.

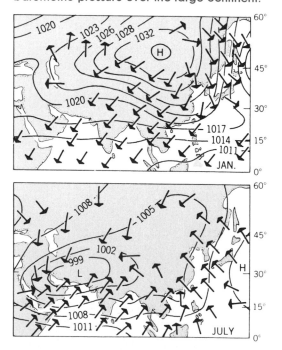

passing over India, Indochina, and China. This air flow constitutes the *summer monsoon* and is accompanied by heavy rainfall in south-eastern Asia.

In winter, Asia is dominated by a strong center of high pressure from which there is an outward flow of air reversing that of the summer monsoon (January map, Figure 5.13). Blowing southward and southeastward toward the equatorial oceans, this *winter monsoon* brings dry weather for a period of several months.

North America does not have the remarkable extremes of monsoon winds experienced by southeastern Asia but, even so, there is a distinct alternation of average temperature and pressure conditions between winter and summer. Wind records show that in summer there is a prevailing tendency for air originating in the Gulf of Mexico to move northward across the central and eastern part of the United States, whereas in winter there is a prevailing tendency for air to move southward from high-pressure sources in Canada. Wind arrows in Figure 5.9 show this seasonal alternation in the airflow pattern.

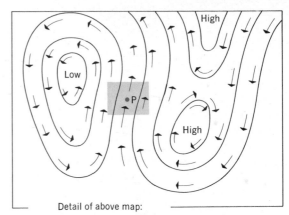

Detail of above map:

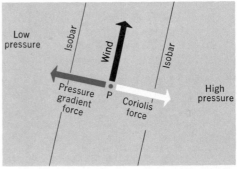

figure 5.14

At high levels above the earth's surface, the wind blows parallel with the isobars. For the northern hemisphere, shown here, lower pressure is always toward the left of the direction of motion, while higher pressure is toward the right.

Winds aloft

The surface wind systems we have examined represent only a shallow basal air layer a few thousands of feet deep, whereas the troposphere is several miles deep. How does air move at these higher levels? Large, slowly moving high-pressure and low-pressure systems are found aloft, but these are generally simple in pattern, with smoothly curved isobars.

Winds high above the earth's surface are not affected by friction with the ground or water over which they move. The Coriolis force turns the flow of air until it becomes parallel with the isobars, as Figure 5.14 shows. In this position the pressure gradient force and Coriolis force are exactly opposed and are exactly balanced.

The upper part of Figure 5.14 is a simplified map of pressure and winds high in the troposphere. Notice how the wind arrows run parallel with the isobars, forming circular flow patterns around lows and highs. Our rules for winds in cyclones and anticyclones need to be slightly modified, as compared with surface winds. Figure 5.15 shows the flow patterns for both hemispheres. If you were an airline pilot flying in the northern hemisphere and keeping a tailwind at all times, your rule would be "keep the highs on your right and the lows on your left."

The global circulation at upper levels

The general pattern of upper-air flow is sketched in Figure 5.16. Two systems dominate. One is the system of *upper-air westerlies* blowing in a complete circuit about the earth from about lat. 25° almost to the poles. At high latitudes these westerlies form a huge circumpolar vorotex, coinciding with a great polar low-pressure center. Toward lower latitudes the pressure rises steadily at a given altitude, forming high-pressure ridges at lat. 15° to 20° N and S. These are the high-altitude parts of the surface subtropical highs, but they are shifted somewhat equatorward.

Between the high-pressure ridges is a trough of weak low pressure in which the winds are easterly. These winds comprise the second major circulation system of the troposphere. They are called the *tropical easterlies*.

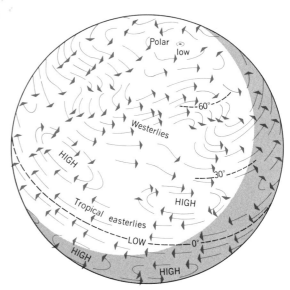

figure 5.16
A generalized plan of global winds high in the troposphere.

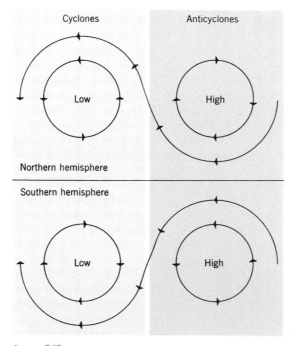

figure 5.15
Upper-level winds form circular patterns around cyclones and anticyclones, but directions are reversed in the two hemispheres.

Seen in meridional cross section (Figure 5.11), circulation in equatorial and tropical latitude zones resolves itself into two circuits, one in each hemisphere. Heated air rises over the equatorial zone but subsides in the subtropical cells, forming the *Hadley cell*. Some of the subsiding air escapes poleward into the westerlies. This cellular circulation of low latitudes is basically a heat engine of the type described earlier in our explanation of sea and land breezes.

Rossby waves and the jet stream

The uniform flow of the upper-air westerlies is frequently disturbed by the formation of large undulations, called *Rossby waves*. As shown in detail in Figure 5.17, these waves grow in amplitude and finally are cut off. The waves develop in a zone of contact between cold, polar air and warm, tropical air.

It is by means of the upper-air waves that warm air of low latitudes is carried far north at the

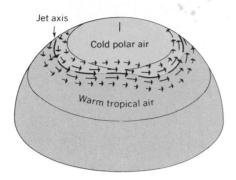

A. Jet stream begins to undulate

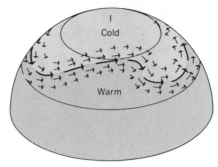

B. Rossby waves begin to form

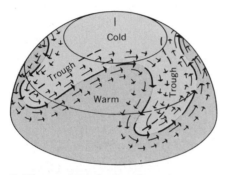

C. Waves strongly developed

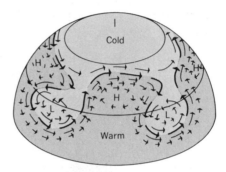

D. Cells of cold and warm air are formed

figure 5.17
Development of upper-air waves in the westerlies. (Data of J. Namias, National Weather Service. From A. N. Strahler, 1971, The Earth Sciences, 2nd ed., Harper & Row, New York.)

same time that cold air of polar regions is brought equatorward. In this way horizontal mixing develops on a vast scale and provides heat exchange between the equatorial region of energy surplus and the polar regions of energy deficit.

Associated with the development of such upper-air waves at altitudes of 30,000 to 40,000 ft (10 to 12 km) are narrow zones in which wind streams attain velocities up to 200 to 250 mi (350 to 450 km) per hour. This phenomenon is named the *jet stream*. It consists of pulselike movements of air following broadly curving tracks (Figure 5.18).

In cross section the jet resembles a stream of water moving through a hose. The center line of highest velocity is surrounded by concentric zones of less rapidly moving fluid, as pictured in Figure 5.19.

The jet stream is an important factor in the operation of jet aircraft in the range of their normal cruising altitudes. In addition to strongly increasing or decreasing the ground speed of the aircraft, the jet stream carries a form of air turbulence that at times reaches hazardous levels. This is clear air turbulence (CAT); it is avoided when known to be severe.

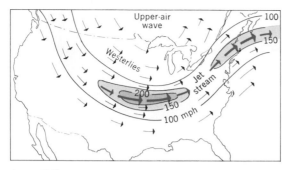

figure 5.18
The jet stream is shown on this map by lines of equal wind speed. (After National Weather Service.)

Local winds

In certain localities, *local winds* are generated by immediate influences of the surrounding terrain rather than by the large-scale pressure systems that produce global winds and large traveling storms. Local winds are of environmental importance in various ways. They can exert a powerful stress on animals and plants when the air is dry and extremely hot or cold. Local winds are also important in affecting the movement of atmospheric pollutants.

One class of local winds — sea breezes and land breezes — was explained earlier in this chapter. The cooling sea breeze (or lake breeze) of summer is an important environmental resource of coastal communities, since it adds to the attraction of the shore zone as a recreation facility.

Mountain winds and *valley winds* are local winds following a daily alternation of direction in a manner similar to the land and sea breezes. During the day, air moves from the valleys, upward over rising mountain slopes, toward the summits. At this time hill slopes are intensely heated by the sun. At night the air then moves valleyward, down the hill slopes, which have been cooled at night by radiation of heat from ground to air. These winds are

responding to local pressure gradients set up by heating or cooling of the lower air.

Still another group of local winds are known as *drainage winds,* in which cold air flows under the influence of gravity from higher to lower regions. Such cold, dense air may accumulate in winter over a high plateau or high interior valley. When general weather conditions are favorable, some of this cold air spills over low divides or through passes to flow out on adjacent lowlands as a strong, cold wind.

Drainage winds occur in many mountainous regions of the world and go by various local names. The bora of the northern Adriatic coast and the mistral of southern France are well-known examples. In southern California there issues on occasion from the Santa Ana

figure 5.19
A pictorial representation of the jet stream. (National Weather Service.)

Valley a strong, dry, east wind, the Santa Ana, which blows across the coastal lowland. This air is of desert origin and may carry much dust in suspension. On the ice sheets of Greenland and Antarctica powerful drainage winds move down the gradient of the ice surface and are funneled through coastal valleys to produce powerful blizzards lasting for days at a time.

Ocean currents

An ocean current is any persistent, dominantly horizontal flow of ocean water. Ocean currents are important regulators of thermal environments at the earth's surface. On a global scale, the vast current systems aid in exchange of heat between low and high latitudes and are essential in sustaining the global energy balance. On a local scale warm water currents bring a moderating influence to coasts in arctic latitudes; cool currents greatly alleviate the heat of tropical deserts along narrow coastal belts.

Practically all of the important surface currents of the oceans are set in motion by prevailing surface winds. Energy is transferred from wind to water by the frictional drag of the air blowing over the water surface. Because of the Coriolis force, the water drift is impelled toward the right of its path of motion (northern hemisphere): therefore the current at the water surface is in a direction about 45° to the right of the wind direction.

To illustrate global surface water circulation, we can refer to an idealized ocean extending across the equator to latitudes of 60° or 70° on either side (Figure 5.20). Perhaps the most outstanding features are the circular movements, called gyres, around the subtropical highs. The gyres are centered about lat. 25° to 30° N and S. An equatorial current with westward flow marks the belt of the trades. Although the trades blow to the

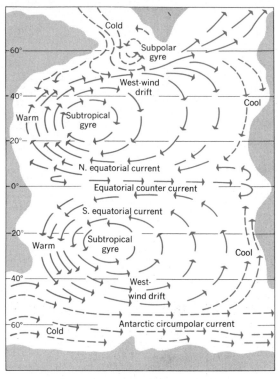

figure 5.20
Two great gyres, one in each hemisphere, dominate the circulation of shallow ocean waters.

southwest and northwest, obliquely across the parallels of latitude, the water movement follows the parallels. The equatorial currents are separated by an equatorial countercurrent. A slow, eastward movement of water over the zone of the westerlies is named the west-wind drift. It covers a broad belt between lat. 35° and 45° in the northern hemisphere and between lat. 30° and 60° in the southern hemisphere. A world map, Figure 5.21, shows these ocean currents in greater detail.

Along the west sides of the oceans in low latitudes the equatorial current turns poleward, forming a warm current paralleling the coast. Examples are the Gulf Stream (Florida stream or Caribbean stream) and the Japan current (Kuroshio). These currents bring higher than average temperatures along these coasts.

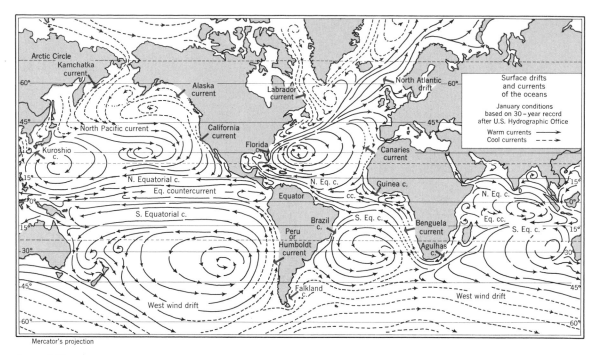

figure 5.21

Surface drifts and currents of the oceans in January. (U.S. Navy Oceanographic Office.)

The west-wind drift, upon approaching the east side of the ocean, is deflected both south and north along the coast. The equatorward flow is a cool current, accompanied by upwelling of colder water from greater depths. It is well illustrated by the Humboldt current (Peru current) off the coast of Chile and Peru; by the Benguela current off the southwest African coast; by the California current off the west coast of the United States; and by the Canaries current, off the Spanish and North African coast.

In the northeastern Atlantic Ocean, the west-wind drift is deflected poleward as a relatively warm current. This is the North Atlantic current, which spreads around the British Isles, into the North Sea, and along the Norwegian coast. The Russian port of Murmansk, on the arctic circle, has year-round navigability by way of this coast.

In the northern hemisphere, where the polar sea is largely landlocked, cold water flows equatorward along the west side of the large straits connecting the Arctic Ocean with the Atlantic basin. One example of a cold current is the Labrador current, moving south from the Baffin Bay area through Davis Strait to reach the coasts of Newfoundland, Nova Scotia, and New England.

In both the North Atlantic and North Pacific oceans the Icelandic and Aleutian lows coincide in a very rough way with centers of counterclockwise circulation involving the cold arctic currents and the west-wind drifts. This type of center is labeled "subpolar gyre" on Figure 5.20.

The antarctic region has a relatively simple current scheme consisting of a single antarctic circumpolar current moving clockwise around the antarctic continent in lat. 50° to 65° S, in a continuous expanse of open ocean.

Oceanic circulation also involves the complex motions of water masses of different temperature and salinity characteristics. Sinking and upwelling are both important motions in certain areas of the oceans.

The global circulation and Man's environment

We have outlined the major mechanical circuits within which sensible heat is transported from low latitudes to high latitudes. In low latitudes the Hadley cell operates like a simple heat engine to transport heat from the equatorial zone to the subtropical zone. Upper-air waves take up the transport and move warm air poleward in exchange for cold air. Ocean currents perform a similar function through the turning of the great gyres.

The global atmospheric circulation also transports heat in the latent form held by water vapor. This heat is released by condensation, a process we shall examine in the next chapter. The movement of water vapor also represents a mass transport of water and is a part of the world water balance. This topic is the subject of Chapter 9.

Winds of the lower troposphere have great environmental significance. Arriving at a mountainous coastal zone after a long travel path over a great ocean, these winds carry a large amount of water vapor, and this is deposited as precipitation on the coast. In this way the distribution of our water resources is partly determined by the atmospheric circulation patterns. Winds also transport atmospheric pollutants, carrying them tens and hundreds of miles from the source of pollution. These are environmental topics related to winds; we can return to them in later chapters.

Atmospheric Moisture and Precipitation

Chapter 6

WE HAVE STRESSED that heat and water are vital ingredients of the environment of the biosphere, or life layer. Plant and animal life of the lands, on which humans depend for much of their food, require fresh water. People use fresh water in many ways. The only basic source of fresh water is from the atmosphere through condensation of water vapor. In this chapter we are concerned mostly with water in the vapor state in the atmosphere and the processes by which it passes into the liquid or solid state and ultimately arrives at the surface of the ocean and the lands through the process of precipitation.

Water also leaves the land and ocean surfaces by evaporation and so returns to the atmosphere. Evidently, the global pathways of movement of water form a complex network. There is a global water balance, just as there is an energy balance; the water balance deals with flow of matter and so complements the energy balance.

Humidity

The amount of water vapor that may be present in the air at a given time varies widely from place to place. It ranges from almost nothing in the cold, dry air of arctic regions in winter to as much as 4 or 5 percent of a given volume of the atmosphere in the humid equatorial zone.

The general term *humidity* refers to the amount of water vapor present in the air. For any specified temperature there is a definite limit to the quantity of moisture that can be held by the air. This limit is known as the *saturation point*. The proportion of water vapor present relative to the maximum quantity is the *relative humidity*, expressed as a percentage. At the saturation point, relative humidity is 100 percent; when half of the total possible quantity of vapor is present, relative humidity is 50 percent, and so on.

A change in relative humidity of the atmosphere can be caused in one of two ways. If an exposed water surface is present, the humidity can be increased by evaporation. This is a slow process requiring that the water vapor diffuse upward through the air. The other way is through a change of temperature. Even though no water vapor is added, a lowering of temperature results in a rise of relative humidity. This change is automatic because the capacity of the air to hold water vapor is lowered by cooling. After cooling, the existing amount of vapor represents a higher percentage of the total capacity of the air. Similarly, a rise of air temperature results in decreased relative humidity, even though no water vapor has been taken away. The principle of relative humidity change caused by temperature change is illustrated by a graph of these two properties throughout the day (Figure 6.1). As air temperature rises, relative humidity falls, and vice versa.

A simple example may help to illustrate these principles (Figure 6.2). At 10 a.m., the air temperature is 60° F (16° C), and the relative humidity is 50 percent. By 3 p.m., the air has become warmed by the sun to 90° F (32° C). The relative humidity has automatically dropped to 20 percent, which is very dry air. Next, the air becomes chilled during the night and, by 4 a.m., its temperature has fallen to 40° F (5° C). Now the relative humidity has automatically risen to 100 percent, the saturation point. Any further cooling will cause *condensation* of the excess vapor into liquid or solid form. As the air temperature continues to fall, the humidity remains at 100 percent, but condensation continues. This may take the form of minute droplets of dew or fog. If the temperature falls below freezing, condensation occurs as frost on exposed surfaces.

Dew point is the critical temperature at which the air is fully saturated. Below the dew point

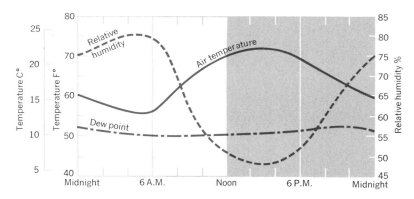

figure 6.1
Relative humidity, temperature, and dew point for May in Washington, D.C. Data of National Weather Service. (From A. N. Strahler, 1971, The Earth Sciences, 2nd ed., Harper & Row, New York.)

condensation normally occurs. An excellent illustration of condensation caused by cooling is seen in the summer, when beads of moisture form on the outside surface of a pitcher or glass filled with ice water. Air immediately adjacent to the cold glass or metal surface is chilled enough to fall below the dew point temperature, causing moisture to condense on the surface of the glass.

Relative humidity is measured by an instrument called a *hygrometer*. Figure 6.3 illustrates a simple, homemade hygrometer in which a strand of human hairs is attached to the end of a pointer. The hairs change length in response to changes in relative humidity and, in this way, move the pointer up or down. In a more sophisticated instrument, the hygrograph,

the strand of hair or other fiber operates a pen point and makes a continuous record on graph paper wrapped around a slowly turning drum.

Specific humidity

Although relative humidity is an important indicator of the state of water vapor in the air, it is a statement only of the relative quantity present compared to a saturation quantity. The

figure 6.2
Relative humidity changes with temperature because capacity of warm air is greater than for cold air.

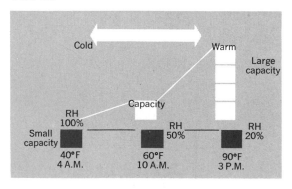

figure 6.3
A simple hygrometer.

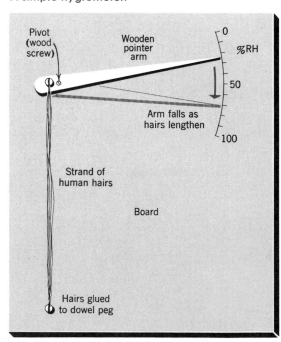

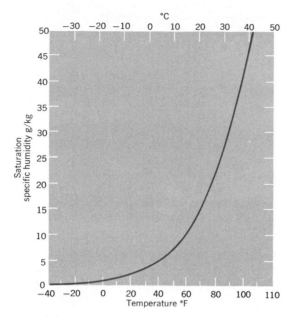

figure 6.4

✳ **Maximum specific humidity of a mass of air increases sharply with rising temperature.**

actual quantity of moisture present is denoted by *specific humidity*, defined as the mass of water vapor contained in a given mass of air. Mass of water is given in grams; the unit mass

of air is the kilogram. Specific humidity is stated in terms of grams per kilogram (gm/kg). For any specified air temperature, there is a maximum weight of water vapor that a kilogram of air can hold (the saturation quantity). Figure 6.4 is a graph showing this maximum moisture content of air for a wide range of temperatures.

Specific humidity is often used to describe the moisture characteristics of a large mass of air. For example, extremely cold, dry air over arctic regions in winter may have a specific humidity of as low as 0.2 gm/kg, whereas extremely warm moist air of equatorial regions often holds as much as 18 gm/kg. The total natural range on a worldwide basis is such that the largest values of specific humidity are from 100 to 200 times as great as the least.

Figure 6.5 is a graph showing how relative humidity and specific humidity vary with latitude. Notice that the relative humidity curve has two saddles, one over each of the subtropical high-pressure belts where the world's tropical deserts are found. Humidity is high in equatorial and arctic zones. In contrast,

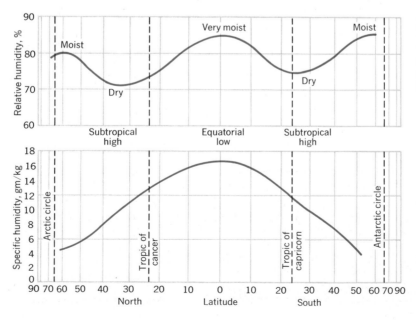

figure 6.5

Pole-to-pole profiles of average relative humidity (above) and of specific humidity (below). (Data of Haurwitz and Austin, 1944.)

the specific humidity curve has a single peak, near the equator, and declines toward high latitudes.

In a real sense, specific humidity is a geographer's yardstick of a basic natural resource — water — to be applied from equatorial to polar regions. It is a measure of the quantity of water than can be extracted from the atmosphere as precipitation. Cold air can supply only a small quantity of rain or snow; warm air is capable of supplying very large quantities.

Condensation and the adiabatic process

Falling rain, snow, sleet, or hail are referred to collectively as _precipitation_. Only where large masses of air are experiencing a steady drop in temperature below the dew point can precipitation occur in appreciable amounts. Precipitation cannot be brought about by the simple process of chilling of the air through loss of heat by longwave radiation during the night. Precipitation requires that a large mass of air be rising to higher elevations.

One of the most important laws of meteorology is that rising air experiences a drop in temperature, even though no heat energy is lost to the outside. The drop of temperature is a result of the decrease in air pressure at higher elevations, permitting the rising air to expand. Because individual molecules of the gas are more widely diffused and do not strike one another so frequently, the sensible temperature of the expanding gas is lowered. This process is described by the adjective _adiabatic_, which simply means "occurring without any gain or loss of heat." In the adiabatic process heat energy as well as matter remain within the system. The process is thus completely reversible. Expansion always results in cooling; compression always results in warming.

Within a rising body of air the rate of drop of temperature, termed the _dry adiabatic lapse_

rate, is about 5½F° per 1000 ft of vertical rise. In metric units the rate is 1 C° per 100 m. The dry rate applies only when no condensation is taking place. The dew point also declines gradually with rise of air: the rate is 1 F° per 1000 ft (0.2 C° per 100 m).

Adiabatic cooling rate should not be confused with the environmental lapse rate, explained in Chapter 2. The environmental lapse rate applies only to nonrising air whose temperature is measured at successively higher levels.

Lapse rates can be shown on a simple graph in which altitude is plotted on the vertical scale and temperature on the horizontal scale (Figure 6.6). The chain of dots connected by arrows represents air rising as if it were a bubble. Suppose that a body of air near the ground has a temperature of 70°F (21° C) and that its dew point temperature is 52° F (11° C); these are the conditions shown in Figure 6.6. If, now, a bubble of air undergoes a steady ascent, its air temperature will decrease much faster than its dew point temperature. Consequently, the two lines on the graph are rapidly converging. At

figure 6.6

Adiabatic decrease of temperature in a rising mass of air leads to condensation of water vapor and the formation of a cloud.

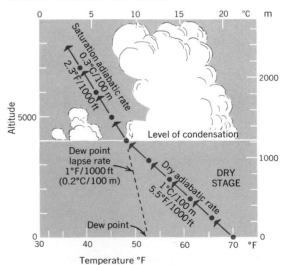

an altitude of 4000 ft (1200 m), the air temperature will have met the dew point temperature. The air bubble has now reached the saturation point. Further rise results in condensation of water vapor into minute liquid particles, and so a cloud is produced. The flat base of the cloud is a common visual indicator of the level of condensation.

As the bubble of saturated air continues to rise, further condensation takes place, but now a new principle comes into effect. When water vapor condenses, heat in the latent form is transformed into sensible heat, which is added to the existing heat content of the air. *Latent heat* represents energy of rapid motions of the gas molecules of water vapor. This heat amounts to about 600 calories for each gram of water. Originally, this stored energy was obtained during the process of *evaporation* at the time the water vapor entered the atmosphere.

You know that evaporation of water cools the liquid surface from which evaporation is taking place. Evaporation of perspiration cools the skin. The desert water bag also works on this principle. Evaporation of water seeping through the coarse flax cloth of the bag cools the remaining water. The heat drawn off by evaporation enters the atmosphere along with the water vapor in the form of the rapid motion of the gaseous water molecules. When these same molecules come to rest on a liquid surface as water, the energy of their motion turns into sensible heat.

As condensation continues within a rising mass of air, the latent heat liberated by that condensation partly offsets the temperature drop by adiabatic cooling. As a result, the adiabatic rate is substantially reduced. The reduced rate, which ranges between 2 and 3 F° per 1000 ft (0.3 and 0.6 C° per 100 m), is termed the *wet adiabatic lapse rate*. On the graph, this reduced rate is expressed by the more steeply inclined section of line above the level of condensation.

Clouds

A *cloud* is a dense mass of suspended water or ice particles in the diameter range of 20 to 50 microns. Each cloud particle has formed on a *nucleus* of solid matter, which is originally on the order of $1/10$ to 1 micron in diameter. Nuclei of condensation must be present in large numbers, and they must be of such a composition as to attract water vapor molecules. In Chapter 2 we referred to these minute suspended particles collectively as atmospheric dust and noted that one source is the surface of the sea. Droplets of spray from the crests of waves are carried rapidly upward in turbulent air. Evaporation of the water leaves a solid residue of crystalline salt, which strongly attracts water molecules. All of you are familiar with the way in which ordinary table salt becomes moist when exposed to warm humid air. Another example is calcium chloride, a commercial salt, widely used to control dust on dirt roads; it attracts atmospheric water vapor and keeps the soil damp, even in dry weather.

Although we may campaign vigorously for "clean air," the term is only relative, since all air of the troposphere is charged with dust. As a result, there is no lack of suitable condensation nuclei. As we shall find in the discussion of air pollution, the heavy load of dust carried by polluted air over cities substantially aids in condensation and the formation of clouds and fog.

We are accustomed to finding that liquid water turns to ice when the surrounding temperature falls to the freezing point (32° F, 0° C) or below. However, water in such minute particles as those comprising clouds remains in the liquid state at temperatures far below freezing. Such water is described as *supercooled*. Clouds consist entirely of water droplets at temperatures down to about 10° F (−12° C). Between 10 and −20° F (−12 to −30° C), the cloud is a mixture of water droplets and ice crystals. Below −20° F (−30°

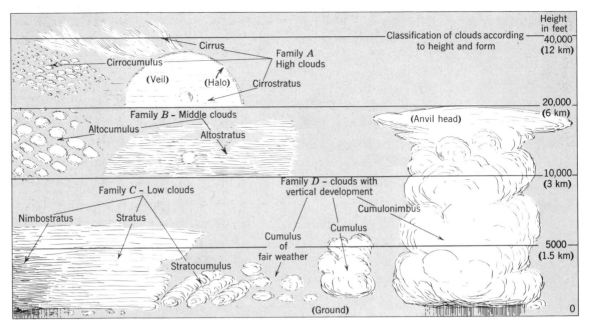

figure 6.7
Clouds are grouped into families on the basis of height. Individual cloud types are named according to their form.

C) the cloud consists <u>predominantly of ice</u> crystals; <u>below</u> −40° F (−40° C) <u>all</u> of the cloud particles are ice crystals. Very high, thin clouds, formed at altitudes of 20,000 to 40,000 ft (6 to 12 km) are formed of ice particles.

The common cloud forms are pictured in Figure 6.7. Clouds are classified into families, arranged by height; these are named on Figure 6.7 along with individual types within each family. On the basis of form there are two major classes of clouds: *stratiform,* or <u>layered</u> clouds, and *cumuliform,* or <u>globular</u> clouds.

The <u>stratiform</u> clouds are blanketlike and <u>cover</u> <u>large</u> areas. A common type is called *stratus*. The important point about stratiform clouds is that they represent <u>air layers being forced to</u> <u>rise gradually over stable underlying air layers</u> of greater density. As forced rise continues, the rising layer of air is adiabatically cooled, and condensation is sustained over a large area. Stratiform clouds can yield substantial amounts of rain or snow.

The <u>cumuliform</u> clouds are globular masses representing <u>bubblelike bodies</u> of warmer air spontaneously rising because they <u>are less</u> <u>dense than the surrounding air</u>. These clouds go by the name *cumulus* (Figure 6.8). Some

figure 6.8
Cumulus of fair weather. (National Weather Service.)

cumuliform clouds are small masses no higher than wide; others develop tall, stalklike shapes and penetrate high into the troposphere. The tall, dense form is called *cumulonimbus* (Figure 6.9). It comprises the thunderstorm, a weather disturbance with heavy rain and sometimes hail, to the accompaniment of lightning and thunder.

In later paragraphs we shall explain the conditions under which both stratiform and cumuliform clouds are produced and the ways in which they yield their precipitation.

Fog

Fog is simply a cloud layer in contact with the land or sea surface or lying very close to the surface. Fog is a major environmental hazard to Man in an industrialized world. Dense fog on high-speed highways is a cause of terrifying chain-reaction accidents, sometimes involving dozens of vehicles and often taking a heavy toll in injuries and deaths. Landing delays and shutdowns at major airports because of fog bring economic losses to airlines and in-

convenience to thousands of travelers in a single day. For centuries fog at sea has been a navigational hazard. Today, with huge supertankers carrying oil, fog adds to the probability of ship collisions capable of creating enormous oil spills. Polluted fogs are a health hazard to urban dwellers and in some occurrences have taken a heavy toll in lives.

One type of fog, known as a *radiation fog*, is formed at night when temperature of the stagnant basal air falls below the dew point. This kind of fog is associated with a low-level temperature inversion (Figure 4.7). Another fog type, *advection fog*, results from the movement of warm, moist air over a cold or snow-covered ground surface. Losing heat to the ground, the air layer undergoes a drop of temperature below the dew point, and condensation sets in. A similar type of advection fog is formed over oceans where air from over a warm current blows across the cold surface of an adjacent cold current. Fogs of the Grand Banks off Newfoundland are largely of this origin because here the cold Labrador current comes in contact with warm waters of Gulf Stream origin.

figure 6.9
Cumulonimbus, an isolated thunderstorm showing rain falling from base.

Precipitation forms

During the rapid ascent of a mass of air in the saturated state, cloud particles grow rapidly and attain a diameter of 50 to 100 microns. They then coalesce through collisions and grow quickly into droplets of about 500 microns diameter (about $1/50$ in.). Droplets of this size reaching the ground constitute a drizzle, one of the recognized forms of precipitation. Further coalescence increases drop size and yields *rain*. Average raindrops have diameters of about 1000 to 2000 microns ($1/25$ to $1/10$ in.), but they can reach a maximum diameter of about 7000 microns (¼ in.). Above this value they become unstable and break into smaller drops while falling. One kind of rain forms directly by liquid condensation and droplet coalescence in warm clouds of the equatorial and tropical zones. However, rain of middle and high latitudes is largely the product of the melting of snow as it makes its way to lower, warmer levels.

Snow is produced in clouds that are a mixture of ice crystals and supercooled water droplets. The falling crystals serve as nuclei to intercept water droplets. As these adhere, the water film freezes and is added to the crystalline structure. The crystals readily clot together to form larger snowflakes, and these fall more rapidly from the cloud. When the underlying air layer is below the freezing temperature, snow reaches the ground as a solid form of precipitation; otherwise, it will melt and arrive as rain. A reverse process, the fall of raindrops through a cold air layer, results in freezing of rain and produces pellets or grains of ice. These are commonly referred to in North America as *sleet* but, among the British, sleet refers to a mixture of snow and rain.

figure 6.10
Heavily coated wires and branches caused heavy damage in eastern New York State in January 1943 as a result of this icing storm. (National Weather Service.)

Hail, another form of precipitation, consists of large pellets or spheres of ice. The formation of hail will be explained in our discussion of the thunderstorm.

When rain falls on a frozen ground surface that is covered by an air layer of below-freezing temperature, the water freezes into clear ice after striking the ground or other surfaces such as trees, houses, or wires (Figure 6.10). The coating of ice that results is called a *glaze,* and an *ice storm* is said to have occurred. Actually, no ice falls, so that ice glaze is not a form of precipitation. Ice storms cause great damage, especially to telephone and power wires and to tree limbs. Roads and sidewalks are made extremely hazardous.

How precipitation is measured

Precipitation is measured in units of depth of fall per unit of time; for example, inches per hour or per day. One inch of rainfall is a quantity sufficient to cover the ground to a depth of 1 in., provided that none is lost by runoff, evaporation, or sinking into the ground. A simple form of rain gauge can be operated merely by setting out a straight-sided, flat-bottomed pan and measuring the depth to which water accumulates during a particular period. Unless this period is short, however, evaporation seriously upsets the results.

Very small amounts of rainfall, such as 0.1 in. (0.25 cm), make too thin a layer to be accurately measured. To avoid this difficulty as well as to reduce evaporation loss, *rain gauges* are made in the form of a cylinder whose base is a funnel leading into a narrow tube (Figure 6.11). A small amount of rainfall will fill the narrow pipe to a considerable height, thus making it easy to read accurately, once a simple scale has been provided for the pipe. This gauge requires frequent emptying unless it is equipped with automatic devices for this purpose.

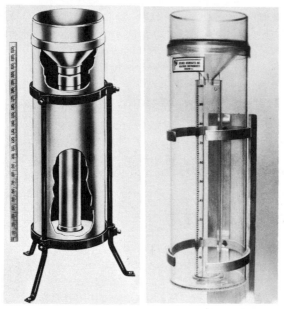

figure 6.11
Standard 8-in. rain gauge used by the National Weather Service (left). A plastic 4-in. gauge showing funnel and graduated tube inside (right). (Science Associates, Inc., Princeton, N.J.)

Snowfall is measured by melting a sample column of snow and reducing it to an equivalent in water. In this way rainfall and snowfall records may be combined for purposes of comparison. Ordinarily, a 10-in. layer of snow is assumed to be equivalent to 1 in. of rainfall, but this ratio may range from 30 to 1 in very loose snow to 2 to 1 in old, partly melted snow.

How precipitation is produced

Precipitation in substantial quantities is induced by two basic types of mechanisms. One is spontaneous rise of moist air; the other is forced rise of moist air.

Spontaneous rise of moist air is associated with *convection,* a form of atmospheric motion consisting of strong updrafts taking place within a *convection cell.* Air rises in the cell because it is less dense than the surrounding air. Perhaps a fair analogy is the updraft of

heated air in a chimney but, unlike the steady air flow in a chimney, air motion in a convection cell takes place in pulses as bubblelike masses of air rise in succession.

To illustrate the convection process, let us suppose that on a clear, warm summer morning the sun is shining on a landscape consisting of patches of open fields and woodlands. Certain of these types of surfaces, such as the bare ground, heat more rapidly and transmit radiant heat to the overlying air. Air over a warmer patch becomes warmed more than adjacent air and begins to rise as a bubble, much as a hot-air balloon rises after being released. Vertical movements of this type are often called "thermals" by sailplane pilots who use them to obtain lift.

As the air rises, it is cooled adiabatically so that eventually it is cooled below the dew point. At once condensation begins, and the rising air column appears as a cumulus cloud. The flat base shows the critical level above which condensation is occurring (Figure 6.12). The bulging "cauliflower" top of the cloud represents the top of the rising warm air column, pushing into higher levels of the atmosphere. Usually, the small cumulus cloud dissolves after drifting some distance downwind. However, should this convection continue to develop, the cloud may grow to a dense cumulonimbus mass, or thunderstorm, from which heavy rain will fall.

figure 6.12
The rise of a bubble of heated air forms a small cumulus cloud. (From A. N. Strahler, 1971, <u>The Earth Sciences</u>, 2nd ed., Harper & Row, New York.)

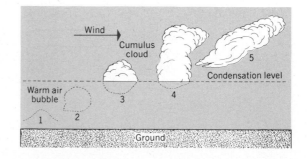

Why, you ask, does such spontaneous cloud growth take place and continue beyond the initial cumulus stage, long after the original input of heat energy is gone? Actually, the unequal heating of the ground served only as a trigger effect to release a spontaneous updraft, fed by latent heat energy liberated from the condensing water vapor. Recall that for every gram of water formed by condensation, 600 calories of heat are released. This heat acts like fuel in a bonfire.

Air capable of rising spontaneously during condensation is said to be *unstable air*. In such air the updraft tends to increase in intensity as time goes on, much as a bonfire blazes with increasing ferocity as the updraft draws in greater supplies of oxygen. Of course, at very high altitudes, the bulk of the water vapor has condensed and fallen as precipitation, so that the energy source is gone. When this happens the convection cell weakens and air rise finally ceases.

Unstable air, given to spontaneous convection in the form of heavy showers and thunderstorms, is most likely to be found in warm, humid areas such as the equatorial and tropical zones throughout the year, and the midlatitude regions during the summer season.

Thunderstorms

Convection activity manifests itself in the *thunderstorm,* an intense local storm associated with a tall, dense cumulonimbus cloud in which there are very strong updrafts of air (Figure 6.9). Thunder and lightning normally accompany the storm, and rainfall is heavy, often of cloudburst intensity, for a short period. Violent surface winds may occur at the onset of the storm.

A single thunderstorm consists of individual cells. Air rises within each cell as a succession of bubblelike air bodies, instead of as a continuous updraft from bottom to top (Figure

6.13). As each bubble rises, air in its wake is brought in from the surrounding region. Precipitation in the thunderstorm cell can be in the form of rain in the lower levels, mixed water and snow at intermediate levels, and snow at high levels.

Upon reaching high levels, which may be about 20,000 to 40,000 ft (6 to 12 km) or even higher, the rising rate diminishes, and the cloud top is dragged downwind to form an anvil top. Ice particles falling from the cloud top act as nuclei for condensation at lower levels, a process called *cloud seeding*. The rapid fall of raindrops adjacent to the rising air bubbles exerts a frictional drag on the air and sets in motion a downdraft. Striking the ground where precipitation is heaviest, this downdraft forms a local squall wind, which is sometimes strong enough to fell trees and do severe structural damage to buildings.

Recent years have seen a sharp increase in Man's efforts to modify weather phenomena.

figure 6.13
A schematic diagram of the interior of a thunderstorm cell.

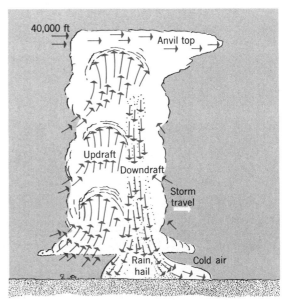

One goal is to increase precipitation in areas experiencing drought; another is to lessen the severity of storms. One method of inducing convectional precipitation is by artificial cloud seeding, the introduction of minute particles into dense cumulus clouds. The particles, which may be of silver iodide smoke, serve as nuclei for added intensity of condensation and can lead to formation of cumulonimbus clouds with heavy rainfall. Some measure of success by cloud seeding was obtained in 1970 and 1971 over drought-ridden southern and central Florida.

Hail and lightning — environmental hazards

In addition to high-wind gusts and torrential falls of rain important environmental hazards connected with the thunderstorm are hail, lightning, and tornadoes. Hailstones are formed by the accumulation of ice layers on an ice pellet suspended in the strong thunderstorm updrafts. The phenomenon is much like that of the icing of aircraft flying through a cloud of supercooled water droplets. After the hailstones have grown to diameters that often reach 1 to 2 in., they escape from the updraft and fall to the ground (Figure 6.14).

figure 6.15
A severe hail storm devastated this corn crop. (NCAR photograph.)

An important aspect of planned weather modification is that of reducing the severity of hailstorms. Annual losses from crop destruction by hailstorms are estimated to approach $300 million (Figure 6.15). Damage to wheat crops is particularly severe in a north-south belt of the High Plains, running through Nebraska, Kansas, and Oklahoma. A much larger region of somewhat less hailstorm frequency extends eastward generally from the Rockies to the Ohio Valley. Corn is a major crop over much of this area. Scientists are studying the isolated cumulonimbus clouds that produce hail. Research will lead to development of cloud-seeding techniques by means of which the severity of hailstorms can be reduced.

Another effect of convection cell activity is to generate lightning, one of the environmental hazards that results annually in the death of many persons and livestock and in the setting of many forest and building fires. Lightning is a great electric arc — a gigantic spark — passing between cloud and ground, or between parts of a cloud mass. During lightning discharge a

figure 6.14
These hailstones are larger than hens' eggs (arrow). (National Weather Service.)

current of as much as 60,000 to 100,000 amperes may develop. Rapid heating and expansion of the air in the path of the lightning stroke sends out intense sound waves, which we recognize as a thunder clap. In the United States lightning causes a yearly average of about 150 human deaths and property damage of about $100 million, including loss by fires set by lightning.

Orographic precipitation

Forced ascent of large masses of air occurs under two quite different sets of controlling conditions. Where prevailing winds encounter a mountain range, the air layer as a whole is forced to rise in order to surmount the barrier. Precipitation produced in this way is described as *orographic*, meaning "related to mountains." A layer of cold air may act in much the same way as a mountain barrier. Warm air in motion often encounters a cold air layer. The cold air is denser than the warm and

will remain close the the ground, acting as a barrier to progress of the warm air. The warm air is then forced to rise over the barrier. In a related type of mechanism, the cold air layer is in motion and forces the warm air to rise over it. Precipitation resulting from these activities is explained in Chapter 7.

Figure 6.16 shows the steps associated with production of orographic precipitation. Moist air arrives at the coast after passing over a large ocean surface. As the air rises on the windward side of the range, it is cooled at the adiabatic rate. When cooling is sufficient, precipitation sets in. After passing over the mountain summit, the air begins to descend the lee side of the range. Now it undergoes a warming through the same adiabatic process and, having no source from which to draw up moisture, becomes very dry. Upon reaching sea level, it is much warmer than at the start. A belt of dry climate, often called a *rainshadow*, exists on the lee side of the range. Several great deserts of the earth are of this type.

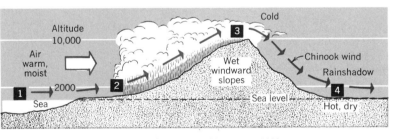

figure 6.16
Forced ascent of oceanic air masses produces precipitation and a rainshadow desert. (From A. N. Strahler, 1971, The Earth Sciences, 2nd ed., Harper & Row, New York.)

Dry, warm _chinook winds_ often occur on the lee side of a mountain range in the western United States. These winds cause extremely rapid evaporation of snow or soil moisture. Chinook winds result from turbulent mixing of lower and upper air in the lee of the range. The upper air, which has little moisture to begin with, is greatly dried and heated when swept down to low levels.

An excellent illustration of orographic precipitation and rainshadow occurs in the far west of the United States (see Figure 15.22). Prevailing westerly winds bring moist air from the Pacific Ocean over the coast ranges of central and northern California and the great Sierra Nevada, whose summits rise to 14,000 ft (4000 m) above sea level. Heavy rainfall is experienced on the windward slopes of these ranges, nourishing rich forests. Passing down the steep eastern face of the Sierra Nevada, air must descend nearly to sea level, even below sea level in Death Valley. The resulting adiabatic heating lowers the humidity, producing part of America's great desert zone, covering a strip of eastern California and all of Nevada.

Much orographic rainfall at low latitudes is actually of the convectional type in that it takes the form of heavy showers and thunderstorms. The convectional storms are set off by the forced ascent of unstable air as it passes over the mountain barrier.

Latent heat and the global balances of energy and water

Let us now take another look at the global energy balance so as to include the mechanism of heat transport in latent form. Figure 6.17 is a schematic diagram showing how evaporation and condensation are involved in energy exchange in each of the major latitude zones. We can now also visualize the pattern of transport of water vapor across the parallels of

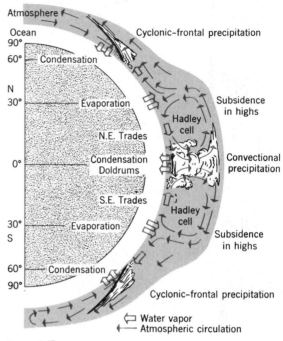

figure 6.17
This schematic diagram summarizes the exchanges of water vapor from pole to pole.

latitude; this is a transport of matter and is part of the global water balance.

The equatorial zone is characterized by a rise of moist, warm air in innumerable convection cells reaching to the upper limits of the troposphere. As condensation and precipitation occur here, enormous amounts of latent heat energy are liberated. This zone has been called the "fire box" of the globe in recognition of this intense production of sensible heat by condensation. Moving equatorward to replace the rising air are the trade winds, part of the Hadley cell circulation. Water evaporates from the ocean surface over the subtropical highs and is carried equatorward in vapor form.

Next, we move into the midlatitudes, where upper-air waves constantly form and dissolve in the westerlies. Here, cyclones and anti-cyclones exchange cold, polar air for warm, tropical air across the parallels of latitude. Water vapor is also transported poleward, carrying with it much latent heat. Condensation in cyclonic storms removes water vapor from the troposphere toward high latitudes, so that the movement of water vapor declines and reaches zero at the poles.

Air pollution

We can make good use of an understanding of the processes of condensation and precipitation when probing into man-made changes in the atmospheric environment. These are inadvertent changes, largely the result of urbanization and the growth of industrial processes, which have enormously increased the rate of combustion of hydro-carbon (fossil) fuels in the past half century or so. We have laready considered the thermal effects of Man's injecting carbon dioxide into the atmosphere. Now we turn to consider the kinds of foreign matter, or *pollutants,* injected by Man into the lower atmosphere, and the effects of that pollution on air quality and urban climates.

We recognize two classes of atmospheric pollutants. First, there are solid and liquid particles, which are designated collectively as *particulate matter.* Dusts found in smoke of combustion, as well as droplets naturally occurring as cloud and fog, fall into the category of particles. Second, there are compounds in the gaseous state, included under the general term *chemical pollutants* in that they are not normally present in measurable quantities in clean air remote from densely populated, industrialized regions. (Excess carbon dioxide produced by combustion is not usually classed as a pollutant.)

One group of chemical pollutants of industrial and urban areas is of primary origin, for example, produced directly from a source on the ground. Gases included in this group are carbon monoxide (CO), sulfur dioxide (SO_2), oxides of nitrogen (NO, NO_2, NO_3), and hydrocarbon compounds. However, these chemical pollutants cannot be treated separately from particulate matter, since they are often combined within a single suspended particle. We have already seen that certain dusts, particularly the sea salts, have an affinity for water and easily take on a covering water film. The water film in turn absorbs the chemical pollutants.

When particles and chemical pollutants are present in considerable density over an urban area, the resultant mixture is known as *smog.* Almost everyone living in large cities is familiar with smog through its irritating effects on the eyes and respiratory system and its ability to obscure distant objects. When concentrations of suspended matter are less dense, obscuring visibility of very distant objects but not otherwise objectionable, the atmospheric condition is referred to as *haze.* Atmospheric haze builds up quite naturally in stagnant air masses as a result of the infusion of various surface materials. Haze is normally present whenever air reaches high relative humidity, because water films grow on suspended nuclei. Nuclei of natural atmospheric haze particles consist of mineral dusts from the soil, crystals of salt blown from the sea surface, hydro-carbon compounds (pollens and terpenes) exuded by plant foilage, and smoke from forest and grass fires. Dusts from volcanoes may, on occasion, add to atmospheric haze.

It is evident at this point that what we are calling atmospheric pollutants are of both natural and man-made origin, and that Man's activities can supplement the quantities of natural pollutants present. Table 6.1 illustrates this complexity by listing the primary pollutants according to sources.

Not all man-made pollution comes from the cities. Isolated industrial activities can produce pollutants far from urban areas. Particularly

table 6.1

Sources of primary atmospheric pollutants

Pollutants from Natural Sources	Sources of Man-made Pollutants
Volcanic dusts	Fuel combustion (CO_2, SO_2, lead)
Sea salts from breaking waves	Chemical processes
Pollens and terpenes from plants	Nuclear fusion and fission
(Aggravated by Man's activities):	Smelting and refining of ores
Smoke of forest and grass fires	Mining, quarrying
Blowing dust	Farming
Bacteria, viruses	

Data from Assn. of Amer. Geographers, 1968, *Air Pollution*, Commission on College Geography, Resource Paper No. 2, Figure 3, p. 9.

important are smelters and manufacturing plants in small towns and rural areas. Sulfide ores (metals in combination with sulfur compounds) are processed by heating in smelters close to the mine. Here, sulfur compounds are sent into the air in enormous concentrations from smokestacks. Fallout over the surrounding area is destructive to vegetation.

Mining and quarrying operations send mineral dusts into the air. For example, asbestos mines (together with asbestos processing and manufacturing plants) send into the air countless threadlike mineral particles, some of which are so small that they can be seen only with the electron microscope. These particles travel widely and are inhaled by humans, lodging permanently in the lung tissue. Nuclear test explosions inject into the atmosphere a wide range of particles, including many radioactive substances capable of traveling thousands of miles in the atmospheric circulation.

Man-induced forest and grass fires add greatly to smoke palls in certain seasons of the year. Plowing, grazing, and vehicular traffic raise large amounts of mineral dusts from dry soil surfaces. Bacteria and viruses, which we have not as yet mentioned, are borne aloft in air turbulence when winds blow over contaminated surfaces such as farmlands, grazing lands, city streets, and waste disposal sites.

Figure 6.18 shows major atmospheric pollutants in percentage form as to both component matter and source in recent years. Much of the carbon monoxide, half of the hydrocarbons, and about a third of the oxides of nitrogen comes from exhausts of gasoline and diesel engines in vehicular traffic. Generation of electricity and various industrial processes contribute most of the sulfur oxides, because the coal and lower-grade fuel oil used for these purposes are comparatively rich in sulfur. These same sources also supply most of the particulate matter. Fly ash consists of the coarser grades of soot particles emitted from smokestacks of generating plants. These particles settle out quite quickly within close range of the source. Combustion used for heating of buildings is a comparatively minor contributor to pollution, because the higher grades of fuel oil are low in sulfur and are usually efficiently burned. Very finely divided carbon comprises much of the smoke of combustion and is capable of remaining in suspension almost indefinitely because of its colloidal size. Forest fires comprise a secondary contributor of particles. Burning of refuse is a minor contributor in all categories of pollutants.

In the smog of cities there are, in addition to the ingredients mentioned above, certain chemical elements contained in particles contributed by exhausts of automobiles and trucks. Included are particles that contain lead, chlorine, and bromine. About half of the particles in automobile exhaust contain lead.

ATMOSPHERIC MOISTURE

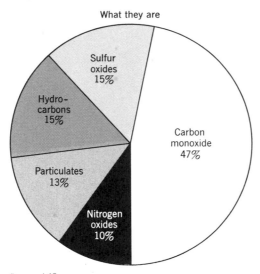

What they are

Sulfur
oxides
15%

Hydro–
carbons
15%

Particulates
13%

Nitrogen
oxides
10%

Carbon
monoxide
47%

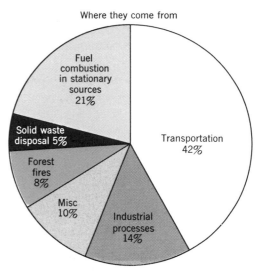

Where they come from

Fuel
combustion
in stationary
sources
21%

Solid waste
disposal 5%

Forest
fires
8%

Misc
10%

Industrial
processes
14%

Transportation
42%

figure 6.18
Air pollution emissions in the United States in percentage by weight.
(Data from National Air Polution Control Administration, HEW.)

Primary pollutants are conducted upward from the emission sources by rising air currents that are a part of the normal convectional process. The larger particles settle under gravity and return to the surface as *fallout*. Particles too small to settle out are later swept down to earth by precipitation, a process called *washout*. By a combination of fallout and washout the atmosphere tends to be cleaned of pollutants. In the long run a balance is achieved between input and output of pollutants, but there are large fluctuations in the quantities stored in the air at a given time. Pollutants are also eliminated from the air over their source areas by winds that disperse the particles into large volumes of cleaner air in the downwind direction. Strong winds can quickly sweep away most pollutants from an urban area but, during periods when a stagnant anticyclone is present, the concentrations rise to high values.

In polluted air certain chemical reactions take place among the components injected into the atmosphere, generating a secondary group of pollutants. For example, sulfur dioxide may combine with oxygen and then react with water of suspended droplets to yield sulfuric acid. This acid is irritating to organic tissues and corrosive to many inorganic materials. In another typical reaction, the action of sunlight on nitrogen oxides and organic compounds produces ozone, a toxic and destructive gas. Reactions brought about by the presence of sunlight are described as *photochemical*. One toxic product of photochemical action is ethylene, produced from hydrocarbon compounds.

Ozone in urban smog has a most deleterious effect on plant tissues and, in some cases, has caused the death or severe damage of ornamental trees and shrubs. Sulfur dioxide is injurious to certain plants and is a cause of loss of productivity in truck gardens and orchards in polluted air. Atmospheric sulfuric acid in cities has in some places largely wiped out lichen growth.

Inversion and smog

Concentration of pollutants over a source area rises to its highest levels when the vertical mixing (convection) of the air is inhibited by a stable configuration of the air mass. Recall from Chapter 4 that at night, when the air is calm and the sky clear, rapid cooling of the ground surface typically produces a low-level temperature inversion (see Figure 4.7). In cold air the reversal of the temperature gradient may extend hundreds of feet into the air. A low-level temperature inversion represents an unusually stable air structure. When this type of inversion develops over an urban area, conditions are particularly favorable for entrapment of pollutants to the degree that heavy smog or highly toxic fog can develop, as shown in Figure 6.19. The upper limit of the inversion layer coincides with the cap, or lid, below which pollutants are held. The lid may be situated 500 to 1000 ft (150 to 300 m) above the ground.

Although situations dangerous to health during prolonged low-level inversion have occurred a number of times over European cities since the Industrial Revolution began, the first major tragedy of this kind in the United States occurred at Donora, Pennsylvania, in late October 1948. The city occupies a valley floor hemmed in by steeply rising valley walls that prevent the free mixing of the lower air layer with that of the surrounding region (Figure 6.19). Industrial smoke and gases from factories poured into the inversion layer for five days, increasing the pollution level. Since

humidity was high, a poisonous fog formed and began to take its toll. Twenty persons died and several thousand persons were made ill before a change in weather pattern dispersed the smog layer.

Related to the low-level inversion, but caused in a somewhat different manner, is the *upper-level inversion*, illustrated in Figure 6.20. Recall that anticyclones are cells of subsiding air that diverges at low levels. Within the center of the cell, winds are calm or very gentle. As the air subsides, it is adiabatically warmed, so that the normal temperature lapse rate is displaced to the right in the temperature-altitude graph, as shown in Figure 6.20 by the diagonal arrows. Below the level at which subsidence is occurring, the air layer remains stagnant. The temperature curve consequently develops a kink in which a part of the curve shows an inversion. For reasons already explained, the layer of inverted temperature structure strongly resists mixing and acts as a lid to prevent the continued upward movement and dispersal of pollutants. Upper-level inversions develop occasionally over various parts of the United States when an anticyclone stagnates for several days at a stretch.

For the Los Angeles Basin of southern California and, to a lesser degree, the San Francisco Bay area and over other west coasts in these latitudes generally, special climatic conditions produce prolonged upper-air inversions favorable to persistent smog accumulation. The Los Angeles Basin is a low,

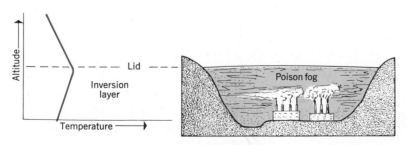

figure 6.19
A low-level inversion was the predisposing condition for poison fog accumulation at Donora, Pennsylvania in October 1948. (From A. N. Strahler, 1972, Planet Earth, Harper & Row, New York.)

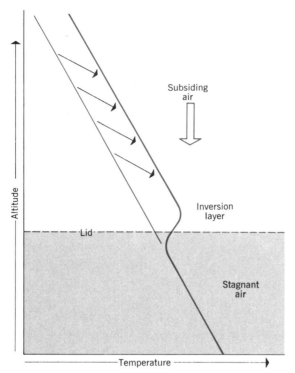

figure 6.20
**An upper-air inversion caused by subsidence.
(From A. N. Strahler, 1972, Planet Earth, Harper &
Row, New York.)**

strongest. The subsiding air over the Los
Angeles Basin is warmed adiabatically as well
as heated by direct absorption of solar
radiation during the day, so that it is markedly
warmer than the cool stagnant air below the
inversion lid. Pollutants accumulate in the cool
air layer, producing the characteristic smog.
The upper limit of the smog stands out sharply
in contrast to the clear air above it, filling the
basin like a lake and extending into valleys in
the bordering mountains.

One important physical effect of urban air
pollution is that it reduces visibility and
illumination. A smog layer can cut illumination
by 10 percent in summer and 20 percent in
winter. Ultraviolet radiation is absorbed by
smog which, at times, completely prevents
these wavelengths from reaching the ground.
Reduced ultraviolet radiation may prove to be
important in permitting increased bacterial
activity at ground level. City smog cuts
horizontal visibility to one fifth to one tenth of
the normal distance for clean air. Over cities,
winter fogs are much more frequent than over
the surrounding countryside. Coastal airports,
such as La Guardia (New York), Newark (New
Jersey), and Boston suffer severely from a high
incidence of fogs augmented by urban air
pollution.

A related effect of the urban heat island
(Chapter 4) is the general increase in cloudiness and precipitation over a city as compared
with the surrounding countryside. This
increase results from intensified convection
generated by heating of the lower air. For
example, it has been found that thunderstorms
over London, England, produce 30 percent
more rainfall over the city than over the
surrounding country. Increased precipitation
over an urban area is estimated to average
from 5 to 10 percent over the normal for the
region in which it lies.

sloping plain lying between the Pacific Ocean
and a massive mountain barrier on the north
and east sides. Cool air is carried inland over
the basin on weak winds from the south and
southwest, but it cannot move farther inland
because of the mountain barrier. It is a
characteristic of these latitudes that the air on
the eastern sides of the subtropical high
pressure cells is continually subsiding, creating
a more-or-less permanent upper-air inversion
that dominates the western coasts of the
continents and extends far out to sea (see
Figure 13.20). The effect is particularly marked
in the summer season, when the Azores and
Bermuda highs are at their largest and

Air Masses and Cyclonic Storms

Chapter 7

THE ATMOSPHERE EXERTS stress, often severe, on Man and other life forms through weather disturbances involving hazards from high winds, cold, and precipitation. As severe storms, these phenomena generate environmental hazards indirectly through their attendant phenomena — storm waves and storm surges on the seas and river floods, mudflows, and landslides on the lands.

Weather disturbances of lesser magnitude are among the beneficial environmental phenomena, since they bring precipitation to the land surfaces and so recharge the vital supplies of fresh water on which Man and all other terrestrial life forms depend.

An understanding of weather disturbances of all intensities enables Man to predict their times and places of occurrence and therefore to give warnings and allow protective measures to be taken. This function of the atmospheric scientist can be placed under the heading of environmental protection. It is also possible to a limited degree for Man to modify atmospheric processes in a deliberate way in order to lessen the wind speeds of storms or to increase precipitation over drought areas. These activities come under the heading of planned weather modification. Here, again, we find that the relationship of Man with the atmosphere is one of interaction.

Traveling cyclones

Much of the unsettled, cloudy weather we experience in midlatitudes is associated with traveling cyclones. Convergence of masses of air toward these centers is accompanied by lift of air and adiabatic cooling which, in turn, produces cloudiness and precipitation. By contrast, much of our fair sunny weather is associated with traveling anticyclones in which the air subsides and spreads outward, causing adiabatic warming. This process is unfavorable to the development of clouds and precipitation.

Many cyclones are very mild in intensity, passing with little more than a period of cloud cover and light rain or snow. On the other hand, when pressure gradients are strong, winds ranging in strength from moderate to gale force accompany the cyclone. In such a case, the disturbance can be called a *cyclonic storm*. Moving cyclones fall into three general classes: (1) the wave cyclone of midlatitude and arctic zones; this variety ranges in severity from a weak disturbance to a powerful storm; (2) the tropical cyclone of tropical and subtropical zones over ocean areas, it ranges from a mild disturbance to the terribly destructive hurricane, or typhoon; and (3) the tornado; although a very small storm, the tornado is an intense cyclonic vortex of enormously powerful winds; it is on a very much smaller scale of size than other types of cyclones and is related to severe convectional activity.

Air masses

Cyclones of middle and high latitudes depend for their development on the coming together of large bodies of air of contrasting physical properties. A body of air in which the upward gradients of temperature and moisture are fairly uniform over a large area is known as an *air mass*. In horizontal extent a single air mass may be as large as a part of a continent; in vertical dimension it may extend through the troposphere. A given air mass is characterized by a distinctive combination of temperature, environmental lapse rate, and specific humidity. Air masses differ widely in temperature — from very warm to very cold — and in moisture content — from very dry to very moist.

A given air mass usually has a sharply defined boundary between itself and a neighboring air mass. This discontinuity is termed a *front*. We found an example of a front in the contact between polar and tropical air masses below the axis of the jet in upper-air waves, as shown in Figure 5.17. This feature we called the *polar front;* it represents the highest degree of global generalization. Fronts may be nearly vertical, as in the case of air masses having little motion relative to one another. Fronts may be inclined at an angle nor far from the horizontal in cases where one air mass is sliding over another. A front may be almost stationary with respect to the earth's surface but, nevertheless, the adjacent air masses may be moving rapidly with respect to each other along the front.

The properties of an air mass are derived partly from the regions over which it passes. Because the entire troposphere is in more or less continuous motion, the particular air-mass properties at a given place may reflect the composite influence of a travel path covering thousands of miles and passing alternately over oceans and continents. This complexity of influences is particularly important in middle and high latitudes in the northern hemisphere, within the flow of the global westerlies.

However, there are vast tropical and equatorial areas over which an air mass reflects quite simply the properties of an ocean or a land surface over which it moves slowly or tends to stagnate. Over a warm equatorial ocean surface the lower levels of the overlying air mass develop a high water vapor content. Over a large tropical desert, slowly subsiding air forms a warm air mass with low relative humidities. Over cold, snow-covered land surfaces in the arctic zone in winter, the lower layer of the air mass remains very cold with a very low water vapor content. Meteorologists have designated as *source regions* those land or ocean surfaces that strongly impress their temperature and moisture characteristics on overlying air masses.

Air masses move from one region to another following the patterns of barometric pressure. During these migrations, lower levels of the air mass undergo gradual modification, taking up or losing heat to the surface beneath, and perhaps also taking up or losing water vapor.

Air masses are classified according to two categories of source regions: (1) the latitudinal position on the globe, which primarily determines thermal properties; and (2) the underlying surface, whether continent or ocean, determining the moisture content. With respect to latitudinal position, five types of air masses are as follows.

Air Mass	Symbol	Source Region
Arctic	A	Arctic ocean and fringing lands
Antarctic	AA	Antarctica
Polar	P	Continents and oceans, lat. 50-60° N and S
Tropical	T	Continents and oceans, lat. 20-35° N and S
Equatorial	E	Oceans close to equator

With respect to type of underlying surface, two further subdivisions are imposed on the preceding.

Air Mass	Symbol	Source Region
Maritime	m	Oceans
Continental	c	Continents

By combining types based on latitudinal position with those based on underlying surface a list of six important air masses results (Table 7.1). Figure 7.1 shows the global distribution of source regions of these air masses. Table 7.1 also gives some typical values of temperature and specific humidity at the surface, although a wide range in these properties may be expected, depending on season.

The equatorial air mass (mE) holds about 200 times as much water vapor as the extremely

table 7.1
Properties of typical air masses

Air Mass	Symbol	Properties	Temperature °F	(°C)	Specific Humidity (g/kg)
Continental-arctic (and continental antarctic)	cA (cAA)	Very cold, very dry (winter)	−50°	(−46°)	0.1
Continental polar	cP	Cold, dry (winter)	12°	(−11°)	1.4
Maritime polar	mP	Cool, moist (winter)	39°	(4°)	4.4
Continental tropical	cT	Warm, dry	75°	(24°)	11
Maritime tropical	mT	Warm, moist	75°	(24°)	17
Maritime equatorial	mE	Warm, very moist	80°	(27°)	19

cold arctic and antarctic air masses, cA and cAA. The maritime tropical air mass (mT) and maritime equatorial air mass (mE) are quite similar in temperature and water vapor content. Both are capable of very heavy yields of precipitation. The continental tropical air mass (cT) has its source region over subtropical deserts of the continents. Although it has a substantial water vapor content, it has low relative humidity when highly heated during the daytime. The polar maritime air mass (mP) originates over midlatitude oceans. Although the quantity of water vapor it holds is not large compared with the tropical air masses, the mP

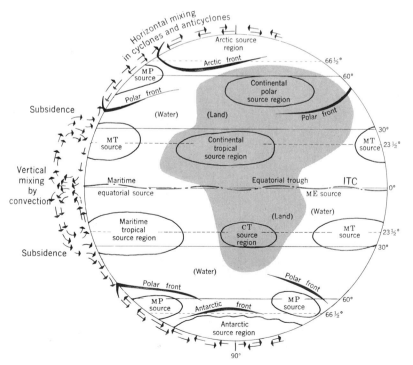

figure 7.1
This schematic global diagram shows the source regions of the major air mass types.

air mass can yield heavy precipitation. Much of this precipitation is of the orographic type, over mountain ranges on the western coasts of continents.

Cold and warm fronts

Figure 7.2 shows the structure of a front in which cold air is invading the warm-air zone. A front of this type is called a *cold front*. The colder air mass, being the denser, remains in contact with the ground and forces the warmer air mass to rise over it. The slope of the cold front surface is greatly exaggerated in the figure, being actually of the order of slope of 1 in 40 (meaning that the slope rises 1 mi vertically for every 40 mi of horizontal distance). Cold fronts are associated with strong atmospheric disturbance. As the unstable warm air is lifted, it may break out in severe thunderstorms. Thunderstorms can be seen on the radar screen (Figure 7.3).

Figure 7.4 illustrates a *warm front* in which warm air is moving into a region of colder air. Here, again, the cold air mass remains in contact with the ground and the warm air mass is forced to rise, as if it were ascending a long ramp. Warm fronts have lower slopes than cold fronts — on the order of 1 in 80, to as low as 1 in 200. Moreover, warm fronts commonly represent stable atmospheric conditions and

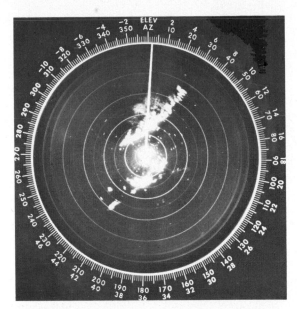

figure 7.3
Photograph of a radar screen on which lines of thunderstorms show as bright patches. The heavy circles are spaced 50 nautical miles (80 km) apart. (National Weather Service.)

lack the turbulent air motions of the cold front. Of course, if the warm air is unstable, it will develop convection cells, and there will be heavy showers or thunderstorms.

Cold fronts normally move along the ground at a faster rate than warm fronts. So, when both types are in the same neighborhood, the cold

figure 7.2
A cold front.

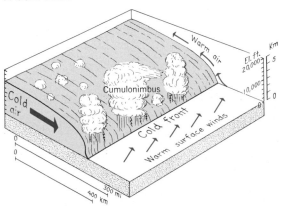

figure 7.4
A warm front.

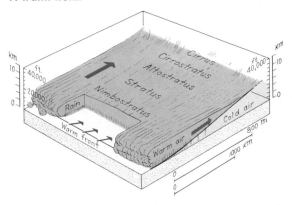

front overtakes the warm front. An _occluded front_ then results (Figure 7.5). The colder air of the fast-moving cold front remains next to the ground, forcing both the warm air and the less cold air to rise over it. The warm air mass is lifted completely free of the ground.

Wave cyclones

The dominant type of weather disturbance of middle and high latitudes is the _wave cyclone_, a vortex that repeatedly forms, intensifies, and dissolves along the frontal zone between cold and warm air masses. The Norwegian meteorologist, J. Bjerknes, at the time of World War I, recognized the existence of atmospheric fronts and developed his _wave theory_ of cyclones.

The term "front," used by Bjerknes, was particularly apt because of the resemblance of this feature to the fighting fronts in western Europe, then active. Just as vast armies met along a sharply defined front that moved back and forth, so masses of cold polar air meet in conflict with warm, moist tropical air. Instead of mixing freely, these unlike air masses remain clearly defined, but they interact along the polar front in great spiralling whorls.

A situation favorable to the formation of a wave cyclone is shown by a surface weather map (Figure 7.6). A trough of low pressure lies between two large cyclones (highs); one is made up of a cold dry polar air mass, the other of a warm moist maritime air mass. Air flow is converging from opposite directions on the two sides of the front, setting up an unstable situation.

A series of individual blocks (Figure 7.7) shows the sequence of stages in the life history of a wave cyclone. At the start of the cycle, the polar front is a smooth boundary along which air is moving in opposite directions. In Figure 7.7A, the polar front shows a wave beginning to form. Cold air is turned in a southerly direction and warm air in a northerly direction so that each invades the domain of the other.

In Figure 7.7B the wave disturbance along the polar front has deepened and intensified. Cold air is now actively pushing southward along a cold front; warm air is actively moving

figure 7.5
An occluded warm front.

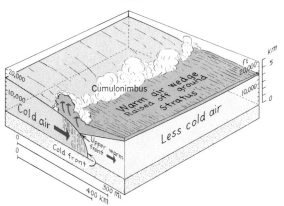

figure 7.6
The trough between two high-pressure regions is a likely zone for development of a wave cyclone.

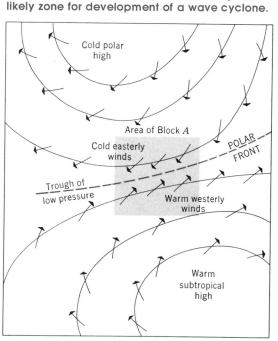

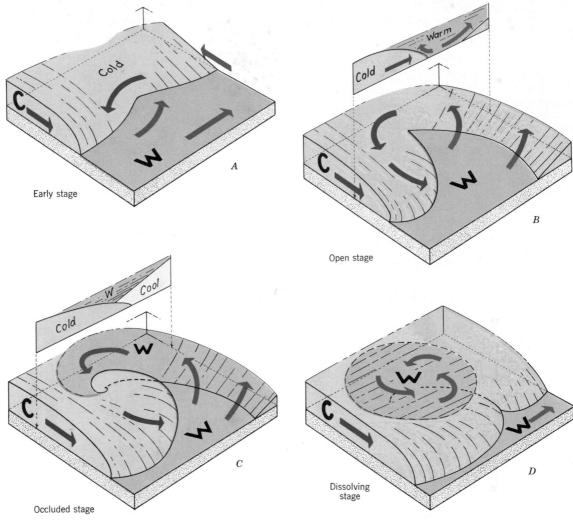

figure 7.7
Stages in the development of a wave cyclone.

northeastward along a warm front. Each front is convex in the direction of motion. The zone of precipitation is now considerable, but wider along the warm front than along the cold front. In a still later stage the more rapidly moving cold front has reduced the zone of warm air to a narrow sector.

In Figure 7.7C, the cold front has overtaken the warm front, producing an occluded front. The warm air mass has been forced off the ground and is now isolated from the parent region of warm air to the south. Because the source of moisture and energy has been cut off, the cyclonic storm gradually dies out. The polar front is now reestablished as originally (Figure 7.7D).

Weather changes within a wave cyclone

We can predict the changing weather conditions accompanying the passage of a wave cyclone in midlatitudes by examining some details of the internal structure of the storm. Figure 7.8 shows simplified surface weather maps depicting conditions on

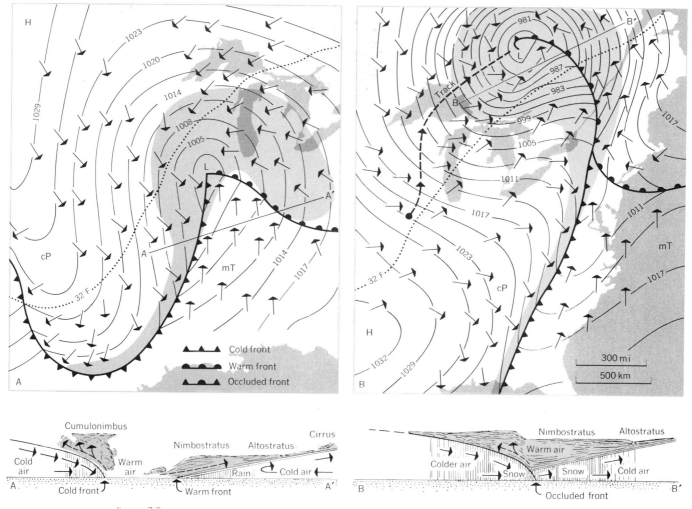

figure 7.8
Simplified surface weather maps and cross sections through a wave cyclone. (A) Open stage. (B) Occluded stage.

successive days. The structure of the storm is defined by the isobars, labeled in millibars. Three kinds of fronts are shown by specialized line symbols. Areas in which precipitation is occurring are shown in color.

Map *A* shows the cyclone in an open stage, similar to that in Figure 7.7B. Note the following points. (1) Isobars of the low are closed to form an oval-shaped pattern. (2) Isobars make a sharp V where crossing cold

and warm fronts. (3) Wind directions, indicated by arrows, are at an angle to the trend of the isobars and form a pattern of counterclockwise inspiralling. (4) In the warm-air sector there is northward flow of warm, moist tropical air toward the direction of the warm front. (5) There is a sudden shift of wind direction accompanying the passage of the cold front. This fact is indicated by the widely different wind directions at stations close to the cold front, but on opposite sides. There is also a sharp drop in temperature accompanying the passage of the cold front. (6) Precipitation is occurring over a broad zone near the warm front and in the central area of the cyclone, but extends as a thin band down the length of the cold front. Cloudiness prevails generally over the entire cyclone. (7) The low is followed on the west by a high (anticyclone) in which low temperatures and clear skies prevail. (8) The 32° F (0° C) isotherm crosses the cyclone diagonally from northeast to southwest, showing that the southeastern part is warmer than the northwestern part.

A cross section through Map A along line AA shows how the fronts and clouds are related. Along the warm front is a broad layer of stratiform clouds. These take the form of a wedge with a thin leading edge of cirrus. Westward this thickens to altostratus, then to stratus, and finally to nimbostratus with steady rain. Within the warm air mass sector the sky may partially clear with scattered cumulus. Along the cold front are cumulonimbus clouds associated with thunderstorms. These yield heavy rains, but only along a narrow belt.

The second weather map, Map B, shows conditions 24 hours later. The cyclone has moved rapidly northeastward, its track shown by a dashed line. The center has moved about 1000 mi (1600 km) in 24 hours, a speed of just over 40 mi (65 km) per hour. The cyclone has occluded. An occluded front replaces the separate warm and cold fronts in the central part of the disturbance. The high-pressure area, or tongue of cold polar air, has moved in to the west and south of the cyclone, and the cold front has pushed far south and east. Within the anticyclone the skies are clear and winds are weak. A cross section below the map shows conditions along line BB, cutting through the occluded part of the storm. Notice that the warm air mass is being lifted higher off the ground and is giving heavy precipitation.

Wave cyclones in all stages of development are seen in weather satellite photographs. A good example is pictured in Figure 7.9. The two photographs have been placed side by side in such a way that a strong three-dimensional effect results when you view them together with a stereoscope. Try it — the effect is quite stunning. Since 1960, specialized orbiting satellites have been in operation, circling the globe continuously and sending back data. In addition to transmitting sharp TV pictures, these satellites record infrared radiation. They also record the vertical profile of temperatures in the atmosphere and the global distributions of ozone, water vapor, and precipitation.

Cyclone tracks and cyclone families

Long observation of the movements of cyclones and anticyclones has shown that certain tracks are most commonly followed. Figure 7.10 is a map of the United States showing the common cyclone paths. Notice that some cyclonic storms travel across the entire United States from a place of origin in the North Pacific; others originate in the Rocky Mountain region, the central states, or the Gulf Coast. Most tracks converge toward the northeastern United States and pass out into the North Atlantic, where they tend to concentrate in the region of the Icelandic low.

The general world distribution of paths of cyclonic storms is shown in Figure 7.11. Notice the heavy concentration in the neighborhood of the Aleutian and Icelandic lows. Wave cyclones commonly form in succession to travel in a chain across the North Atlantic and North Pacific oceans. Figure 7.12, a world weather map, shows several such cyclone families. As each cyclone moves northeastward, it deepens and occludes, becoming an intense upper-air vortex. For this

figure 7.9
An occluded cyclone over the eastern Pacific Ocean shows on this satellite image as a tight cloud spiral. The cold front makes a dense, narrow cloud band sweeping to the south and southwest of the cyclone center. View these frames with a stereoscope for strong three-dimensional effect. (NOAA-2 satellite image, courtesy of National Environmental Satellite Service.)

reason, intense cyclones arriving at the western coasts of North America and Europe are usually occluded.

In the southern hemisphere, storm tracks are more nearly along a single lane, following the parallels of latitude. This is the result of uniform ocean surface circling the globe at these latitudes; only the southern tip of South America breaks the monotonous oceanic expanse. Also, the polar-centered ice sheet of Antarctica provides a centralized source of very cold air.

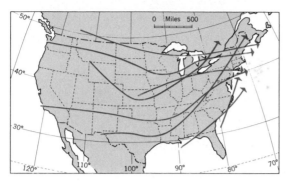

figure 7.10
Common tracks taken by wave cyclones passing across the United States.

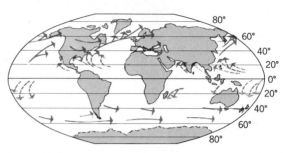

figure 7.11
Principal tracks of wave cyclones are shown by solid lines; those of tropical cyclones are shown by dashed lines.

This brief survey of wave cyclones will help you to understand the physical environment of a broad belt of the earth, including the midlatitude zone and the subarctic zone, in which the industrial nations of North America and Eurasia are largely concentrated. Wide fluctuations of weather conditions within short periods of time characterize this belt, together with strong seasonality of climate.

Tropical and equatorial weather disturbances

Weather systems of the tropical and equatorial zones show some basic differences from those of midlatitudes. The Coriolis force is weak close to the equator, and there is a lack of strong contrast between air masses. Consequently, clearly defined fronts and large,

figure 7.12
A daily weather map of the world for a given day during July or August might look like this map, which is a composite of typical weather conditions. (After M. A. Garbell.)

intense wave cyclones are missing. On the other hand, there is intense atmospheric activity in the form of convection cells because of the high moisture content of the maritime air masses in these low latitudes. In other words, there is an enormous reservoir of energy in the latent form. This same energy source powers the most formidable of all storms — the tropical cyclone.

The world map (Figure 7.12) shows a typical set of weather conditions in the subtropical and equatorial zones. Fundamentally the arrangement consists of two rows of high-pressure cells, one or two cells to each land or ocean body. The northern row lies approximately along the tropic of cancer; the southern row lies along the tropic of capricorn. Between the subtropical highs lies the equatorial trough of low pressure. Toward this trough the northeast and southeast trades converge along the intertropical convergence zone (ITC). At higher levels in the troposphere, the air flow is almost directly from east to west in the form of persistent tropical easterlies, described in Chapter 5.

One of the simplest forms of weather disturbance is an *easterly wave*, a slowly moving trough of low pressure within the belt of tropical easterlies (trades). These waves occur in lat. 5° to 30° N and S over oceans, but not over the equator itself. Figure 7.13 is a simplified weather map of an easterly wave showing isobars, winds, and the zone of rain. The wave is simply a series of indentations in the isobars to form a shallow pressure trough. The wave travels westward at a rate of 200 to 300 mi (325 to 500 km) per day. Air flow converges on the eastern, or rear, side of the wave axis. This convergence causes the moist air to be lifted and to break out into scattered showers and thunderstorms. The rainy period may last for a day or two.

Another related disturbance is the *weak equatorial low*, which forms near the center of the equatorial trough. Moist equatorial air masses converge on the center of the low, causing rainfall from many individual convectional storms. Several such weak lows are shown on the world weather map (Figure 7.12) lying along the ITC. Because the map is for a day in July or August, the ITC is shifted well north of the equator. At this season the rainy monsoon is in progress in Southeast Asia.

Another distinctive feature of tropical-zone weather is the occasional penetration of powerful tongues of cold polar air from the midlatitudes into very low latitudes. These tongues are known as *polar outbreaks*; they bring unusually cool, clear weather with strong, steady winds moving behind a cold front with squalls. The polar outbreak is best developed in the Americas. Outbreaks that move southward from the United States over the Caribbean Sea and Central America are called "northers" or "nortes"; those that move north from Patagonia into tropical South America are called "pamperos." One such outbreak is shown over South America on the world weather map (Figure 7.12).

figure 7.13
An easterly wave passing over the West Indies. (After Riehl.)

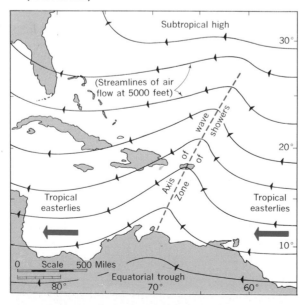

The tropical cyclone – an environmental hazard

One of the most powerful and destructive types of cyclonic storms is the *tropical cyclone*, otherwise known as the *hurricane or typhoon*. The storm develops over oceans in lat. 8° to 15° N and S, but not close to the equator, where the Coriolis force is extremely weak. In many cases an easterly wave simply deepens and intensifies, growing into a deep, circular low. High sea-surface temperatures that are over 80° F (27° C) in these latitudes are important in the environment of storm origin because warming of air at low level creates instability. Once formed, the storm moves westward through the trade wind belt. It may then curve northwest and north, finally penetrating well into the belt of westerlies.

The tropical cyclone is an almost circular storm center of extremely low pressure into which winds are spiralling at high speed, accompanied by very heavy rainfall (Figures 7.14 and 7.15). Storm diameter may be 100 to 300 mi (150 to 500 km). Wind speeds range from 75 to 125 mi (120 to 200 km) per hour, and sometimes much higher. Barometric pressure in the storm center commonly falls to 950 mb or lower.

figure 7.15

Hurricane Gladys, photographed on October 8, 1968, from Apollo 7 spacecraft at an altitude of about 110 mi (180 km). The storm center was near Tampa, Florida. (NASA photograph.)

A characteristic feature of the tropical cyclone is its *central eye*, in which calm prevails (Figure 7.16). The eye is a cloud-free vortex produced by the intense spiralling of the storm. In the eye air descends from high altitude and is adiabatically warmed. Passage of the central eye may take about half an hour, after which the storm strikes with renewed ferocity, but with winds in the opposite direction.

World distribution of tropical cyclones is limited to six regions, all of them over tropical and subtropical oceans (see Figure 7.12): (1) West Indies, Gulf of Mexico, and Caribbean Sea; (2) western North Pacific, including the Philippine Islands, China Sea, and Japanese Islands; (3) Arabian Sea and Bay of Bengal; (4) eastern Pacific coastal region off Mexico and Central America; (5) south Indian Ocean, off Madagascar; and (6) western South Pacific, in the region of Samoa and Fiji Islands and the east coast of Australia. Curiously, these storms are unknown in the South Atlantic. Tropical cyclones never originate over land, although they often penetrate well into the margins of continents.

figure 7.14

A hurricane of the West Indies.

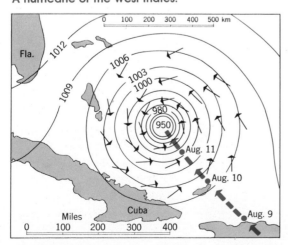

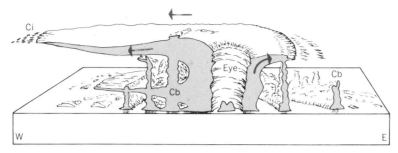

figure 7.16
Schematic diagram of a hurricane. Cumulonimbus clouds in concentric rings rise through dense stratiform clouds. Width of the diagram represents about 600 mi (1000 km). (Redrawn from NOAA, National Weather Service, R. C. Gentry, 1964.)

Tracks of tropical cyclones of the North Atlantic are shown in Figure 7.17. Most of the storms originate at 10° to 20° latitude, travel westward and northwestward through the trades, then turn northeast at about 30° to 35° latitude into the zone of the westerlies. Here the intensity lessens and the storms change into typical midlatitude wave cyclones. In the trade wind belt the cyclones travel 6 to 12 mi (10 to 20 km) per hour.

The occurrence of tropical cyclones is restricted to certain seasons of year, and these vary according to the global location of the storm region. For hurricanes of the North Atlantic the season runs from May through November, with maximum frequency in late summer or early autumn. The general rule

is that tropical cyclones of the northern hemisphere occur in the season during which the ITC has moved north; those of the southern hemisphere occur when it has moved south.

The environmental importance of tropical cyclones lies in their tremendously destructive effect on inhabited islands and coasts (Figure 7.18). Wholesale destruction of cities and their inhabitants has been reported on several occasions. A terrible hurricane that struck Barbados in the West Indies in 1780 is reported to have torn stone buildings from their foundations, destroyed forts, and carried cannons more than a 100 ft from their locations. Trees were torn up and stripped of their bark. More than 6000 persons perished there.

Coastal destruction by storm waves and greatly raised sea level is perhaps the most serious effect of tropical cyclones. Where water level is raised by strong wind pressure, great storm surf attacks ground ordinarily far inland of the limits of wave action. A sudden rise of water level, known as *storm surge*, may take place as the hurricane moves over a coastline. Ships are lifted bodily and carried inland to become stranded. If high tide accompanies the storm the limits reached by inundation are even higher. The terrible hurricane disaster at Galveston, Texas in 1900 was wrought largely by a sudden storm surge, inundating the low coastal city and drowning about 6000 persons. At the mouth of the Hooghly River on the Bay

figure 7.17
Tracks of some typical hurricanes occurring during August.

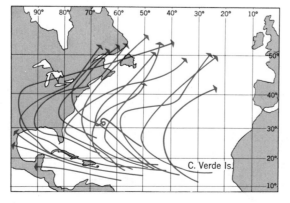

of Bengal, 300,000 persons died as a result of inundation by a 40-ft (12 m) storm surge that accompanied a severe tropical cyclone in 1737. Low-lying coral atolls of the western Pacific may be entirely swept over by wind-driven seawater, washing away palm trees and houses and drowning the inhabitants.

Important, too, is the large quantity of rainfall produced by tropical cyclones. A considerable part of the summer rainfall of some coastal regions can be traced to a few such storms. While this rainfall is a valuable water resource, it may prove a menace as a producer of unwanted river floods and, in areas of steep mountain slopes, gives rise to disastrous mudslides and landslides.

An attempt to reduce the severity of a hurricane by cloud seeding methods was made in 1969 by scientists of the National Oceanic and Atmospheric Administration (NOAA). The experiment, known as Project Stormfury, apparently produced significant results. A 30 percent reduction in wind speed is claimed to have been achieved for a brief period.

The tornado — an environmental hazard

The smallest but most intense of all known storms is the _tornado._ It seems to be a typically American storm, since it is most frequent and violent in the United States. Tornadoes also occur in Australia in substantial numbers and are reported occasionally in other places in midlatitudes.

The tornado is a small, intense cyclone in which the air is spiralling at tremendous speed. It appears as a dark _funnel cloud_ (Figure 7.19) hanging from a cumulonimbus cloud. At its lower end the funnel may be 300 to 1500 ft (90 to 460 m) in diameter. The funnel appears dark because of the density of condensing moisture, dust, and debris swept up by the wind.

Wind speeds in a tornado exceed anything known in other storms. Estimates of wind speed run as high as 250 mi (400 km) per hour. As the tornado moves across the country, the funnel writhes and twists. The end of the funnel cloud may alternately sweep the ground, causing complete destruction of

figure 7.18
This devastation along the south coast of Haiti was caused by Hurricane Flora on October 3, 1963. (Miami News photo.)

figure 7.19
This funnel cloud was photographed by William L. Males at Cheyenne, Oklahoma on May 4, 1961. The tornado was less than 1 mi from the observer when the picture was taken.

anything in its path, and rise in the air to leave the ground below unharmed. Tornado destruction occurs both from the great wind stress and from the sudden reduction of air pressure in the vortex of the cyclonic spiral. Closed houses literally explode. Even corks will pop out of empty bottles, so great is the drop in air pressure.

Tornadoes occur as parts of powerful cumulonimbus clouds traveling in advance of a cold front. They seem to originate where turbulence is greatest. They are most common in the spring and summer, but occur in any month. Where maritime polar air lifts warm, moist tropical air on a cold front, conditions may become favorable for tornadoes. They occur in greatest numbers in the central and southeastern states and are rare over mountainous and forested regions. They are almost unknown from the Rocky Mountains westward and are relatively fewer on the eastern seaboard (Figure 7.20).

Devastation from a tornado is complete within the narrow limits of its path (Figure 7.21). Only the strongest buildings constructed of concrete and steel can resist major structural damage. A tornado can often be seen or heard approaching, but those that come during the night may give no warning. The National Weather Service maintains a tornado forecasting and warning system. Whenever weather conditions conspire to favor tornado development, the danger area is alerted, and systems for observing and reporting a tornado are set in readiness.

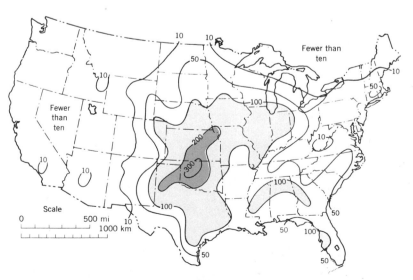

figure 7.20
Distribution of tornadoes in the United States, 1955 to 1967. Figures tell number of tornadoes observed within two-degree squares of latitude and longitude. (Data of M. E. Pautz, 1969, ESSA.)

figure 7.21
This wide swath of destruction through Xenia, Ohio, was cut by a tornado on the afternoon of April 4, 1974. The storm was one of 148 tornadoes that struck in 12 central states on two successive days, altogether killing over 300 persons and injuring over 5000 — the largest single tornado outbreak in the world's history. (Wide World Photos.)

Cyclonic activity and the global environment

Reviewing the major concepts covered in this chapter, we can recognize first the important role that cyclonic storms and their fronts play in the global balances of heat and water. Along with convectional activity cyclonic storms liberate enormous quantities of latent heat carried as water vapor. Moist maritime air masses migrate poleward from the low-latitude belt of surplus energy. These air masses carry heat as well as water, and this heat is released in higher latitudes.

Cyclonic storms supply precipitation vital to native and cultivated plants of the midlatitude and tropical zones. But these same storms also produce environmental hazards to life. As yet, Man has been able to exert no far-reaching or lasting control over the enormous releases of energy in cyclonic storms. However, looking into the future, we can anticipate much more effective methods of achieving weather modification than those now being tried. With such success may come side effects that could be unwanted and undesirable, and these will require close watching.

The Global Scope of Climate

Chapter 8

A GOAL OF physical geography is to recognize and describe environmental regions of significance to Man and to all life forms generally. Which are the vital components of the environment of the life layer with respect to Man in particular and to the biosphere in general? Keep in mind that the basic food supply of animals is organic matter synthesized by plants. Plants are the primary producers of organic matter, which sustains other life.

In view of Man's dependence on the primary producers, it seems logical that we should emphasize those components of the environment that are vital to plant growth. Plants occupy both marine and terrestrial environments. Because plant life of the lands is most directly exposed to the atmosphere for exchanges of energy and matter, and because most of Man's food is derived from terrestrial plants, we will want to concentrate on the land-atmosphere interface.

Knowing our broad goals in assessing the environment of the land-atmosphere interface is an important first step. To repeat, our concern is with those environmental components essential to plant life on the lands. Basically, there are two vital, atmospherically derived components that vary greatly from place to place and from season to season: energy and water; both must be in available forms. Plants require exposure to solar radiation in order to carry on photosynthesis. (Photosynthesis is the process by which carbohydrate molecules are synthesized.) Plants need carbon dioxide as well as water to synthesize carbohydrate compounds. But carbon dioxide is nearly uniform in its atmospheric concentration over the globe at all seasons, so we can omit it as a variable factor in the environment.

Plants require the presence of sensible heat as measured by air and soil temperatures within specified limits. Plants also require water in the root zone of the soil. Although land animals are consumers they, too, require fresh water and a tolerable range of surrounding temperatures. Plants also release energy and water to the atmosphere through the process of respiration, in which the carbohydrate molecules are broken down into carbon dioxide and water. (Photosynthesis and respiration are discussed in Chapter 11.)

Besides energy, carbon dioxide, and water, plants need nutrients, which they derive from the soil. Most important of the nutrient elements are nitrogen, calcium, potassium, silicon, magnesium, sulfur, aluminum, phosphorus, chlorine, sodium, iron, and manganese. Many other elements are used in lesser quantities, but are nonetheless vital. Except for nitrogen, which is derived from the atmosphere, plant nutrients found in the soil are mostly derived from the parent mineral matter. The nutrients are released by the plants when the plant tissue dies. In this way the nutrients are returned, or recycled, to the soil.

To summarize, land plants require light energy, carbon dioxide, water, and nutrients. The input of these ingredients comes from two basic sources: (1) the adjacent atmosphere, and (2) the soil. Input of light energy, heat energy, and water from the atmosphere is encompassed by the concept of climate. In short, plants depend on climate and soil. Of these two sources of energy and matter the soil is the more strongly affected by the plants it serves through the recycling of matter. Climate, when broadly defined as a source of energy and water, is an independent agent of control. Climate is determined by latitude and by large-scale air motions and air-mass interactions within the troposphere. If we are to establish a pyramid of priorities and interactions, climate occupies the apex as the independent control (Figure 8.1). Below it and forming the base of the pyramid are (1) the organic process of plants, and (2) the soil process. Plants and soil interact with one another at the basal level.

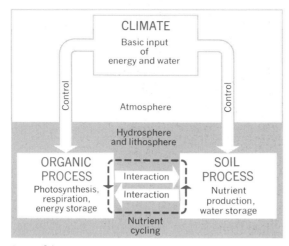

figure 8.1
The role of climate in environmental processes of the life layer.

Climate and climate classification

Climate has always been a keystone in physical geography and has formed the basis for defining physical regions of the globe. Let us first examine the content of traditional _climatology,_ the science of climate. In the broadest sense, _climate_ is the characteristic condition of the atmosphere near the earth's surface at a given place or over a given region. Components that enter into the description of climate are mostly the same as weather components used to describe the state of the atmosphere at a given instant. If weather information deals with the specific event, then climate represents a generalization of weather. A statement of the climate of a given observing station, or of a designated region, is described through the medium of weather observations accumulated over many years time. Not only are mean, or average, values taken into account, but also the departures from those means and the probabilities that such departures will occur.

Earlier chapters contained much information falling within the definition of climate. For example, world maps showing average January and July barometric pressures, winds, and air temperatures are expressions of climate. The physical components of climate are many; they include measurable quantities such as radiation, sensible heat, barometric pressure, winds, relative and specific humidity, dew point, cloud cover and type, fog, precipitation type and intensity, evaporation and transpiration, incidence of cyclones and anticyclones, and frequency of frontal passages. Which of these components are really significant in our analysis of climate? Approaching climatology as geographers, we shall take only what we need.

Our problem in devising a meaningful system of climate classes (i.e., a classification of climates) is to select categories of available information that correlate closely with the needs of life on the lands. We then use those categories to define and delineate regional classes of climates. If we have done our work well, the regional units within the climate system will reflect strongly the atmosphere's control on terrestrial life. This same system will give some indication of the opportunities and constraints that the atmospheric environment imposes on humans as they seek to increase their food supplies and water resources at the same time that they extend the areas of urban and industrial land use. The climate classification we seek must have utility in guiding land-use planning and population growth as well as describing the natural environment. This role is particularly crucial in guiding the progress of the developing nations as they seek to augment inadequate food resources.

Let us now select those categories of information that will produce a meaningful system of climate classification.

Air temperature as a basis for climate classification

All species of organisms, whether plants or animals, are subject to limiting temperatures of the surrounding air, water, or soil; above or below these limits survival is not possible. Few growing plants can survive temperatures over 120° F (50° C) for more than a few minutes. Many species of plants native to the tropical and equatorial latitude zones die if they experience even a short period of below-freezing temperature (32° F; 0° C). Freezing of water in the plant tissues causes physical disruption in plants not adapted to such conditions. Alternate freezing and thawing of the soil can have a disruptive effect on plant roots and is an important factor in limiting plant growth in the arctic and alpine zones.

Apart from extreme limits of temperature tolerance, plants react to an increase in air and water temperature through an increase in the intensity of physical and chemical activity. The optimum rates at which both photosynthesis and respiration take place increase with rising temperature in the range from near-freezing to 70 to 80° F (20 to 25° C), after which the rate of photosynthesis begins to decline. Respiration, however, continues to increase in rate to considerably higher temperatures.

Knowing that plants are strongly influenced by the factor of sensible air and soil temperatures that surround them, we include air temperature as one of the essential climate factors.

Besides being an important environmental factor in plant physiology and reproduction, air temperature also enters into many activities of animal life, for example, hibernation and migration. For Man, air temperature is an important physiological factor and relates directly to the quantity of energy expended in space heating and air conditioning of buildings. Nevertheless, air temperature alone does not define meaningful climate classes, because the ingredient of water availability is missing.

Temperature of the lower air layer, as measured in the standard thermometer shelter (Chapter 4), has long provided one essential variable quantity in leading systems of climate classification. Monthly data based on daily readings of the maximum-minimum thermometer have been accumulated for many decades at thousands of observing stations the world over. Consequently, availability has been an important factor in favoring the use of air temperature in climate classification.

Thermal regimes

In Chapter 4 we studied the annual air temperature cycle and its relationship to the annual net radiation cycle. Figure 4.9 showed annual temperature cycles for four stations, ranging from the equatorial zone to the subarctic zone. Let us now carry this investigation of annual temperature cycles a step further to explore the global range of annual cycles.

A comparison of annual air temperature cycles for many observing stations over the globe allows us to recognize a number of distinctive types, which can be called thermal regimes. A few of these are shown in Figure 8.2. Each regime has been labeled according to the latitude zone in which it lies: equatorial, tropical, midlatitude, subarctic, and polar. The label also includes words descriptive of the location in terms of position with respect to land mass. "Continental" refers to a continental interior location; "west coast" and "marine" refer to a location close to the ocean, usually on the western side of the continent.

The equatorial regime is uniformly very warm; temperatures are close to 80° F (27° C) year around, and there are no temperature seasons.

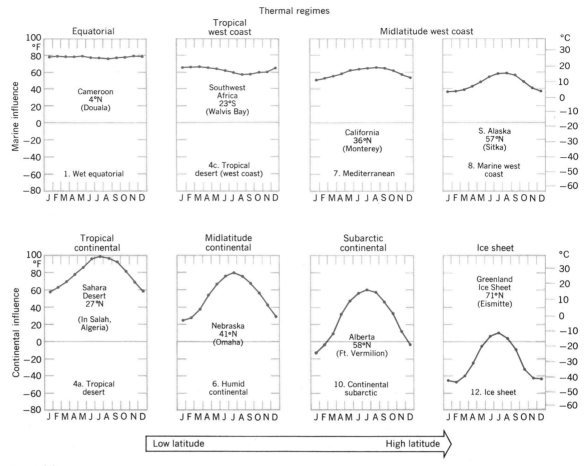

figure 8.2
(above and right) Some important thermal regimes, represented by annual cycles of air temperature (Based on Goode Base Map.)

The tropical continental regime ranges from very hot when the sun is high, near one solstice, to mild at the opposite solstice. However, close to the ocean we find the tropical west coast regime with only a weak annual cycle and no extreme heat. The same weak annual cycle can be traced into the midlatitude west coast regime, and it persists even much farther poleward. In the continental interiors the strong annual cycle prevails through the midlatitude continental regime into the subarctic continental regime, where the annual range is enormous. The ice sheet regime of Greenland is in a class by itself, with severe cold all year.

Other regimes can be identified and, because they grade into one another, the list could be expanded indefinitely. Instead of memorizing the eight regimes shown in Figure 8.2, you should simply get a feel for the concepts they illustrate. One concept is that of *continentality*, the tendency of a large landmass to impose a large annual range on the annual cycle. Continentality becomes stronger with higher latitude, because the insolation cycle shows stronger seasonal extremes with increasing latitude. Another concept is that of the marine influence, tending to weaken the annual cycle and to hold temperatures in a moderate range. This effect is due, of course, to the capacity of

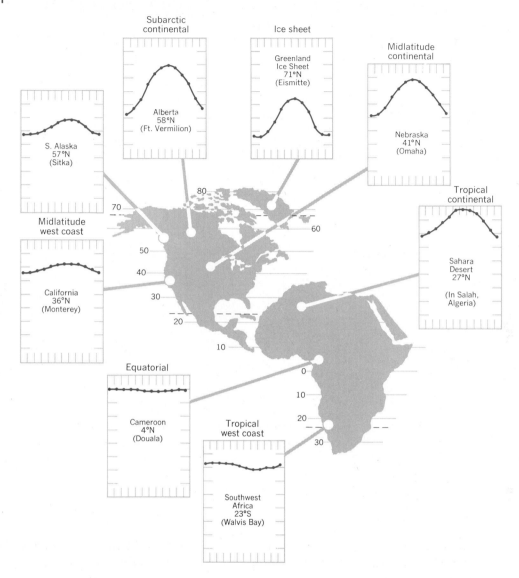

the oceans to hold a large amount of heat in storage and to take up and give off that heat very slowly, as compared with land areas.

Precipitation as a basis for climate classification

Precipitation data, obtained by the simple rain gauge (Chapter 6), are abundantly available for long periods of record for thousands of observing stations widely distributed over the globe. Small wonder, then, that monthly and annual precipitation data form the cornerstone of most of the widely used climate classifications.

At this point it is important for you to obtain an overview of the global patterns of precipitation and their relationship to air-mass source regions and prevailing movements of air masses.

figure 8.3
World precipitation. Isohyets show mean annual precipitation in inches. (Isohyets modified and simplified from The Times Atlas, John Bartholomew, Editor, The Times Publishing Company, Ltd., London, 1958. Based on Goode Base Map.)

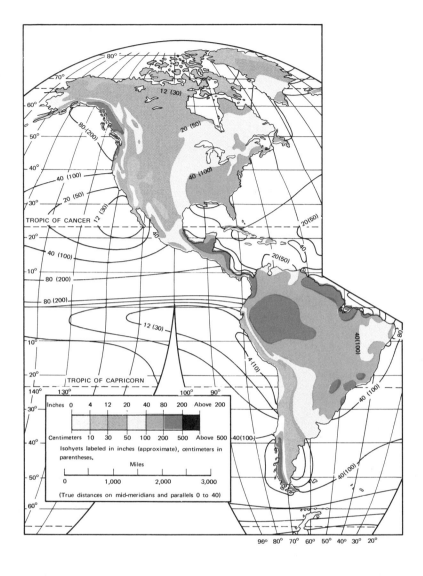

Average annual precipitation is shown on a world map (Figure 8.3). This map uses *isohyets*, lines drawn through all points having the same annual quantity. For regions where all or most of the precipitation is rain, we use the word "rainfall"; for regions where snow is an important part of the annual total, we use the word "precipitation."

Seven precipitation regions can be recognized in terms of annual total in combination with location. Figure 8.4 is a schematic diagram of the global distribution of precipitation, simplifying the picture as much as possible in terms of an imaginary continent. Notice that the adjectives "wet," "humid," "subhumid,"

"semiarid," and "arid" have been applied to each of five levels of annual precipitation ranging from 80 in. (200 cm) down to zero.

1. The wet equatorial belt of heavy rainfall, over 80 in. (200 cm) annually, straddles the equator and includes the Amazon River basin in South America, the Congo River basin of equatorial Africa, much of the African coast from Nigeria west to Guinea, and the East Indies. Here the prevailingly warm temperatures and high moisture content of the mE air masses favor abundant convectional rainfall. Thunderstorms are frequent year around.

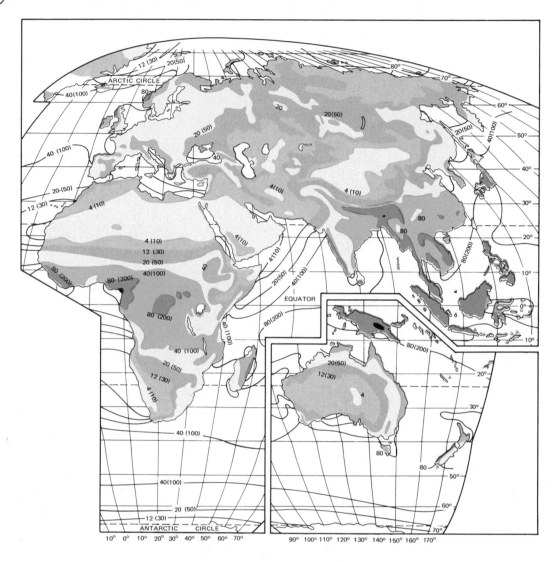

2. Narrow coastal belts of high rainfall, 60 to 80 in. and even more (150 to 200 cm +), extend from near the equator to latitudes of about 25° to 30° N and S on the eastern sides of every continent or large island. For example, see the eastern coasts of Brazil, Central America, Madagascar, and northeastern Australia. These are the trade-wind coasts, where moist mT air masses from warm oceans are brought over the land by the trades. Encountering coastal hills and mountains, these air masses produce heavy orographic rainfall.

3. In striking contrast to the wet equatorial belt astride the equator are the two zones of vast tropical deserts lying approximately on the tropics of cancer and capricorn. These are hot, barren deserts, with less than 10 in. (25 cm) of rainfall annually and in many places with less than 2 in. (5 cm). They are located under and are caused by the subtropical cells of high pressure where the subsiding cT air mass is adiabatically warmed and dried. Note that these deserts extend off the west coasts of the lands and out over the oceans. Rain such as these areas experience is largely convectional and extremely unreliable.

4. Farther northward, in the interiors of Asia and North America between lat. 30° and lat. 50°, are great continental midlatitude deserts

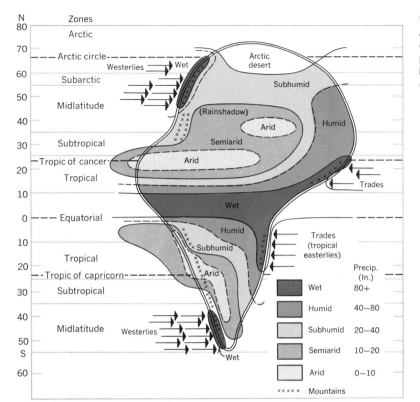

figure 8.4
A schematic diagram of the distribution of annual precipitation over an idealized continent and adjoining oceans.

and expanses of semiarid grasslands known as steppes. Annual precipitation ranges from less than 4 in. (10 cm) in the driest areas to 20 in. (50 cm) in the moister steppes. Dryness here results from remoteness from ocean sources of moisture. Located in a region of prevailing westerly winds, these arid lands occupy the position of rain shadows in the lee of coastal mountains and highlands. For example, the Cordilleran Ranges of Oregon, Washington, British Columbia, and Alaska shield the interior of North America from moist mP air masses originating in the Pacific. Upon descending into the intermontane basins and interior plains, the mP air masses are warmed and dried.

Similarly, mountains of Europe and the Scandinavian peninsula obstruct the flow of moist mP air masses from the North Atlantic

into western Asia. The great southern Asiatic ranges likewise prevent the entry of moist mT and mE air masses from the Indian Ocean.

The southern hemisphere has too little land in the midlatitudes to produce a true continental desert, but the dry steppes of Patagonia, lying on the lee side of the Andean chain, are roughly the counterpart of the North American deserts and steppes of Oregon and northern Nevada.

5. On the southeastern sides of the continents of North America and Asia, in lat. 25° to 45° N, are the humid subtropical regions, with 40 to 60 in. (100 to 150 cm) of rainfall annually. Smaller areas of the same kind are found in the southern hemisphere in Uruguay, Argentina, and southeastern Australia. These regions lie on the moist western sides of the subtropical high-pressure centers in such a position that

humid mT air masses from the tropical ocean are carried poleward over the adjoining land. Commonly, too, these areas receive heavy rains from tropical cyclones.

6. Another distinctive wet location is on midlatitude west coasts of all continents and large islands lying between about 35° and 65° in the region of prevailing westerly winds. These zones have been mentioned in Chapter 6 as good examples of coasts on which abundant orographic precipitation falls as a result of forced lift of mP air masses. Where the coasts are mountainous, as in Alaska and British Columbia, southern Chile, Scotland, Norway, and South Island of New Zealand, the annual precipitation is over 80 in. (200 cm). These coasts formerly supported great valley glaciers that carved the deep bays (fiords), so typically a part of their scenery.

7. A seventh precipitation region is formed by the arctic and polar deserts. Northward of the 60th parallel, annual precipitation is largely under 12 in. (30 cm), except for the west coast belts. Cold cP and cA air masses cannot hold much moisture; consequently, they do not yield large amounts of precipitation. At the same time, however, the relative humidity is high and evaporation rates are low.

As you would expect, zones of gradation lie between these precipitation regions. Moreover, this list does not recognize the fact that seasonal patterns of precipitation differ from region to region.

Seasonal patterns of precipitation

While the annual total precipitation is a useful quantity in establishing the characteristics of a climate type, it can be a misleading statistic because there may be strong seasonal cycles of precipitation. It makes a great deal of difference in terms of native plants and agricultural crops if there are alternately dry and wet seasons instead of a uniform distribution of precipitation throughout the year. It also makes a great deal of difference whether the wet season coincides with a season of higher temperatures or with a season of lower temperatures, because plants need both water and heat.

The seasonal aspects of precipitation can be largely covered by three major patterns: (1) uniformly distributed precipitation; (2) a precipitation maximum during the summer (or season of high sun) in which insolation is at its peak; and (3) a precipitation maximum during the winter or cooler season, when insolation is least. The first pattern can include a wide range of possibilities from little or no precipitation in any month to abundant precipitation in all months.

A study of monthly means of precipitation throughout the year at many observing stations over the globe shows a number of outstanding precipitation types. Some of these are illustrated in Figure 8.5. For each the average monthly precipitation is shown by the height of the bar in the small graphs.

Dominant in very low latitudes is an equatorial rainy type, illustrated by Singapore, near the equator. Rain is abundant in all months, but some months have considerably more than others. The tropical desert type has so little rain in any month that it cannot be shown on the graph. The tropical wet-dry type has a very wet season at time of high sun (summer solstice) and a very dry season at time of low sun (winter solstice). This seasonal alternation is carried to its greatest extreme in the Asiatic monsoon type, with an extremely wet high-sun monsoon season and a very dry period at low sun.

The summer precipitation maximum is carried into higher latitudes on the eastern sides of continents in a subtropical moist type. This same feature persists into the midlatitude

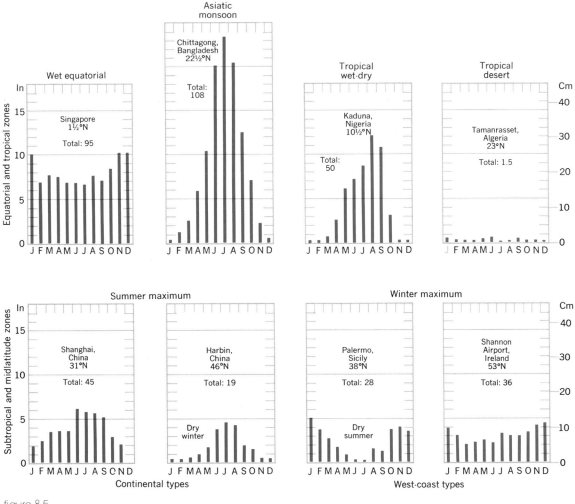

figure 8.5
(above and right) Eight precipitation types selected to show various seasonal patterns. (Based on Goode Base Map.)

continental type, which has a long, dry winter but a marked summer rain period.

A cycle with a winter precipitation maximum is typical of the west sides of continents in midlatitudes. A Mediterranean type, named for its prevalence in the Mediterranean region, has a very dry summer but a moist winter. This same cycle is carried into higher midlatitudes along narrow strips of west coasts. In the midlatitude west-coast type there is precipitation throughout the year, but with a

distinct maximum in winter and a distinct minimum in summer.

We do not intend that you memorize these precipitation types, but that you get a feel for the patterns illustrated. Each will appear in more detail in the descriptions of individual climates in Chapters 13 to 15. The reasons for annual precipitation cycles will have to be developed further in terms of shifting belts of pressure and winds and the seasonal changes in domination by different air masses.

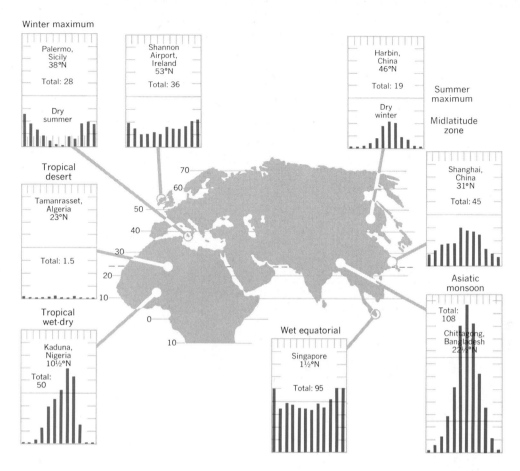

A climate system based on air masses and frontal zones

Recall from Chapter 7 that air masses are classified according to the general latitude of their source regions and the surface qualities of the source region, whether land or ocean. This air-mass classification has built into it a recognition of the factors of temperature and precipitation, because (1) air-mass temperature decreases poleward, and (2) precipitation capability of an air mass is related to sources of moisture. Source regions of air masses change seasonally along with shifting belts of air temperature, pressure, and winds. Frontal zones also migrate seasonally. In this way seasonal cycles of both air temperature and precipitation can be described in terms of changing air-mass activity.

The climate system we shall develop here uses the concepts of thermal regimes and precipitation types, but explains them in terms of air-mass activity. This explanatory system makes use of all of the global climatic information developed systematically in previous chapters. The system allows you to use that information to understand a given climate and to appreciate the reasons for its occurrence in a particular location.

The major climate groups

Our climate system is based on the location of air-mass source regions and the nature and movement of air masses, fronts, and cyclonic storms. A schematic global diagram (Figure 8.6) shows the principles of the classification.

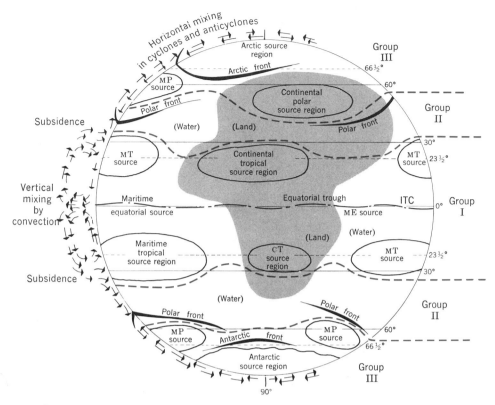

figure 8.6
Three major climate groups are related to air-mass source regions and frontal zones.

Notice that this figure is the same as Figure 7.1. We have placed it here to show how three major climate groups can be superimposed.

I *Low-Latitude Climates*. Group I includes the tropical air-mass source regions and the equatorial trough, or ITC, between. Climates of Group I are controlled by the subtropical high-pressure cells, or anticyclones, which are regions of air subsidence and are basically dry, and by the great equatorial trough of convergence that lies between them. Although air of polar origin occasionally invades the tropical and equatorial zones, the climates of Group I are almost wholly dominated by tropical and equatorial air masses. Tropical cyclones are important in this climate group.

II *Midlatitude Climates*. Climates of Group II are in a zone of intense interaction between unlike air masses: the *polar front zone*. Tropical air masses moving poleward and polar air masses moving equatorward are in conflict in this zone, which contains a procession of eastmoving wave cyclones. Locally and seasonally, either tropical or polar air masses may dominate in these regions, but neither has exclusive control.

III *High-Latitude Climates*. Climates of Group III are dominated by polar and arctic (including antarctic) air masses. The two polar continental air-mass source regions of northern Canada and Siberia fall into this group, but there is no southern hemisphere counterpart to these continental centers. In the arctic belt of the 60th to 70th parallels, air masses or arctic origin meet polar continental air masses along an *arctic front zone,* creating a series of east-moving wave cyclones.

Within each climate group are a number of climate types, or simply climates. There are twelve climates altogether, four in Group I, five in Group II, and three in Group III. For ease in identification the climates are numbered, but the numbers have no significance as a code. Only the climate names need be used, since these are descriptive of global location.

Table 8.1 gives a brief description of each climate. Along with each description is a representative *climograph* showing the cycles of mean air temperature and monthly mean precipitation. The climograph allows you to

superimpose the annual temperature cycle on the annual precipitation cycle. At a single glance, you can integrate the important features of the climate.

The dry climates (4 and 9) are subdivided into (a) desert, and (b) steppe. *Desert* refers to an extremely dry climate, while *steppe* refers to a semiarid climate with sufficient moisture to support grasslands. Generally, the steppe regions are transitional zones with adjoining regions of more humid climate.

Figure 8.7 is a schematic diagram showing the arrangement of the climates on an imagined supercontinent, patterned after a combined Eurasia and Africa. Major mountain barriers, which have strong controls over the placement of rainshadow deserts, are not suggested in this diagram.

figure 8.7

A schematic diagram of the placement of climates on an idealized continent. Compare with the world map, Figure 8.8.

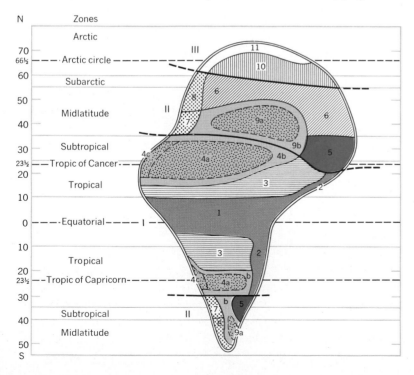

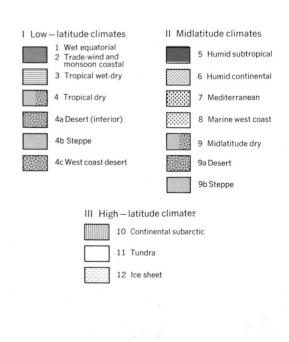

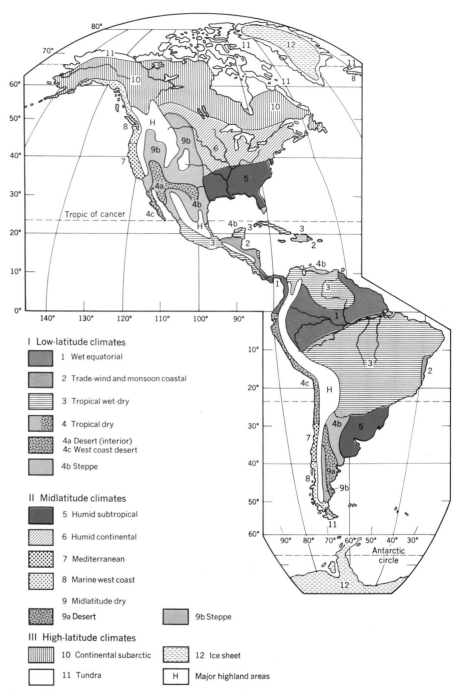

I Low-latitude climates

1   Wet equatorial

2   Trade-wind and monsoon coastal

3   Tropical wet-dry

4   Tropical dry

4a Desert (interior)
4c West coast desert

4b Steppe

II Midlatitude climates

5   Humid subtropical

6   Humid continental

7   Mediterranean

8   Marine west coast

9   Midlatitude dry

9a Desert   9b Steppe

III High-latitude climates

10 Continental subarctic   12 Ice sheet

11 Tundra   H   Major highland areas

figure 8.8
**Generalized and simplified world map showing distribution of twelve
climates. (Based on Goode Base Map.)**

Figure 8.8 is a generalized world map showing the distribution of the twelve climates. Keep in mind that sharp boundaries on the map are only a graphic device and do not exist in nature. Climates grade imperceptibly from one to the next except where a steep mountain barrier exists. The major highland areas are identified with no attempt to show the complex range of climates each contains. Many minor highlands of specialized climate are simply ignored on the map. The concept of a global climate pattern is what this map should convey.

table 8.1
Climate classification based on air masses and frontal zones

Group I: Low-Latitude Climates (controlled by equatorial and tropical air masses)

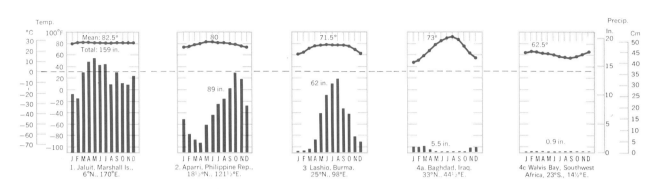

1. Jaluit, Marshall Is., 6°N., 170°E.
2. Aparri, Philippine Rep., 18½°N., 121½°E.
3. Lashio, Burma, 25°N., 98°E.
4a. Baghdad, Iraq, 33°N., 44½°E.
4c. Walvis Bay, Southwest Africa, 23°S., 14½°E.

Climate Name	Dominant Latitude Range	Air Mass Source Regions, Frontal Zones, General Characteristics.
1. Wet equatorial climate	10°N-10°S (Asia 10°-20°N)	Equatorial trough (ITC) climate, dominated by warm, moist maritime tropical (mT) and equatorial (mE) air masses yielding heavy convectional rainfall. Remarkably uniform temperatures prevail throughout the year.
2. Trade-wind and monsoon coastal climate	10°-25°N and S	Trade winds bring maritime tropical (mT) air masses from moist western sides of oceanic high-pressure cells to give narrow east-coast zones of heavy orographic rainfall. Temperatures uniformly warm. Rainfall shows strong seasonal cycle. Climate includes monsoon coasts of Southeast Asia.
3. Tropical wet-dry climate	5°-20°N and S (Asia 10°-30°N)	Seasonal alternation of moist mT or mE air masses with dry cT air masses gives climate with wet season at high-sun, dry season at low-sun. Hot season comes before rains.
4. Tropical dry climates (4a) Desert (4b) Steppe) (4c) West-coast) desert	15°-35°N and S	Source regions of cT air masses in high-pressure cells over tropics of cancer and capricorn. Subsiding air mass is stable, dry. Interior desert (4a) and steppe (4b) have very high maximum temperature and moderate annual range. West-coast desert (4c) relatively cool with small annual range, frequent fog.

Group II: Midlatitude Climates (controlled by both tropical and polar air masses)

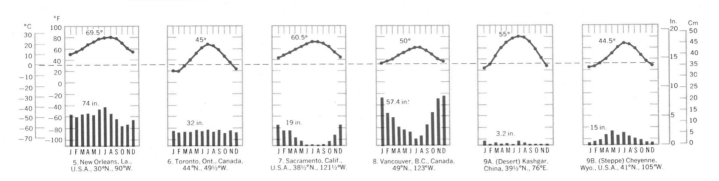

5. New Orleans, La., U.S.A., 30°N., 90°W. — 69.5° — 74 in.
6. Toronto, Ont., Canada. 44°N., 49½°W. — 45° — 32 in.
7. Sacramento, Calif., U.S.A., 38½°N., 121½°W. — 60.5° — 19 in.
8. Vancouver, B.C., Canada. 49°N., 123°W. — 50° — 57.4 in.
9A. (Desert) Kashgar, China, 39½°N., 76°E. — 55° — 3.2 in.
9B. (Steppe) Cheyenne, Wyo., U.S.A., 41°N., 105°W. — 44.5° — 15 in.

Climate Name	Dominant Latitude Range	Air Mass Source Regions, Frontal Zones, General Characteristics.
5. Humid subtropical climate	20°-35°N and S	Subtropical eastern continental margins dominated by moist maritime tropical (mT) air masses flowing from the western sides of the oceanic high-pressure cells. In summer rainfall is copious and temperatures warm. Winters are cool with frequent invasions of continental polar (cP) air masses. Frequent cyclonic storms.
6. Humid continental climate	35°-60°N	Located in central and eastern parts of continents of midlatitudes, this climate is in the polar front zone, the battleground of polar and tropical air masses. Seasonal temperature contrasts are strong, and weather is highly variable. Ample precipitation throughout the year is increased in summer by invading mT air masses. Cold winters are dominated by cP and cA air masses from subarctic source regions.
7. Mediterranean climate	30°-45°N and S	A wet-winter, dry-summer climate resulting from seasonal alternation of conditions causing climates 4 and 8. Moist mP air masses invade often in winter with cyclonic storms and ample rainfall. Subsiding mT and cT air masses dominate in summer with extreme drought. Moderate annual temperature range.
8. Marine west-coast climate	40°-60°N and S	Midlatitude west coasts, windward to westerlies, receive frequent cyclonic storms with cool, moist mP air masses. Precipitation in all months, but with a winter maximum. Annual temperature range is smaller for midlatitudes.

Group II: Midlatitude Climates — continued

Climate Name	Dominant Latitude Range	Air Mass Source Regions, Frontal Zones, General Characteristics.
9. Midlatitude dry climates (9a) Desert (9b) Steppe	35°-50°N and S	Interior midlatitude deserts and steppes distant from sources of mT or mP air masses, but dominated by cT air masses in summer and by cP air masses in winter. Very large annual temperature range; summers warm to hot, winters cold. Precipitation maximum in summer.

Group III: High-Latitude Climates (controlled by polar and arctic air masses)

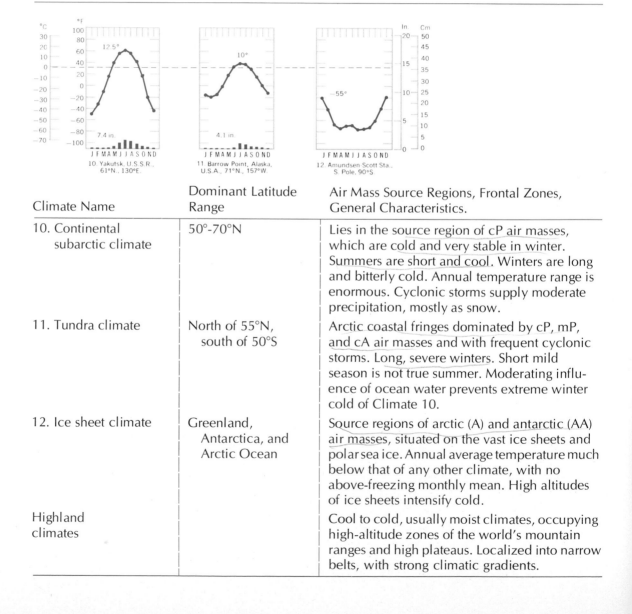

10. Yakutsk, U.S.S.R., 61°N., 130°E.

11. Barrow Point, Alaska, U.S.A., 71°N., 157°W.

12. Amundsen-Scott Sta., S. Pole, 90°S.

Climate Name	Dominant Latitude Range	Air Mass Source Regions, Frontal Zones, General Characteristics.
10. Continental subarctic climate	50°-70°N	Lies in the source region of cP air masses, which are cold and very stable in winter. Summers are short and cool. Winters are long and bitterly cold. Annual temperature range is enormous. Cyclonic storms supply moderate precipitation, mostly as snow.
11. Tundra climate	North of 55°N, south of 50°S	Arctic coastal fringes dominated by cP, mP, and cA air masses and with frequent cyclonic storms. Long, severe winters. Short mild season is not true summer. Moderating influence of ocean water prevents extreme winter cold of Climate 10.
12. Ice sheet climate	Greenland, Antarctica, and Arctic Ocean	Source regions of arctic (A) and antarctic (AA) air masses, situated on the vast ice sheets and polar sea ice. Annual average temperature much below that of any other climate, with no above-freezing monthly mean. High altitudes of ice sheets intensify cold.
Highland climates		Cool to cold, usually moist climates, occupying high-altitude zones of the world's mountain ranges and high plateaus. Localized into narrow belts, with strong climatic gradients.

The Köppen climate system

Among professional geographers, the most widely used climate classification system is that devised originally by Dr. Wladimir Köppen in 1918 and subsequently revised by his students, R. Geiger and W. Pohl. The version we present here was published in 1953 and is sometimes called the Köppen-Geiger-Pohl system. Köppen was both climatologist and plant geographer, so that his main interest lay in finding climate boundaries that coincided approximately with boundaries between major vegetation types. Although he was not entirely successful in achieving this goal, his climate system has appealed to geographers because it is strictly empirical and allows no room for subjective decisions.

Under the Köppen system each climate is defined according to assigned values of temperature and precipitation, computed in terms of annual or monthly values. Any given station can be assigned to its particular climate group and subgroup solely on the basis of the records of temperature and precipitation at that place provided, of course, that the period of record is long enough to yield meaningful averages. To use the Köppen system, it is not necessary to understand how climate controls operate.

The Köppen system features a shorthand code of letters designating major climate groups, subgroups within the major groups, and further subdivisions to distinguish particular seasonal characteristics of temperature and precipitation.

Five major climate groups are designated by capital letters as follows:

A *Tropical rainy climates.* Average temperature of every month is above 64.4° F (18° C). These climates have no winter season. Annual rainfall is large and exceeds annual evaporation.

B *Dry climates.* Evaporation exceeds precipitation on the average throughout the year. No water surplus; hence no permanent streams originate in B climate zones.

C *Mild, humid (mesothermal) climates.* Coldest month has an average temperature under 64.4° F (18° C), but above 26.6° F (−3° C); at least one month has an average temperature above 50° F (10° C). The C climates thus have both a summer and a winter season.

D *Snowy-forest (microthermal) climates.* Coldest month average temperature under 26.6° F (−3° C). Average temperature of warmest month above 50° F (10° C), that isotherm coinciding approximately with poleward limit of forest growth.

E *Polar climates.* Average temperature of warmest month below 50° F (10° C). These climates have no true summer.

Note that four of these five groups (A, C, D, and E) are defined by temperature averages, whereas one (B) is defined by precipitation-to-evaporation ratios. Groups A, C, and D have sufficient heat and precipitation for growth of forest and woodland vegetation. Figure 8.9 shows the boundaries of the five major climate groups.

Subgroups within the five major groups are designated by a second letter according to the following code.

S Semiarid (steppe).
W Arid (desert).

(The capital letters S and W are applied only to the dry B climates.)

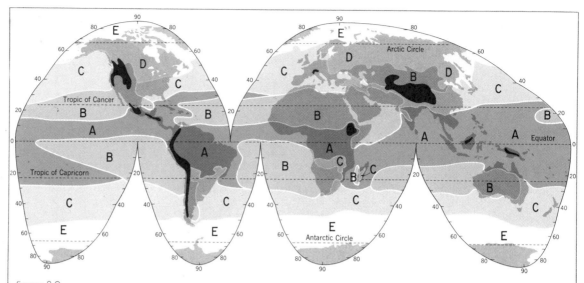

figure 8.9
Highly generalized world map of major climatic regions according to the Köppen classification. Highland areas in black. (Based on Goode Base Map.)

f   Moist. Adequate precipitation in all months. No dry season. This modifier is applied to *A, C,* and *D* groups.

w Dry season in winter of the respective hemisphere (low-sun season).

s   Dry season in summer of the respective hemisphere (high-sun season).

m Rainforest climate despite short, dry season in monsoon type of precipitation cycle. Applies only to *A* climates.

From combinations of the two letter groups, eleven distinct climates emerge.

Af *Tropical rainforest climate.* Rainfall of driest month is 2.4 in. (6 cm) or more.

Am Monsoon variety of *Af.* Rainfall of driest month is less than 2.4 in. (6 cm). Dry period is of short duration.

Aw *Tropical savanna climate.* At least one month has rainfall less than 2.4 in. (6 cm). Dry season is strongly developed.

Figure 8.10 shows the boundaries between *Af, Am,* and *Aw* climates as determined by both annual rainfall and rainfall of the driest month.

BS *Steppe climate.* A semiarid climate characterized by grasslands. It occupies an intermediate position between the desert climate *(BW)* and more humid climates of the *A, C,* and *D* groups. Boundaries are determined by formulas given in Figure 8.11.

figure 8.10
Boundaries of the <u>A</u> climates.

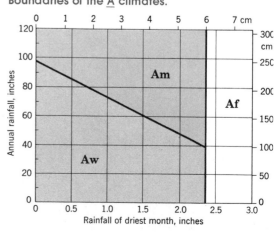

BW *Desert climate.* An arid climate with annual precipitation usually less than 15 in. (40 cm). Boundary with adjacent steppe climate *(BS)* is determined by formulas given in Figure 8.11.

Cf *Mild humid climate with no dry season.* Driest month precipitation average more than 1.2 in. (3 cm).

Cw *Mild humid climate with a dry winter.* The wettest month of summer has at least ten times the precipitation of the driest month of winter. (Alternate definition: 70 percent or more of the mean annual precipitation falls in the warmer 6 months.)

Cs *Mild humid climate with a dry summer.* Precipitation of driest month of summer is less than 1.2 in. (3 cm). Precipitation of the wettest month of winter is at least three times as much as the driest month of summer. (Alternate definition: 70 percent or more of the mean annual precipitation falls in the winter 6 months.)

Df *Snowy-forest climate with a moist winter.* No dry season.

Dw *Snowy-forest climate with a dry winter.*

ET *Tundra climate.* Mean temperature of warmest month is above 32° F (0° C) but below 50° F (10° C).

EF *Perpetual frost climate.* Ice sheet climate. Mean monthly temperatures of all months are below 32° F (0° C).

To denote further variations in climate, Köppen added a third letter to the code group. Meanings are as follows.

a   With hot summer; warmest month over 71.6° F (22° C) (*C* and *D* climates).

b   With warm summer; warmest month below 71.6° F (22° C) (*C* and *D* climates).

c   With cool, short summer; less than four months over 50° F (10° C) (*C* and *D* climates).

figure 8.11
Boundaries of the B climates.

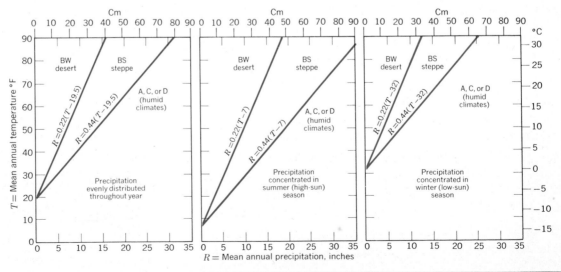

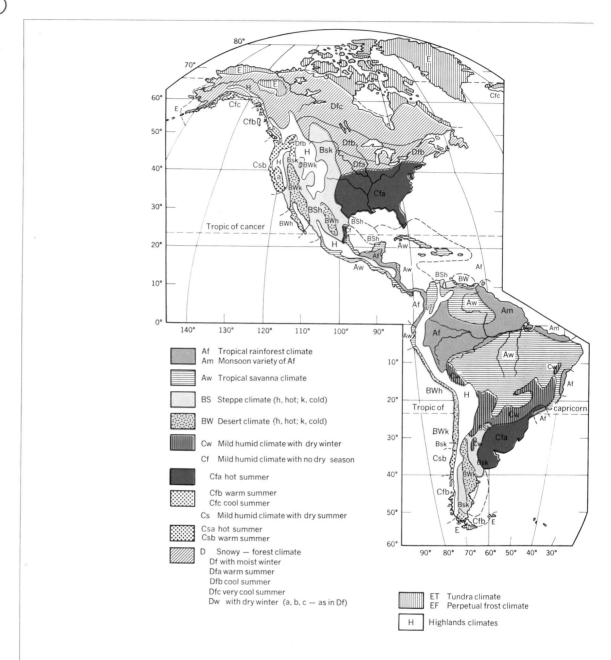

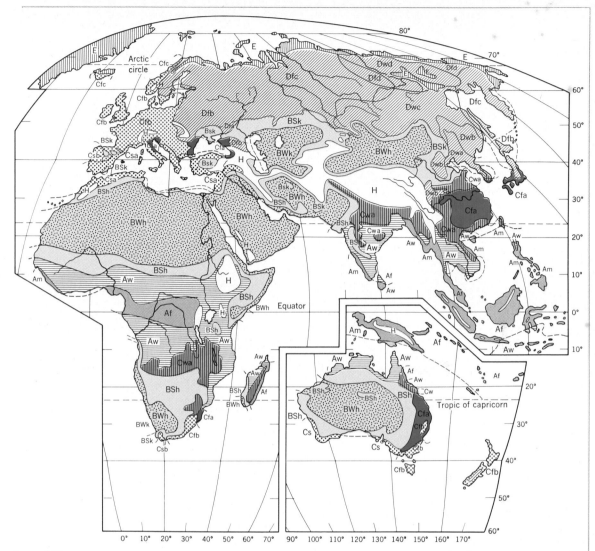

figure 8.12

World map of climates according to the Köppen-Geiger-Pohl system of 1953. (Based on Goode Base Map.)

d With very cold winter; coldest month below −36.4° F (−38° C) (*D* climates only).

h Dry-hot; mean annual temperature over 64.4° F (18° C) (*B* climates only).

k Dry-cold; mean annual temperature under 64.4° F (18° C) (*B* climates only).

As an example of a complete Köppen climate code, *BWk* refers to *cool desert climate,* and *Dfc* refers to *cold, snowy forest climate with cool, short summer.*

Figure 8.12 is a world map of Köppen climates.

Vegetation, soils, and climate

Plant geographers are keenly aware that plants are highly responsive to differences in climate. With each plant species is associated a particular combination of climatic factors most favorable to its growth, as well as certain extremes of heat, cold, or drought beyond which it cannot survive. Plants tend to adapt their physical forms to meet the stresses of climate. We find a wide range of forms taken by assemblages of the dominant plant species. These overall form patterns or habits closely reflect climate controls.

Since the late nineteenth century, soil scientists have recognized that the several fundamental classes of mature soils, seen on a worldwide scope, are more strongly controlled by climatic elements than by any other single factor. But plants also contribute to determining the

properties of the very soils on which they depend.

To develop the concept of an interplay among climate controls, vegetation, and soils, we must make a special study of water within the soil layer. While precipitation is an essential climate ingredient, what really counts in the growth of plants is the amount of water available in the soil zone where a plant's roots are situated. Evaporation from soil and plant surfaces can return a large proportion of the incoming precipitation to the atmosphere. Within the soil there is a balance between incoming and outgoing water. As in the case of the energy balance, there can be seasonal water surpluses and water deficits. In the next chapter we shall investigate the soil-water balance as an important step in increasing our understanding of the varied environments of the life layer.

The Soil-Water Balance

Chapter 9

A BASIC CONCEPT of physical geography is that the availability of water to plants is a more important factor in the environment than precipitation itself. Much of the water received as precipitation is lost in a variety of ways and is not usable for plants. Just as in a fiscal budget, when the monthly or yearly loss of moisture exceeds the precipitation, a budgetary deficit results; when precipitation exceeds the losses, a budgetary surplus results. An analysis of the water budget of a given area of the earth's surface is actually arrived at in much the same way as the understanding of a fiscal budget. The accounting requires only additions and subtractions of amounts within fixed periods of time, such as the calendar month or year.

To understand why such moisture gains and losses occur, we need to study the physical processes affecting water in its vapor, liquid, and solid states not only in the atmosphere but also in the soil and rock and in exposed water

of streams, lakes, and glaciers. The science of *hydrology* treats such relationships of water as a complex but unified system on the earth. The primary concern of this chapter is the hydrology of the soil zone.

Surface and subsurface water

We classify water of the lands according to whether it is *surface water,* flowing exposed or ponded on the land, or *subsurface water,* occupying openings in the soil or rock. Water held in the soil within a few feet of the surface is the *soil water;* it is the particular concern of the plant ecologist, soil scientist, and agricultural engineer. Water held in the openings of the bedrock is referred to as ground water and is studied by the geologist. Surface water and ground water are the subjects of Chapter 19.

The hydrologic cycle and the global water balance

Water of oceans, atmosphere, and lands moves in a great series of continuous interchanges of both geographic position and physical state, known as the *hydrologic cycle* (Figure 9.1). A particular molecule of water might, if we could trace it continuously, travel through any one of a number of possible circuits involving alternately the water vapor state and the liquid or solid state.

The pictorial diagram of the hydrologic cycle given in Figure 9.1 can be quantified for the earth as a whole. Figure 9.2 is a mass-flow diagram relating the principal pathways of the water circuit. We can start with the oceans, which are the basic reservoir of free water. Evaporation from the ocean surfaces totals about 109,000 cu mi per year. (Metric equivalents are shown in table in Figure 9.2). At the same time, evaporation from soil, plants, and water surfaces of the continents totals

figure 9.1

The hydrologic cycle traces the various paths of water from oceans, through atmosphere, to lands, and return to oceans.

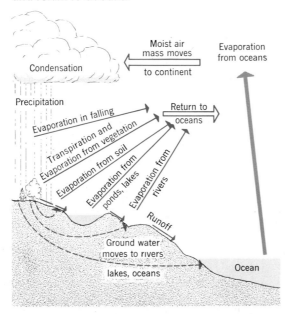

about 15,000 cu mi. Thus the total evaporation term is 124,000 cu mi; it represents the quantity of water that must be returned annually to the liquid or solid state.

Precipitation is unevenly divided between continents and oceans; 26,000 cu mi is received by the land surfaces and 98,000 cu mi by the ocean surfaces. Notice that the continents receive about 11,000 cu mi more water as precipitation than they lose by evaporation. This excess quantity flows over or under the ground surface to reach the sea; it is collectively termed *runoff*.

We can state the *global water balance* as follows.

$$P = E + R$$

where P = precipitation
E = evaporation
R = runoff (positive when out of the continents, negative when into the oceans)

Using the figures given for the continents

$$26,000 = 15,000 + 11,000$$

and for the oceans

$$98,000 = 109,000 - 11,000$$

For the globe as a whole, combining continents and ocean basins, the runoff terms cancel out.

$$26,000 + 98,000 = 15,000 + 109,000$$
$$124,000 = 124,000$$

Global water in storage

The total global water resource is stored in the gaseous, liquid, and solid states. Figure 9.3 gives a breakdown of storage quantities. Water of the oceans constitutes over 97 percent of the total, as we would expect. Next comes water in storage in glaciers, a little over 2 percent. Of the remaining quantity, about two thirds is

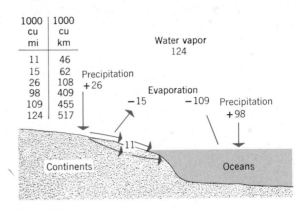

1000 cu mi	1000 cu km
11	46
15	62
26	108
98	409
109	455
124	517

figure 9.2
The global water balance. Figures give average annual water flux in and out of world land areas and world oceans. (Data of M. I. Budyko, 1971.)

ground water, so that surface water in lakes and streams is a very small quantity, indeed. Yet it is this surface water, together with the very small quantity of soil water, that sustains all life of the lands. Some environmentalists consider that fresh surface water will prove to be the limiting factor in the capacity of our planet to support the rapidly expanding human population. The quantity of water vapor held in the atmosphere is also very small — only about ten times greater than that held in streams — yet this atmospheric moisture is the source of all fresh water of the lands.

Infiltration and runoff

Most soil surfaces in their undisturbed, natural states are capable of absorbing the water from light or moderate rains. This absorption process is known as *infiltration* (Figure 9.4). Most soils have natural passageways between poorly fitting soil particles, as well as larger openings such as earth cracks, resulting from soil drying, borings of worms and animals, cavities left from decay of plant roots, or openings made by growth and melting of frost crystals. A mat of decaying leaves and stems breaks the force of falling drops and helps to keep these openings

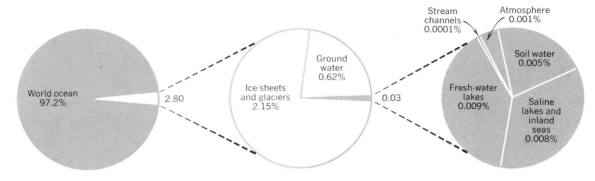

figure 9.3

The total volume of global water in storage is largely held in the world ocean.

clear. If rain falls too rapidly to be passed downward through these soil openings, the excess amount flows as a surface water layer down the direction of ground slope. This surface runoff, called overland flow, is described in Chapter 19.

Evaporation and transpiration

Between periods of rain, water held in the soil is gradually given up by a twofold drying process. First, direct evaporation into the open air occurs at the soil surface and progresses downward. Air also enters the soil freely and may actually be forced alternately in and out of the soil by atmospheric pressure changes. Even if the soil did not "breathe" in this way, there would be a slow diffusion of water vapor surfaceward through the open soil pores. Ordinarily only the first foot (30 cm) of soil is dried by evaporation in a single dry season but, in the prolonged drought of deserts, a dry condition extends to depths of many feet.

Second, plants draw the soil water into their systems through vast networks of tiny rootlets. This water, after being carried upward through the trunk and branches into the leaves, is discharged through leaf pores into the atmosphere in the form of water vapor. The process is termed *transpiration*.

In studies of climatology and hydrology it is convenient to use the term *evapotranspiration* to cover the combined moisture loss from direct evaporation and the transpiration of plants. The rate of evapotranspiration slows down as soil moisture supply becomes depleted during a dry summer period, because plants employ various devices to reduce transpiration. In general, the less moisture remaining, the slower is the loss through evapotranspiration.

Figure 9.5 illustrates the various terms explained up to this point and serves to give a more detailed picture of that part of the

figure 9.4

Rainfall, infiltration, and overland flow. (From A. N. Strahler, 1971, The Earth Sciences, 2nd ed., Harper & Row, New York.)

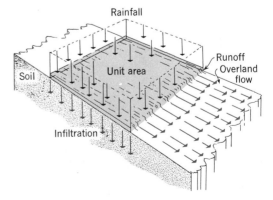

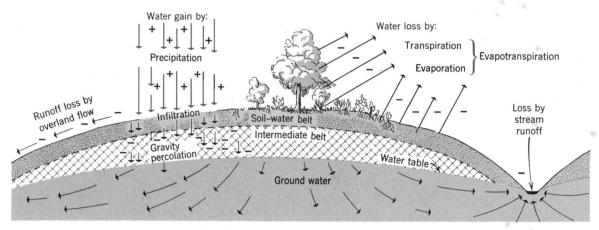

figure 9.5
The soil-water belt occupies an important position in the hydrologic cycle.

hydrologic cycle involving the soil. The soil layer from which plants can draw moisture is the *soil-water belt*. This belt gains water through precipitation and infiltration. As the minus signs show, the soil loses water through transpiration, evaporation, and overland flow. Excess water also leaves the soil by downward *gravity percolation* to the ground water zone below. Between the soil-water belt and the ground water zone is an *intermediate belt*. Here water is held at a depth too great to be returned to the surface by evapotranspiration, since it is below the level of plant roots.

Ground water and stream flow

Gravity percolation carries excess water down to the ground water zone, in which all pore spaces are fully saturated (Figure 9.5). Within this zone water moves slowly in deep paths, eventually emerging by seepage into streams, ponds, lakes, and oceans.

Excess water leaves the area by stream flow. Streams are fed directly by overland flow in periods of heavy, prolonged rain or rapid snowmelt. Streams that flow throughout the year — perennial streams — derive much of

their water from ground water seepage. Streams fed only by overland flow run intermittently, and their channels are dry much of the time between rain periods. We shall investigate ground water flow and stream flow in detail in Chapter 19.

Water in the soil

When infiltration occurs, the water is drawn downward by gravity through the soil pores, wetting successively lower layers. This activity is called *soil-water recharge*. Soon the soil openings are filled with water moving downward, except for some air entrapped in the form of bubbles. Then the percolation continues downward through the underlying intermediate belt.

Suppose now that the rain stops and several days of dry weather follow. The excess soil water continues to drain downward, but some water clings to the soil particles and effectively resists the pull of gravity because of the force of capillary tension. We are all familiar with the way in which a water droplet seems to be enclosed in a "skin" of surface molecules, drawing the droplet together into a spherical shape, so that it clings to the side of a glass indefinitely without flowing down. Similarly, tiny films of water adhere to the soil grains, particularly at the points of grain contacts, and will stay until disposed of by evaporation or by absorption into plant rootlets.

When a soil has first been saturated by water, then allowed to drain under gravity until no more water moves downward, the soil is said to be holding its *storage capacity* of water. Drainage takes no more than two or three days for most soils. Most excess water is drained out within one day. Storage capacity is measured in units of depth, usually inches or centimeters, just as with precipitation. This means that for a given cube of soil, say 12 in. on a side (1 cu ft), if we were to extract all of the stored water, it might form a layer of water 3 in. deep in a pan of 1 sq ft. This would be equivalent to complete absorption of a 3-in. rainfall by a completely dry 12-in. layer of soil.

Storage capacity of a given soil depends largely on its texture. Sandy soil has a very small storage capacity; clay soil has a large capacity.

The soil-water cycle

We can turn next to consider the annual water budget of the soil. Figure 9.6 shows the annual cycle of soil water for a single year at an agricultural experiment station in Ohio. This example can be considered generally representative of conditions in humid, mid-latitude climates where there is a strong temperature contrast between winter and summer. Let us start with the early spring (March). At this time the evaporation rate is low, because of low air temperatures. The abundance of melting snows and rains has restored the soil water to a surplus quantity. For two months the quantity of water percolating through the soil and entering the ground water keeps the soil pores nearly filled with water. This is the time of year when one encounters soft, muddy ground conditions, whether driving on dirt roads or walking across country. This, too, is the season when runoff is heavy, and major floods may be expected on larger streams and rivers. In terms of the soil-water budget, a *water surplus* exists.

figure 9.6

A typical annual cycle of soil-water change in the Middle West shows a short period of surplus in the spring and a long period of deficit during the summer and fall. (Data of Thornthwaite and Mather.)

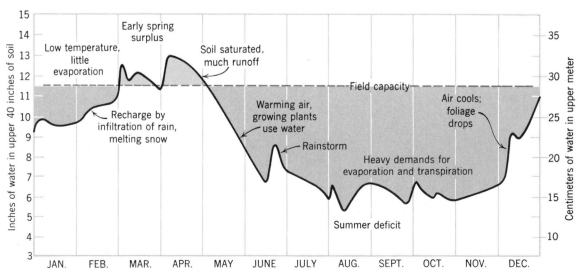

By May, the rising air temperatures, increasing evaporation, and full growth of plant foliage bring on heavy evapotranspiration. Now the soil water falls below the storage capacity, although it may be restored temporarily by unusually heavy rains in some years. By midsummer a large _water deficit_ exists in the water budget. Even the occasional heavy thunderstorm rains of summer cannot restore the water lost by heavy evapotranspiration. Small springs and streams dry up, and the soil becomes firm and dry. By November (and sometimes in September), however, the soil water again begins to increase. This is because the plants go into a dormant state, sharply reducing transpiration losses. At the same time, falling air temperatures reduce evaporation. By late winter, usually in February at this location, the storage capacity of the soil is again fully restored.

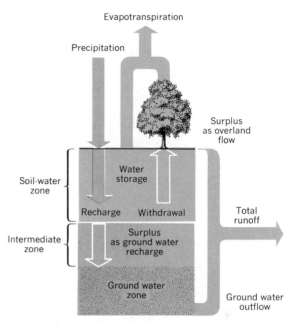

figure 9.7

Schematic diagram of the soil-water balance in a soil column.

The soil-water budget

From the example of soil-water change in a single year, we move forward to a more generalized concept. The gain, loss, and storage of soil water are accounted for in the _soil-water budget._ Figure 9.7 is a flow diagram to illustrate the budget. Water held in storage in the soil-water zone is increased by recharge during precipitation, but decreased by use through evapotranspiration. Surplus water is disposed of by downward percolation or by overland flow.

To proceed, we must recognize two ways to define evapotranspiration. First is _actual evapotranspiration,_ which is the true or real rate of water vapor return to the atmosphere from the ground and its plant cover. Second is _potential evapotranspiration,_ representing the water vapor flux under an ideal set of conditions. One condition is that there be present a complete (or closed) cover of uniform vegetation consisting of fresh green leaves, and no bare ground exposed through that cover.

The leaf cover is assumed to have a uniform height above ground — whether the plants be trees, shrubs, or grasses. A second condition is that there be an adequate water supply, such that the storage capacity of the soil is maintained at all times. This condition can be fulfilled naturally by abundant and frequent precipitation, or artificially by irrigation. To simplify the ponderous terms we have just defined, they may be transformed as follows.

actual evapotranspiration is _water use_
potential evapotranspiration is _water need_

The word "need" signifies the quantity of soil water needed if plant growth is to be maximized for the given conditions of solar radiation and air temperature and the available supply of nutrients.

The difference between water use and water need is the _soil-water shortage._ This is the quantity of water that must be furnished by irrigation to achieve maximum crop growth within an agricultural system.

All of the terms in the soil-water budget can be stated in inches or centimeters of water depth, the same as for precipitation. A particular value of field capacity of the soil is established in advance. In the soil-water budgets presented here, it has been assumed that the soil layer has a storage capacity of 12 in. (30 cm).

A simplified soil-water budget is shown in Figure 9.8, a graph on which average monthly values are plotted as points and connected by smooth lines. In this example, precipitation is much the same in all months, with no strong yearly cycle. In contrast, water need shows a strong seasonal cycle: low values in winter and a high summer peak. For midlatitudes in a humid climate this model is about right.

At the start of the year a large water surplus exists, and this is being disposed of by runoff. By May, conditions have switched over to a water deficit. In this month plants begin to withdraw soil water from storage. *Storage withdrawal* is represented by the difference between the water-use curve and the precipitation curve. As storage withdrawal continues, the plants reduce their water use to

less than the optimum quantity, so that without irrigation the water-use curve departs from the water-need curve. Storage withdrawal continues throughout the summer. The deficit period lasts through September. The area labeled "soil-water shortage" represents the total quantity of water needed by irrigation to insure maximum growth throughout the deficit period.

In October precipitation again exceeds water need, but the soil must first absorb an amount equal to the summer storage withdrawal. So we have a period of *storage recharge*, and it lasts through November. In December the soil has reached its full storage capacity, and a water surplus again sets in, lasting through the winter.

C. Warren Thornthwaite, a distinguished climatologist who was concerned with practical problems of crop irrigation, developed these concepts of a soil-water balance and proposed a system of classification of world climates based on these principles. Associates and students of Thornthwaite extended his work and collected hydrologic data for a vast network of observing stations on all land

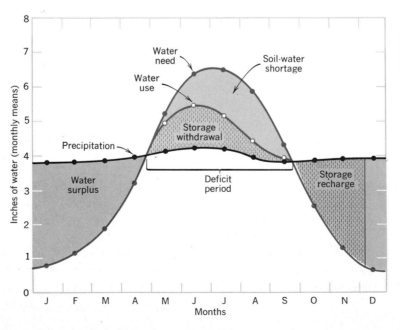

figure 9.8

A simplified soil-water budget typical of a humid midlatitude climate.

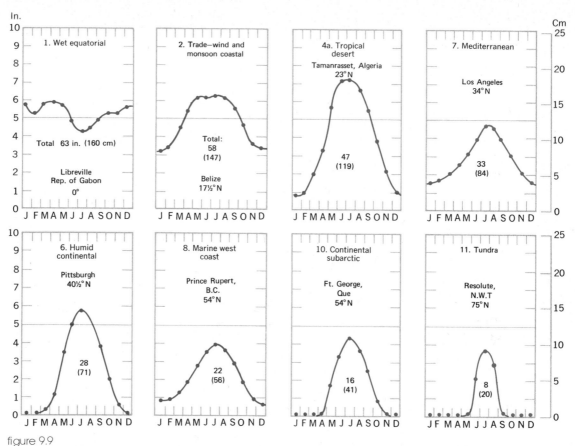

figure 9.9

(above and p. 150) Some representative annual cycles of water need (potential evapotranspiration) over a wide latitude range. (Data of C. W. Thornthwaite Associates.)

areas of the globe. Water need was calculated for thousands of stations by means of a formula using air temperature, precipitation, and latitude. Water need (potential evapotranspiration) is a difficult quantity to measure, and several methods have been developed to estimate its true value.

In a world beset by severe and prolonged food shortages, the Thornthwaite concepts and calculations are of great value in assessing the benefits to be gained by increased irrigation. Only in a few parts of the tropical and midlatitude zones is precipitation ample to fulfill the water need during the growing season. In contrast, the equatorial zone generally has a large water surplus throughout the year.

The global range of water need

Figure 9.9 gives some idea of the effect of latitude on the annual cycle of water need. Both air temperature cycles and insolation cycles are reflected in these cycles. Each chart has been labeled as to climate, according to the system given in Chapter 8.

In the wet equatorial climate (1) water need is high all year and the total is over 60 in. (150 cm). For the trade-wind and monsoon coastal climate (2) in the tropical latitude zone, there is a pronounced annual cycle, but a large annual total water need: almost 60 in. (150 cm). In the tropical desert climate (4a) the annual cycle is very strongly developed and the total water need remains high; nearly 50 in. (127 cm).

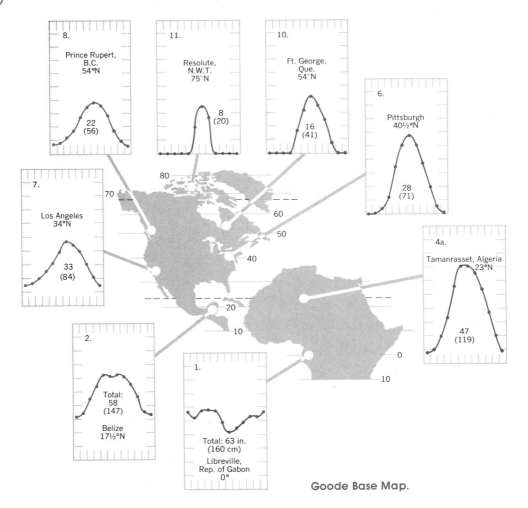

Goode Base Map.

In midlatitudes on the west coast, the Mediterranean climate (7) shows a well-developed annual cycle, but monthly values of water need remain high throughout the mild winter. In the humid continental climate (6) winter months have no water need, but the summer peak is high; the total is 30 in. (75 cm). However, in the marine west-coast climate (8), close to the Pacific Ocean, water need remains substantial throughout the winter because of the mild temperatures. Farther poleward in the continental subarctic climate (10) the number of months of zero water need increases to six, and the annual total water need diminishes to 16 in. (40 cm). North of the arctic circle, the tundra climate station (11)

shows nine consecutive winter months of zero water need and a very narrow summer peak. The total water need here is least of all.

The details of these examples are not important. What counts is the trends they show. Total annual water need diminishes from a maximum in the equatorial zone to a minimum in the arctic zone. At the same time, the annual cycle becomes stronger and stronger in the poleward direction. As the cold of winter makes its effects more pronounced, the number of months of zero water need increase. Of course, plant growth is essentially zero when soil water is frozen, so the growing season is nicely reflected in the graphs of water need. Water need persists through the winter far poleward along the western coastal zone, in contrast to the continental interior regions.

The global spectrum of the soil-water budget

Can the developing nations increase their food production enough to stave off starvation? This is a grave question in our times, and the answers given by well-informed specialists span the range from extreme pessimism to extreme optimism. Another question we hear is, "Will there be enough fresh water to supply the rapidly increasing demands of the energy-consuming industrial nations?" Your appreciation of current problems of agriculture and freshwater supply over the lands of the globe can be greatly increased by a study of some representative soil-water budgets.

Each of the ten soil-water budgets shown in Figure 9.10 is identified by name with one of the global climate types described in Chapter 8. These examples show how the soil-water budget can be meaningfully integrated into our climate system, based as it is on a combination of monthly mean air temperature and monthly mean precipitation. Each climate type has its distinctive soil-water budget.

Wet equatorial climate (1). The soil-water balance of the wet equatorial climate is well illustrated by Figure 9.10A. Januarete, Brazil, is located almost on the equator in the Amazon lowland, deep in the heart of South America. Water need is uniformly high throughout the year, because air temperatures are uniformly very warm and the sun is always high in the sky. Precipitation peaks in a three-month period (April through June), during which time the ITC is migrating northward over this latitude. Precipitation is much greater than water need in every month, so that a water surplus exists in every month. There is a tremendously large annual total water surplus. Soil moisture storage remains close to field capacity in all months. Plant growth can

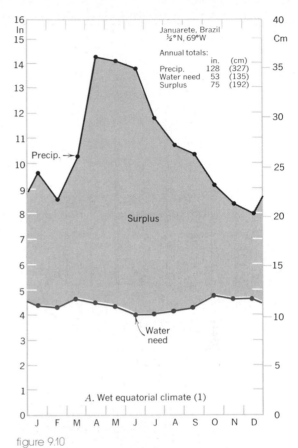

figure 9.10

Ten soil-water budgets, each representative of one of the climate types. (Data from Average Climatic Water Balance Data of the Continents, C. W. Thornthwaite Associates, Laboratory of Climatology, Publ. in Climatology, 1962-1965, Centerton, N.J.)

proceed at the maximum rate in all months; rainforest is the typical natural vegetation. Stream flow in this regime is copious throughout the year. In a mountainous region this large water surplus might be used for the development of hydroelectric power.

Trade-wind and monsoon coastal climate (2). Figure 9.10B illustrates the soil-water balance for a station in the Philippine Islands located in the trade-wind belt in the tropical zone. The precipitation cycle shows a strong annual cycle, peaking strongly following the high-sun season and falling to low values following the low-sun season. Water need also has a marked annual cycle, but it follows closely the

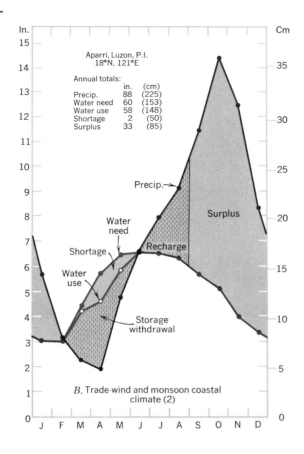

In. / Cm

Aparri, Luzon, P.I.
18°N, 121°E

Annual totals:
	in.	(cm)
Precip.	88	(225)
Water need	60	(153)
Water use	58	(148)
Shortage	2	(50)
Surplus	33	(85)

Precip.→

Water need

Surplus

Shortage

Recharge

Water use

Storage withdrawal

B. Trade-wind and monsoon coastal climate (2)

J F M A M J J A S O N D

insolation cycle with a maximum at summer solstice and a minimum at winter solstice. During the three-month period March through May, water need is greater than precipitation, so that soil water is withdrawn from storage and a small soil-water shortage is developed. However, this shortage is not severe in its effect on forest plants. Then, in June, as rainfall begins to increase, the soil-water storage is recharged and, by September, a surplus sets in. The total surplus is substantial — some 33 in. (85 cm) — and, during this period, streams flow full to their banks. There may also be heavy flooding, particularly when a typhoon strikes. In this climate rice is an important crop, growing during the wet season and maturing for harvest in the short dry season.

Tropical wet-dry climate (3). Figure 9.10C shows the soil-water budget for Calcutta, India, a great city located on the deltaic plain of the

Ganges River. This location is close to the tropic of cancer. Here is a budget of great contrasts. Precipitation is slight in the low-sun months (November to February), but rises to huge values in the rainy season (June to September). The annual curve of water need also shows a strong seasonality, being moderately high in coincidence with the rainy season. As a result, there is a large water surplus (July to September), but a severe soil-water shortage preceding the rains (March to May). Agriculture depends on these monsoon rains; if they do not materialize, the shortage can extend through the year, bringing crop failure. When the monsoon rains are exceptionally heavy, river floods can be an equally serious disaster, driving people from their homes and inundating their crops. This "feast-or-famine" budget is typical of those tropical countries of Asia and Africa most

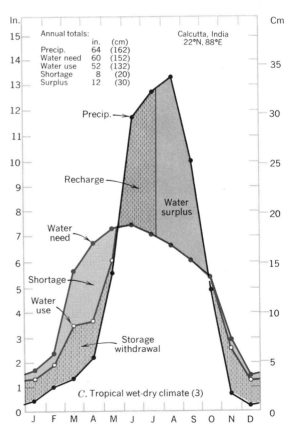

In. / Cm

Annual totals:
	in.	(cm)
Precip.	64	(162)
Water need	60	(152)
Water use	52	(132)
Shortage	8	(20)
Surplus	12	(30)

Calcutta, India
22°N, 88°E

Precip.

Recharge

Water surplus

Water need

Shortage

Water use

Storage withdrawal

C. Tropical wet-dry climate (3)

J F M A M J J A S O N D

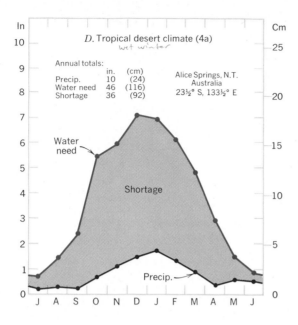

D. Tropical desert climate (4a)
wet winter

Annual totals:
	in.	(cm)
Precip.	10	(24)
Water need	46	(116)
Shortage	36	(92)

Alice Springs, N.T.
Australia
23½° S, 133½° E

Water need

Shortage

Precip.

peak and a marked drier period in the autumn. In contrast, the cycle of water need is strongly developed, with low values throughout the mild winter and a high peak in summer when insolation is at the maximum and air temperatures are high. The water surplus, which is large in the winter, gives way in summer to a four-month period of withdrawal of soil water from storage. However, the soil-water shortage is very small and plants have almost the full amount needed for maximum growth. There is little need for irrigation. Among the crops that are particularly well suited to this water budget are sweet potatoes, sugar cane, and rice.

Humid continental climate (6). Figure 9.10F is the soil-water budget of Pittsburgh, Pennsylvania, representative of the humid continental climate. Precipitation runs quite uniformly through the year, while water need has a strong annual cycle. Three cold winter months show little or no water need, because plants are dormant and soil water is frozen. In summer the water need exceeds precipitation by a moderate amount, but the accumulated soil-water shortage is not severe. A substantial water surplus occurs in winter and spring, so that the larger streams flow throughout the

severely beset with food shortages and famine.

Tropical desert climate (4a). Figure 9.10D illustrates the soil-water budget for a tropical desert station in the heart of the great Australian Desert. Notice that the sequence of months begins with July, so as to keep the cycles in the same phase as in the northern hemisphere. While rainfall is measurable in small amounts in all months, the values of water need are always larger. Consequently, a soil-water deficit prevails throughout all months of the year and racks up a very large total. Soil-water storage is close to zero at all times. It is easy to understand why vegetation is sparse in a desert such as this. Great quantities of irrigation water would be needed to support field crops, and much of this would be lost directly by evaporation from the soil. The cost benefits of such a program would need to be carefully weighed before launching a large water importation program.

Humid subtropical climate (5). Figure 9.10E is the soil-water budget of Baton Rouge, Louisiana, a typical station in the humid subtropical climate. Precipitation is substantial in all months, but there is a small midsummer

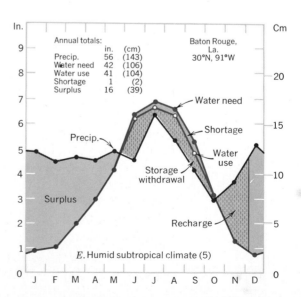

Annual totals:
	in.	(cm)
Precip.	56	(143)
Water need	42	(106)
Water use	41	(104)
Shortage	1	(2)
Surplus	16	(39)

Baton Rouge, La.
30°N, 91°W

Precip.

Water need

Shortage

Water use

Storage withdrawal

Surplus

Recharge

E. Humid subtropical climate (5)

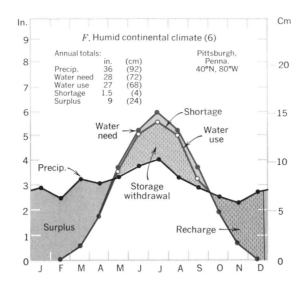

shore. Both precipitation and water need show strong annual cycles, but they are exactly out of phase. As a result, the summer peak of water need develops at just that season when precipitation drops close to zero for four consecutive months. The withdrawal of soil water from storage begins in spring and extends over an eight-month period. The soil-water shortage is severe during this time. Here is a region of summer-dry climate in which irrigation will be extremely valuable; in fact, summer crop cultivation will not be possible here without irrigation. The native plants are adapted to a long, warm summer with little or no rainfall. The winter is a period of soil-water recharge, but it is insufficient to generate a surplus.

Marine west coast climate (8). Figure 9.10*H* shows the soil-water budget for Cork, Ireland, a coastal city strongly influenced by moist air masses from the Atlantic Ocean. Precipitation is moderate in all months, but shows the winter maximum and summer minimum typical of midlatitude west coasts. The cycle of water need is in opposite phase, peaking in the summer when precipitation is least. Soil water is withdrawn from storage in June, July, and August, but the shortage is small, and irrigation would not bring an important advantage. There is a long period of water surplus throughout winter and spring.

year. Some water is held in reserve as snow, to be released rapidly in early spring. Forests thrive in this climate, drawing the necessary moisture from storage during the summer. Irrigation is not generally required for field crops such as corn and for pasture land. This soil-water budget is well suited to diversified farming and dairying.

Mediterranean climate (7). Figure 9.10*G* is the soil-moisture budget for Los Angeles, California, located in a semiarid coastal zone strongly influenced by a persistent subtropical high-pressure cell and a cool current just off

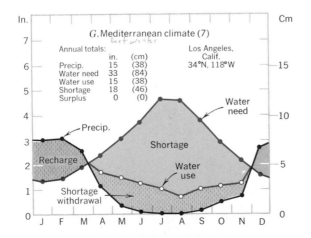

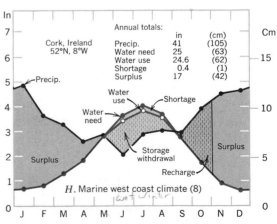

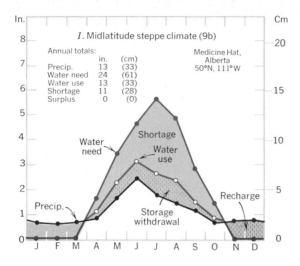

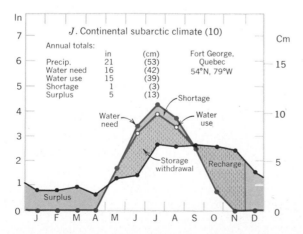

Midlatitude steppe climate (9b). Figure 9.10*I* is the soil-water budget for Medicine Hat, a station on the plains of Alberta. Located to the east of the Rocky Mountains, this is a rainshadow zone with respect to the prevailing westerly winds. Precipitation is small in all months, but there is a well-developed summer peak. Water need, which is zero during five months of cold winter, rises to a high peak in summer, when a large soil-water shortage is accumulated. Soil-water recharge through the winter months is not enough to produce a surplus. Grasslands (steppes) are typical of areas having this soil-water budget, since grasses can enter a dormant state when

soil-moisture declines to low levels or becomes frozen. Certain of the cereals, which are of the grass family, are well adapted to this climate. Of these, wheat is a principal crop. On the other hand, garden crops require heavy irrigation to survive.

Continental subarctic climate (10). Figure 9.10*J* shows the soil-water budget for Fort George, Quebec, a station on the east shore of James Bay at lat. 54° N. For six consecutive winter months the water need is zero. Small amounts of snow fall in these months, and it recharges the soil-water storage when spring melting occurs in May. The curve of water need is sharply peaked within a six-month period, and there is a very small soil-water shortage at the height of summer. A forest of needleleaf evergreen trees is native to this environment, but low air temperatures and shortness of the growing season rule out production of food crops.

Soil water and Man's food resources

This brief survey of soil-water budgets spans the full range of major budget types encountered over the habitable lands of the globe. Perhaps the most important lesson is that few of these budgets provide the full water need of food plants during the growing season, with neither too great a shortage nor too great a surplus. Severe soil-water shortages require irrigation to optimize crop production. On the other hand, large year-around surpluses of water lead to removal of plant nutrients, so that costly applications of fertilizers are needed. Irrigation in a dry climate has serious side effects, such as the accumulation of salts in the soil and the rise of ground water to saturate the soil. The soil-water budget sets stringent limitations on Man's expansion of agricultural resources. Intelligent planning for environmental management depends heavily on an understanding of all phases of the soil-water budget.

The Soil Layer

Chapter 10

THE SOIL IS the very heart of the life layer on the lands. Returning to the key concepts developed in our introductory statements about climate (Chapter 8), we identify the soil layer as a place in which plant nutrients are produced and held. As we emphasized in Chapter 9, the soil layer also holds water in storage for plants to use. The role of climate is to vary the input of water and heat into the soil. This same heat energy and water are responsible for the breakup and chemical change of rock to produce the parent mineral body of the soil. It is from this mineral matter that many of the plant nutrients are derived.

But climate acting on rock cannot make a soil layer capable of sustaining a rich plant cover. Plants themselves, together with many forms of animal life, play a major role in determining the qualities of the soil layer. Those qualities have evolved through centuries of time by the interaction of organic processes with physical and chemical soil processes. The organic processes include the synthesis of organic compounds, and these are eventually added to the body of the soil. Plants use the mineral nutrients to build complex organic molecules. Upon death of plant tissues, these nutrients are released and reenter the soil, where they are reused by living plants. So the concept of nutrient cycling by plants is one of the keys to understanding the development of the soil layer.

The soil is a dynamic layer in that many complex physical and chemical activities are going on simultaneously within it. These activities maintain the soil layer in a certain equilibrium condition over long periods of time. Because climate and plant cover vary greatly from place to place over the globe, the equilibrium condition of the soil is expressed very differently from place to place. It does not take an expert to realize that the soil beneath a spruce forest in Maine is quite different in color and structure from the soil beneath the prairie farmlands of Iowa. Yet, in each of these

localities, the soil has reached an equilibrium suitable to the climate controls and organic processes prevailing there.

The geographer is keenly interested in the differences in soils from place to place over the globe. The capability of a given soil to furnish food crops largely determines which areas of the globe support the bulk of the human population. Despite changes in population distribution made possible by technology and industrialization, most of the world's inhabitants still live where the soil furnishes them food. Many of those same humans die prematurely because the soil does not furnish enough food for all.

The soil

We use the word *soil* to mean a surface layer having distinctive zones, called *soil horizons,* and a set of physical, chemical, and organic properties enabling it to support plant growth. In this strict sense, the surface layer of large expanses of the continents cannot be called soil. For example, dunes of moving sand, bare rock surfaces of deserts and high mountains, and surfaces of fresh lava near active volcanoes do not have a soil layer. In time these areas may develop true soils but, in that process, the surface layers will be greatly altered.

The substance of the soil exists in all three states — solid, liquid, and gas. The solid portion consists of both inorganic (mineral) and organic substances. The liquid present in the soil is a complex solution capable of engaging in a multitude of important chemical reactions. Gases present in the open pores of the soil consist not only of the atmospheric gases, but also gases liberated by biological activity and chemical reactions within the soil. Soil science, often called *pedology,* is obviously a highly complex body of knowledge. We shall do no more than cover a few of the high points of this science.

The mineral soil

For most classes of soils, the bulk of the solid fraction of the soil consists of finely divided mineral matter. As Figure 10.1 shows, this mineral matter can be derived from solid rock (bedrock) through the process of rock weathering. Rock weathering involves two major changes: (1) physical breakdown into smaller and smaller particles; and (2) chemical alteration of the rock-forming minerals. These weathering processes are described in detail in Chapter 18. Weathering reduces the solid rock into *regolith*, a layer of soft mineral matter incapable of supporting plant growth, but capable of being transformed into soil. Other forms of regolith consist of mineral particles transported to a place of rest by action of streams, glaciers, and waves. In all of its varied forms and origins, regolith is the parent matter of which the soil is formed.

The mineral fraction of the soil usually spans a very wide range of particle sizes. The term *soil*

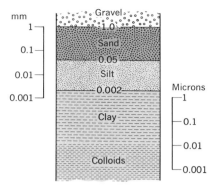

figure 10.2
Size grades used in the description of soil textures. (U.S. Dept. of Agriculture system.)

texture refers to the proportion of particles falling into each of several size grades. Metric units of length are used in describing particle grades. In measuring soil texture, gravel and larger particles (diameter over 1 mm) are eliminated, since these play no important role in soil processes. The remaining grades are classified as *sand, silt,* and *clay*. The diameter range of each of these grades is shown in Figure 10.2. Millimeters are the standard units down to 0.001 mm, when the scale shifts to microns. Each unit on the scale represents a power of ten, so that clay particles of 0.001 microns diameter are one ten-millionth as large as sand grains 1 mm in diameter.

Soil texture is described by a series of names emphasizing the dominant constituent, whether sand, silt, or clay. Figure 10.3 gives examples of five soil textures with typical percentage compositions. A *loam* is a mixture containing a substantial proportion of each of the three grades. Loams are classified as sandy, silty, or clay-rich when one of the grades is dominant.

Texture is important because it largely determines the ability of the soil to retain water or to transmit water to the intermediate belt below. Recall that storage capacity of a soil is

figure 10.1
The true soil is characterized by distinct horizons and the capability to support plants. Below is infertile regolith, derived from the underlying bedrock.

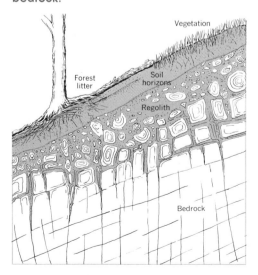

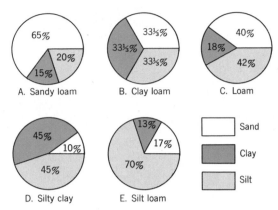

figure 10.3
Typical compositions of five soil texture classes.

its capacity to hold water against the pull of gravity. Figure 10.4 shows how storage capacity varies with soil texture. Pure sand holds the least water, while pure clay holds the most. Loams hold intermediate amounts. Sand transmits the water downward most rapidly; clay most slowly. When planning the quantity of irrigation water to be applied, these factors must be taken into account. Sand reaches its

figure 10.4
Storage capacity and wilting point vary according to soil texture.

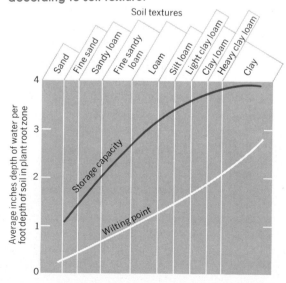

capacity very rapidly, and added water is wasted. Clay-rich loams take up water very slowly and, if irrigation is too rapid, water will be lost by surface runoff. By the same token, sandy soils require more frequent watering than clay-rich soils. (The organic content of a soil also strongly affects its water-holding capacity.)

Soil texture is largely an inherited feature of a given soil and depends on the composition of the parent regolith and underlying rock. Some rocks furnish a large spread in particle sizes, others yield mostly sand or mostly clay. The intermediate loam textures are generally best as agricultural soils because they drain well, but also have favorable water retention properties.

Soil colloids

Colloids play a vital role in the soil. _Soil colloids_ consist of particles smaller than about 0.1 micron. These particles include not only the mineral colloids, which are inorganic, but also organic colloids. _Humus_ is finely divided, partially decomposed organic matter found resting on the soil surface and disseminated through the upper horizons. Humus particles of colloidal dimensions are gradually carried down by percolating soil water to reach lower layers. These particles give the soil its brown and black coloration.

If you examine mineral colloids under the electron microscope, you will find that they consist of thin, platelike bodies (Figure 10.5). When well mixed in water, particles of this small size remain suspended indefinitely, giving the water a murky appearance.

Soil water contains a wide variety of ions. An _ion_ is an electrically charged atom or group of atoms. Compounds that dissolve in water break up into ions. A familiar example is ordinary table salt, sodium chloride, with the formula NaCl. One atom of sodium (Na) is united with

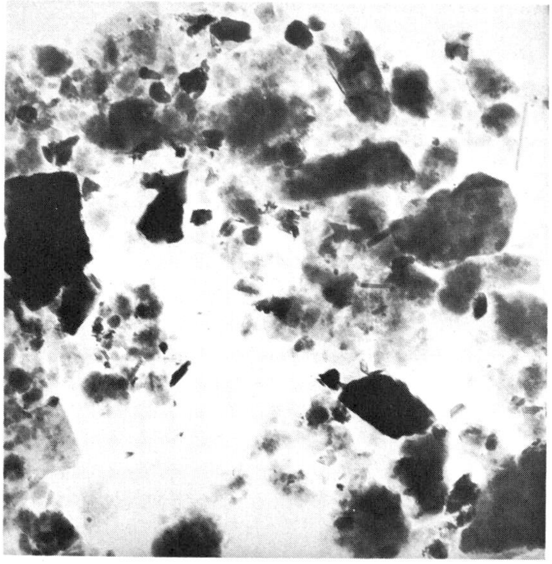

figure 10.5

Seen here enlarged about 20,000 times are tiny flakes of the clay minerals of colloidal dimensions. The minerals are illite (sharp outlines) and montmorillonite (fuzzy outlines). These particles have settled from suspension in San Francisco Bay. (San Francisco District Corps of Engineers, U.S. Army.)

one atom of chlorine (Cl) in a single molecule of salt. When the salt is dissolved, the two atoms are separated and become ions. The sodium ion bears a positive charge and is referred to as a _cation_. The chlorine ion bears a negative charge.

Getting back to the soil colloids, we find that each particle holds a surface electrical charge, as shown in Figure 10.6. Colloids of mineral origin have negatively charged outer surfaces. These negative charges attract and hold such cations as may be present in the soil-water solution. The numerous plus signs in the diagram designate cations held by the colloidal particle. Organic colloids — humus particles, that is — are also capable of holding ions.

Among the many ions in the soil solution, one important group consists of plant nutrients. Known as _base cations_, or simply _bases_, these are principally ions of four metallic elements:

calcium, magnesium, potassium, and sodium. Colloids hold these ions, but also give them up to plants; they enter the plant tissue through root membranes. Were it not for the ion-holding ability of soil colloids, most of these vital nutrients would be carried out of the soil by percolating water and would be taken out of the region in streams, eventually reaching the sea. This leaching process goes on continually in climates having a substantial water surplus, but loss is greatly retarded by the ion-holding capacity of soil colloids.

Soil acidity and alkalinity

The soil solution also contains hydrogen ions, and these are also positively charged ions, or cations. The presence of hydrogen ions in the soil solution tends to make the solution acid in chemical balance. On the other hand, an abundance of the base cations tends to make the soil solution alkaline in balance. An important principle of soil chemistry is that the hydrogen ion has the power to replace the base cations clinging to the surfaces of the soil colloids. As hydrogen ions accumulate, the base cations are carried away by leaching and lost. When this happens, the soil acidity is increased. At the same time, loss of nutrient base cations reduces the fertility of the soil.

The degree of acidity or alkalinity of a solution is designated by the pH number. The lower the pH number, the greater is the degree of acidity.

figure 10.6

A colloidal particle with negative surface charges and a layer of positively charged ions.

Layer of positively charged ions
+ + + + + + + + + + + + + + +
– – – – – – – – – – – – – – –
Clay particle
– – – – – – – – – – – – – – –
+ + + + + + + + + + + + + + +

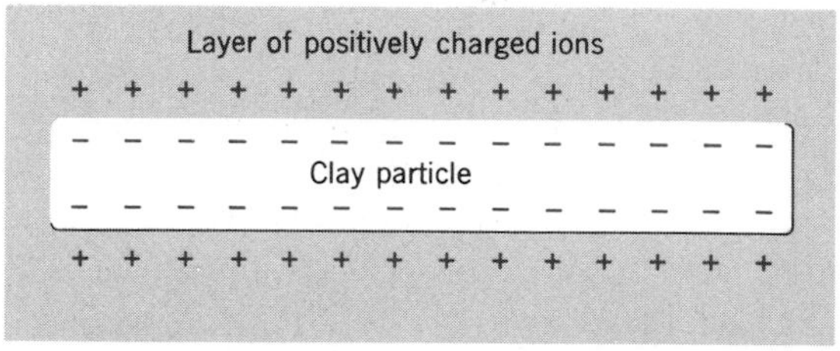

A pH number of 7 represents a neutral state; higher values are in the alkaline range. Table 10.1 shows the natural range of acidity and alkalinity found in soils. High soil acidity is typical of cold, humid climates; soil alkalinity is typical of dry climates. Acidity can be corrected by the application of lime, an oxide of calcium. This is an important agricultural practice because it reduces the hydrogen ion content of the soil. Then, when fertilizers are applied, the base cations can reoccupy positions on the soil colloids.

Soil aggregates

Soil structure refers to the way in which soil grains are grouped together into larger masses, or aggregates. Soil aggregates range in size from small grains to large clods or blocks. Small aggregates of roughly spherical shape give the soil a granular structure or crumb structure, illustrated in Figure 10.7. Blocky structure is also illustrated in Figure 10.7. One factor contributing to formation of soil aggregates is the binding effect of soil colloids. Colloid-rich clays shrink greatly in volume as they dry out, and shrinkage results in formation of soil cracks, defining the surfaces of the aggregates. Other aggregates commonly found in soils take the form of platelike bodies and elongate prisms.

Soils with well-developed granular or blocky structure are easy to till; this is an important agricultural factor in lands where primitive plows, drawn by animals, are still widely used. Soils in which the clay colloids are in a dispersed state lack aggregates. These soils are sticky and heavy when wet and are difficult to cultivate. When dry, they become too hard to be worked.

table 10.1
Soil acidity and alkalinity*

pH	4.0 4.5 5.0	5.5	6.0 6.5 6.7	7.0	8.0	9.0 10.0	11.0
Acidity	Very strongly acid	Strongly acid / Moderately acid	Slightly acid	Neutral	Weakly alkaline / Alkaline	Strongly alkaline	Excessively alkaline
Lime requirements	Lime needed except for crops requiring acid soil	Lime needed for all but acid-tolerant crops	Lime generally not required	No lime needed			
Occurrence	Rare / Frequent	Very common in cultivated soils of humid climates			Common in sub-humid and arid climates	Limited areas in deserts	
Soil groups	Podzols	Gray-brown podzolic soils / Tundra soils	Brown forest soils / Prairie soils / Latosols		Chestnut and brown soils / Tropical black soils	Desert alkali soils	

*Based on data of C. E. Millar, L. M. Turk, and H. D. Foth (1958), *Fundamentals of soil science,* third edition, John Wiley and Sons, New York, 526 pp. See Chart 4.

figure 10.7
Granular structure (left) and blocky structure (right). The length of the bar is 1 in. (Division of Soil Survey, U.S. Dept. of Agriculture.)

Soil processes and the water balance

Climate, in a broad sense, is the most important factor in developing the distinctive characteristics of the soil layer. One characteristic is the presence of soil horizons. These horizons are usually explained by either selective removal or accumulation of certain ions, colloids, and chemical compounds. Development of a distinctive color and texture is also, to some extent, controlled by climate, although these are often inherited characteristics derived from the parent mineral matter.

Let us focus on processes of selective removal and accumulation of matter within the soil. Soils formed in regions of large water surplus are subject to selective removal of matter as the excess water percolates through the soil to the ground water zone and is eventually disposed of as stream runoff. However, in many cases, matter carried downward simply accumulates at a lower level in the soil, where it forms a distinctive horizon (Figure 10.8). the downward transport process is called

eluviation, or more simply, _leaching_. Eluviation produces a distinct soil horizon from which the matter has been removed; it is called the _A horizon_. Accumulation of matter in the underlying zone is called _illuviation;_ it forms the _B horizon_.

Upward migration of ions from lower to higher positions in the soil is characteristic of dry climates having a large soil-moisture shortage and little or no water surplus. During dry

figure 10.8
Downward migration of soil constituents is characteristic of cool, humid climates having a large water surplus.

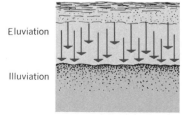

periods or a dry season, water is brought slowly surfaceward by capillary force, much as a cotton wick draws kerosene upward in a lamp. Ions are carried upward in the solution. As evaporation takes place within the soil, the ions are deposited in the solid state. These solids take the form of grains, nodules, plates, or layers. In extreme cases, the deposit is a rocklike slab. Commonest of the mineral deposits of this type is the compound calcium carbonate ($CaCO_3$).

In North America soils with accumulations of excess carbonate minerals are found west of a north-south line running about through central Texas and westernmost Iowa. The eluviation-illuviation process dominates east of this line, but is also active at high altitudes in the western mountain ranges, where there is a humid climate with a water surplus.

Soil processes and temperature

Soil temperatures, like air temperatures, span a wide global range from year-around warmth of the tropical and equatorial zones to the year-around cold of arctic and polar zones. Soil-forming processes respond strongly to the prevailing thermal environment, so that we can recognize a progression of changes in soil composition and horizon development from one thermal environment to the next. Let us begin in the arctic zone and work downward to low latitudes. Figure 10.9 is a generalized diagram of the approximate locations of the soil groups we shall describe.

The world soils map, Figure 10.10, shows those regions in which one might expect to find soils of each of the soil groups. In using this map, keep in mind that it is greatly

figure 10.9

Schematic diagram of ten major soil groups on an imaginary supercontinent. Compare with Figure 10.10.

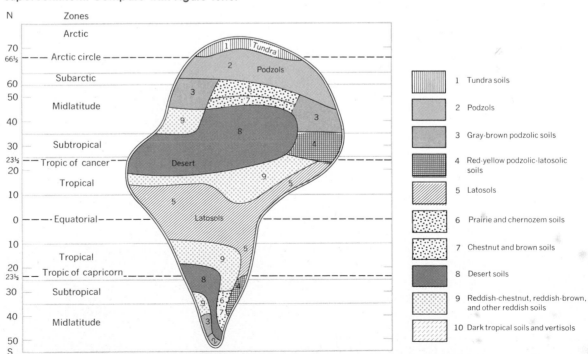

1 Tundra soils

2 Podzols

3 Gray-brown podzolic soils

4 Red-yellow podzolic-latosolic soils

5 Latosols

6 Prairie and chernozem soils

7 Chestnut and brown soils

8 Desert soils

9 Reddish-chestnut, reddish-brown, and other reddish soils

10 Dark tropical soils and vertisols

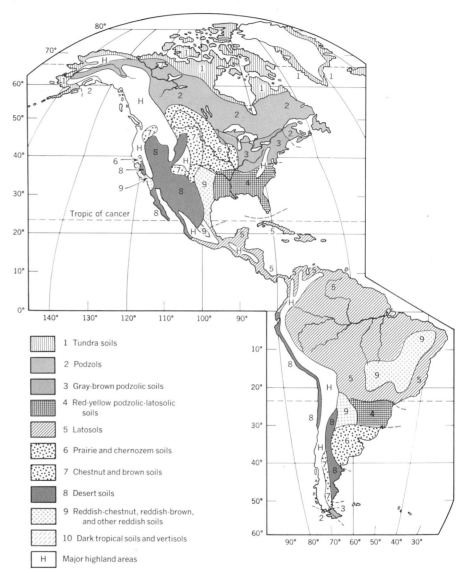

1	Tundra soils
2	Podzols
3	Gray-brown podzolic soils
4	Red-yellow podzolic-latosolic soils
5	Latosols
6	Prairie and chernozem soils
7	Chestnut and brown soils
8	Desert soils
9	Reddish-chestnut, reddish-brown, and other reddish soils
10	Dark tropical soils and vertisols
H	Major highland areas

figure 10.10

A generalized map of world soils. Use this map as a regional diagram only, for comparison with the world climate map, Figure 8.8. (Based on Goode Base Map.)

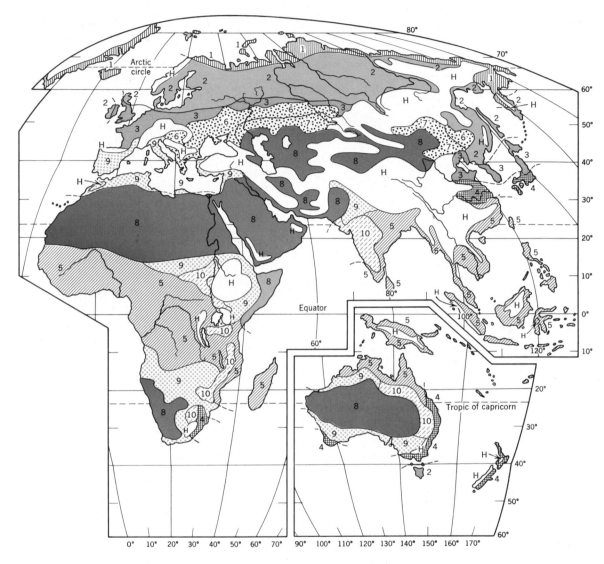

generalized and that in many parts of the world soils are not as yet well mapped.

Tundra soils

In the summerless tundra climate of lands fringing the Arctic Ocean, chemical reactions are greatly inhibited or entirely prevented by soil temperatures below freezing much of the year. Only the upper few inches of the soil are thawed during the brief summer, and water does not drain freely from the saturated soil. Seasonal freezing and thawing of this soil layer disrupt the soil structure, and distinctive soil horizons do not form. The distribution of tundra soils is shown on the world soils map.

Podzolization and the podzolic soils

Traveling next into the continental subarctic climate, we come to a great belt of needleleaf forest, composed of trees such as the spruce, larch, and fir. This climate has bitterly cold winters when all soil moisture is frozen, but there is a short summer during which plant growth and soil-forming processes are active. Fallen plant matter accumulates on the soil

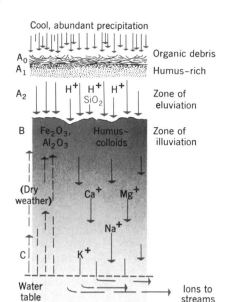

figure 10.11
Schematic diagram of the process of podzolization.

figure 10.12
This podzol profile has formed in sandy parent material in Maine. Depth scale in feet. (Soil Conservation Service, U.S. Dept. of Agriculture.)

surface in a thick layer, where it decays very slowly. Decay is a process of decomposition accomplished by soil bacteria, and these organisms have only a short period each year to do their work. The accumulated plant debris is slowly converted to humus but, in the process, organic acids are produced, and these have an important role in soil development. Percolating down through the zone of eluviation, acids remove the base cations, and they are ultimately exported from the area by runoff. Soil acids also react with compounds of aluminum and iron in the mineral soil and these are carried down to be deposited in a zone of illuviation. Humus particles are also removed from the eluviation zone and accumulate in the illuviation zone. Little remains in the eluviation zone but silica (silicon dioxide, SiO_2).

The result of these processes is to produce strongly defined A and B horizons (Figure 10.11). The A horizon is subdivided into two parts. An upper layer, the A_1 horizon, is a thin acid layer rich in humus and ranging in color from gray through yellowish brown to reddish brown. The zone of eluviation is called the A_2 horizon. It has a soft, friable texture because of removal of colloids. The color of the A_2 horizon is ash-gray or whitish-gray because such coloring agents as iron oxide and humus have been removed and only silica remains behind (Figure 10.12).

The B horizon is a brownish layer, enriched by colloids and bases and therefore having a dense, claylike consistency. Oxides of iron and aluminum are often so heavily concentrated in the B horizon that the soil particles are cemented into stony material called "hardpan."

Russian soil scientists, who were the pioneer investigators of soils, gave the name *podzol* to

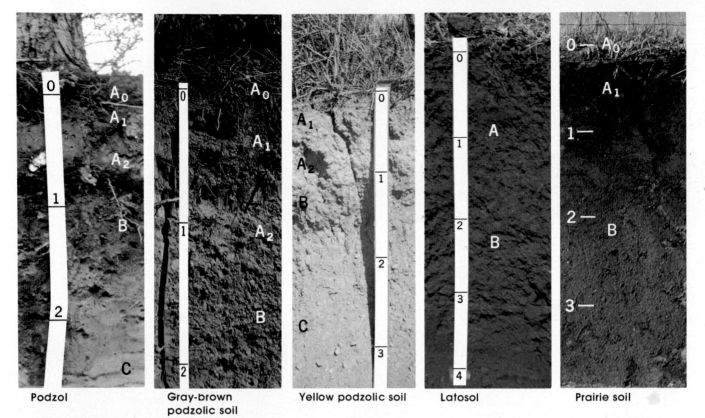

Podzol

Gray-brown
podzolic soil

Yellow podzolic soil

Latosol

Prairie soil

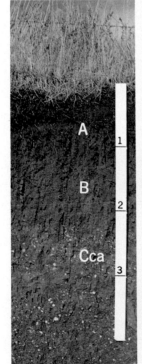

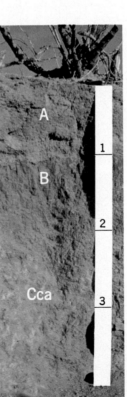

Soil Profiles

Soil Conservation Service,
U.S. Department
of Agriculture

Chernozem

Brown soil

Red desert soil

(below) **Red soil near Asheville, North Carolina. This soil is typical of red-yellow podzolic-latosolic soils widespread over the southeastern United States. (Orlo E. Childs)**

(left) **Sugar cane field on dark red latosol, Island of Kauai, Hawaii. The soil is formed from deeply weathered basaltic lava in a warm, humid climate. (Alan H. Strahler)**

Soils

(above) **Chernozem soil developed on thick loess (wind-deposited silt) near Vermilion, South Dakota. Prairie grassland is the native vegetation. (Arthur N. Strahler)**

(left) **Garden crops cultivated on black, peat-rich soil of a former glacial lake bed, New Jersey. This bog soil, or muck, requires special drainage to remove excess soil water. (Arthur N. Strahler)**

(right) **Gray desert soil ditched for irrigation, near Palm Springs, California. This sandy loam is fertile and easily cultivated. (Ned L. Reglein)**

(below) **Saline desert soil in the Devil's Cornfield, Death Valley, California. Salt bush, a salt-tolerant shrub, grows here on mounds of silt. (June Ryder)**

(above) This air-plant, or epiphyte, is a bromeliad supported on a tree branch. Everglades National Park, Florida. (Arthur N. Strahler)

(right) Rainforest of the western Amazon lowland, northwest of Iquitos, Peru. An exploration well is being drilled from a barge, floated 2300 miles up the Amazon River and its tributary, seen here, the Cuinico River. (Phillips Petroleum Company)

(right) Beech-hemlock forest, Great Smoky Mountains, Tennessee. Lightning has shattered the tree at the left. (Donaldson Koons)

(far right) Spruce-fir forest, Quebec. The ground surface, in deep shade, bears a thick layer of decomposing needles. The soil beneath is a podzol. (Alan H. Strahler)

Natural Vegetation

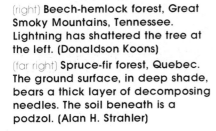

(above) Forest of Old World cedars (Cedrus atlantica) at 5000 ft (1500 m) altitude in the Atlas Mountains, Algeria, North Africa. (Arthur N. Strahler)

(left) Caribou herds grazing upon cotton-grass tundra in northern Alaska. (William R. Farrand)

(left) **Chaparral, a sclerophyll forest of shrubs and small trees, San Gabriel Mountains, southern California. (Arthur N. Strahler)**

(below) **Sawgrass savanna with scattered pines in the dry season (winter), Everglades National Park, Florida. Clumps of tropical forest (left) are known as hammocks. (Arthur N. Strahler)**

(left below) **Cattle and horses grazing on steppe grasslands, southwestern South Dakota. Beneath the grass cover is chestnut soil, developed on wind-deposited loess. (Arthur N. Strahler)**

(left) **Semi-desert vegetation of small shrubs and grasses on sandy soil, Canyon Lands of southeastern Utah. La Sal Mountains in the distance. (Alan H. Strahler)**

Vegetation of Dry Climates

(above) **This prickly-pear cactus grows on the rocky floor of a narrow canyon within the desert zone of vegetation. Havasu Canyon, Grand Canyon, Arizona. (Donald L. Babenroth)**

(right) **Desert shrub vegetation in the Sonoran Desert of northern Mexico. Expanses of barren, stony desert pavement occupy much of the surface. (Mark A. Melton)**

soils of the subarctic forest belt. In Russian the word means "ash soil" and refers to the ashlike appearance of the A_2 horizon. American soil scientists adopted the name podzol for this class of soils, and it has been used widely throughout the English-speaking world. The formative process we have described is often called *podzolization*. This process acts not only in the subarctic continental climate, but also at lower latitudes in the humid continental climate with cold winters. Podzolization acts beneath a cover of deciduous (leaf-shedding) forest over much of the northeastern part of the United States, but the effect is not so strong as in the type region to the north. Soils in this more southerly climate zone are referred to as *gray-brown podzolic soils*. The A_2 horizon is pale, but not intensively leached. These podzolic soils are acid in chemical balance, but not to the extent of the northern podzols.

The world map of soils (Figure 10.10), shows the general distribution of the podzolic soils. Besides the North American region we have already mentioned, the principal occurrence is in an enormous belt extending from western Europe to eastern Siberia. Podzolic soils are also found in Manchuria, Korea, and northern Japan.

The accompanying color plate shows the profiles of these and other soil types discussed in this chapter.

Laterization and the latosolic soils

As we move into lower latitudes, while at the same time staying in a humid climate, the effect of increasing temperature begins to make its impression on soil processes and the soil profile. Soils of the humid subtropical climate can be studied in the southeastern United States in a great belt running from Louisiana to the Carolinas. The most conspicuous feature of soils in this area is their reddish color, which contrasts with the rich greens of the plant cover. In sandy areas the color is yellowish. The red and yellow soil colors are a reflection of another major soil-forming process, called *laterization.* It is the dominant process of the tropical and equatorial zones where the climate is wet year around or has a well-developed wet season. Soils produced by laterization as classed as *latosols,* or *latosolic soils.*

The key to laterization lies in the intense activity of soil bacteria and other micro-organisms (particularly the fungi) in consuming nonliving plant matter. Under the equatorial thermal regime, air and soil temperatures are uniformly high year around. Moisture is also available for long periods. In this environment, the consumer organisms rapidly convert the fallen organic matter to its original components — mostly water and carbon dioxide.

Humus is conspicuously absent both on the soil surface and in the soil horizons. Consequently, the organic acids produced by humus are lacking. The soil is only weakly acid in chemical balance. Under these conditions oxides of iron and aluminum remain in the soil as residual materials and become relatively more abundant as other consituents are removed. Percolating soil water carries down silica (desilication) and most of the base cations, and these are removed from the region by streams (Figure 10.13).

Oxides accumulating in the process of laterization are of a stable type called *sesquioxides,* in which three atoms of oxygen are combined with two atoms of either iron or aluminum (Fe_2O_3, Al_2O_3). These oxides also contain water in their molecular structure. The sesquioxides of iron and aluminum (and sometimes also manganese) are remarkably stable and not easily soluble in the weakly acid soil water. Sesquioxide of iron is the principal cause of the typical reddish color of the latosolic soils.

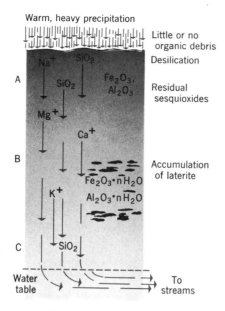

Warm, heavy precipitation

Little or no organic debris

Desilication

Residual sesquioxides

Accumulation of laterite

Water table

To streams

figure 10.13
A schematic diagram of the process of laterization, dominant in wet climates of low latitudes.

An interesting feature of latosols is the local development of accumulations of iron and aluminum sesquioxides into layers that can be cut out as building bricks. The material is called _laterite_. On exposure to the drying effects of the air, the blocks become very hard. In Indochina, particularly, laterite bricks have often been used as a building material.

Valuable mineral deposits occur as laterites. These are thick layers of minerals such as bauxite (hydrous aluminum oxide), limonite (hydrous iron oxide), and manganite (manganese oxide). They are classed as residual ores because they are not soluble in soil water and have continued to accumulate as the parent rock has weathered away and the silica and other soluble constituents have been removed.

Latosols of the wet equatorial climate and the trade-wind and monsoon coastal climate support a luxuriant rainforest, but the trees rely on nutrient bases released by consumption of nonliving plant materials. These nutrients are absorbed close to the soil surface and are not distributed at depth in the profile. As a result, latosols quickly lose their fertility under crop cultivation.

The world soils map (Figure 10.10) shows the distribution of the latosolic soils. They occur over vast areas of Central and South America, Africa, Southeast Asia, and the islands of the East Indies, and northern Australia.

Red-yellow podzolic-latosolic soils

Returning now to the zone of red and yellow soils of the southeastern United States, we recognize these as transitional soil types. They are intermediate between the podzols and the latosols, both in geographical location and in composition. As a group, they can be called _podzolic-latosolic soils_. They show weakly developed horizons but also have accumulations of the sesquioxides of iron and aluminum. Like the latosols, these soils are deficient in humus and are poorly endowed with nutrient bases. Fertilizers are needed when the soil is placed in crop cultivation. Where the parent materials are sandy, leaching of nutrients is severe, and these areas are best suited to cultivation of pine trees or are used for cattle grazing.

Dark tropical soils

Soil scientists who developed their unifying concepts of soils in Europe and North America for many decades gave little attention to soils of the tropical and equatorial zones. It was known that many regions within low latitudes have dark-colored soils, quite unlike the reddish latosols, but these were not well understood. One major area of dark soils lies in western India and seems closely correlated with a large surface exposure of a dark variety of igneous rock (basalt). Other important areas of dark soils occur in eastern Africa, and there is also a wide belt of dark soils in eastern Australia. Most of these soils are in the tropical wet-dry climate in which a very dry season alternates with a very wet season.

Recent studies have led to recognition of a new class of dark soils of the tropical zones. They have been named the *vertisols*. This coined word makes use of the Latin verb *"verto,"* meaning "to turn." Vertisols have a high content of colloidal clay that shrinks and swells greatly with alternate seasonal drying and wetting. During the dry season, the soil develops deep, wide cracks (Figure 10.14). When rains come, some of the surface soil drops into the cracks before they close, so that the soil slowly "swallows itself." Literally, the soil profile inverts itself by this process. The masses of soil between cracks move against one another as they become moist and plastic.

Vertisols show no horizons. Texture is fine, and the soil contains moderate amounts of organic matter. Vertisols are high in base cation content. However, the soil becomes highly plastic and difficult to till when wet, so that crop yields are low when cultivation depends on animal power alone. As a result, large areas of vertisols remain in the natural cover of grasses and scattered trees, and their use is limited to grazing. Application of modern technology and its energy resources could lead to substantial production of food and fiber from these vertisols.

Calcification
and the grassland soils

We turn next to soil-forming processes in the semiarid grasslands of the midlatitudes. These grasslands lie in the midlatitude steppe climate, which is transitional between the humid continental climate and the midlatitude desert climate. A type region for study of grassland

soils is the Great Plains of the United States and Canada. This geographic region extends from about the longitude of western Iowa westward to the Rockies and from Texas northward into the southern parts of Alberta and Saskatchewan.

The dominant soil-forming process of the semiarid grasslands is *calcification,* the accumulation of excess calcium carbonate ($CaCO_3$) in the soil. The evaporation process by which carbonate matter is brought upward by capillary water was explained in earlier paragraphs. The typical soil profile in this environment is illustrated in Figure 10.15. There is a thick, uniform *A* horizon, rich in humus and varying in color from dark brown or almost black to chestnut color or pale brown. Darkness of color corresponds to content of humus. The *A* horizon is fully penetrated by roots of grasses and annual plants (forbs). The horizon is rich in the nutrient bases, and these are taken up by roots and brought into plant tissues. As plant tissues decompose, these nutrients are released and make their way back into the *A* horizon during

figure 10.14

Deep soil cracks in a Texas vertisol, the Houston black clay. The crop shown here is cotton. (Soil Conservation Service, U.S. Dept. of Agriculture.)

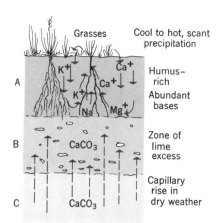

figure 10.15
Calcification is the dominant soil-forming process in midlatitude grasslands in a semiarid climate.

figure 10.16
A chernozem soil profile in North Dakota. (Soil Conservation Service, U.S. Dept. of Agriculture.)

moist periods when there is a water surplus. In this way the nutrients are recycled, and a high level of soil fertility is maintained.

Below the *A* horizon is a thick *B* horizon in which excess carbonate matter accumulates as minute grains or as nodules or platelike masses. The *B* horizon also holds humus and is brown or yellowish brown in color.

Because aridity increases from east to west across the Great Plains, we observe a sequence of grassland soils reflecting a decrease in humus and an increase in excess carbonate as we go from east to west. There is also a decrease in total thickness of the soil profile.

Our traverse should begin in the tall-grass prairies of Illinois. Here the climate is described as subhumid, with insufficient soil moisture for extensive forests. This is a neutral zone, with neither podzolization nor calcification a dominant process. Soils of this region are called *prairie soils*. Here the *A* and *B* horizons are combined into a single thick, dark layer of extremely high fertility. Today, of course, the prairie soils are largely under cultivation as croplands, and almost no vestiges of the original prairie grasslands remain.

Traveling westward across Iowa, we come to the zone of *chernozem soils*. These soils are also sometimes called "black earths." Chernozem is a Russian soil name adopted by Western soil scientists. Chernozem soils have distinct *A* and *B* horizons; both are dark brown in color (Figure 10.16). The *A* horizon has a granular structure such as that illustrated in Figure 10.7. The chernozem soils are among the most fertile of all global soils. Because the summers are hot and dry, with a soil-water deficit, the chernozem soils are best suited to cultivation of wheat and related cereals such as oats, barley, and rye. Enormous grain surpluses are exported from the chernozem areas of North America, the Ukraine, and Argentina. These regions are truly the world's breadbaskets and play a major role in feeding populations of the developing nations during famine periods.

Farther west, the chernozem soils grade into lighter-colored _chestnut soils_; these, in turn grade into _brown soils_ in eastern Colorado and Wyoming. The brown soils typically show a structure of elongate vertical prisms in the _B_ horizon (Figure 10.17).

Traced into lower latitudes — into the sub-tropical and tropical steppes — the chestnut and brown soils grade into _reddish-chestnut_ and _reddish-brown soils_. The reddish colors indicate some accumulation of iron oxide in the soil. These reddish soils also are widespread in the tropical wet-dry climate where the wet season is a short one. Followed into lower latitudes, the reddish soils grade into latosols of tropical wet-dry climates having a well-developed wet season.

figure 10.17

Profile of a brown soil in Colorado. The B horizon has good prismatic structure. (Soil Conservation Service, U.S. Dept. of Agriculture.)

Desert soils

Soils of the deserts — both hot deserts of the tropical desert climate and cold deserts of the midlatitude desert climate — are characterized by pale colors, which range from reddish to gray. Because plants are sparsely distributed and slow growing, very little humus is present in the soil. Horizons are only faintly developed. Calcification is a dominant process in the deserts and often leads to thick accumulations of calcium carbonate. Gypsum, another calcium compound, is also abundant in desert soils.

On the basis of color, desert soils are usually divided into _red desert soils_, tinted by minor amounts of iron oxides, and _gray desert soils_, largely free of coloration. Generally, the red desert soils are found in the tropical deserts, while the gray desert soils are found in the cold midlatitude deserts. Desert soils are rich in nutrient bases and can be highly productive under irrigation where the texture is not too coarse and drainage is good.

The world soils map (Figure 10.10) shows enormous expanses of desert soils in North America, Africa, Asia, and Australia. In South America these soils occupy narrow zones trending north-south, parallel with the Andes mountain chain.

To review the array of soils, Figure 10.18 shows a continuous schematic profile crossing the United States from Maine to Nevada. This representation disregards terrain features and the effects of high altitudes.

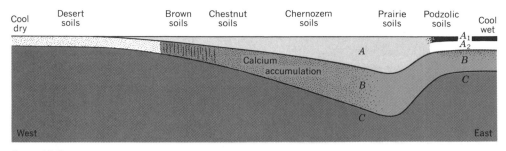

figure 10.18
A schematic soil profile from east to west across the United States in midlatitudes.

Soils of poorly drained locations

Throughout all parts of the continents there are places where soil drainage is poor. Such places include bogs, swamps, and marshes in the humid climates, and basinlike lowlands in the dry climates. Here the water table may lie close to the surface and may actually rise during wet weather to reach the soil layer. The major soil types we have described occupy upland surfaces that are well drained and with land slopes sufficient to carry off overland flow during wet periods. Obviously, special soil-forming conditions prevail in the poorly drained localities, and these conditions will be expressed in special features of the soil profile.

Soils of bogs and low-lying meadows in cold humid climates are characterized by massive accumulations of organic matter in the uppermost soil horizon. _Peat_ is a layer of partly decomposed plant remains, largely lacking in inorganic particles. Peat accumulates in saturated soils because bacterial action is inhibited. Oxygen in the soil water is depleted by bacterial action and, in some cases, only bacteria capable of life without oxygen (anaerobic bacteria) are present. Organic acids also accumulate in the peat layer. Acid-loving plants thrive on the soils of bogs in cold climates. A soil layer consisting of a mixture of organic and inorganic matter is typical of flat meadows adjacent to streams. These meadow soils are richly productive, but may require large additions of lime to neutralize the acidity.

Soils of low-lying basins in dry climates are modified through accumulation of salts. Here the soil-water budget is one of severe water shortage throughout much of the year. Where the water table lies close to the surface, water is drawn to the surface by capillary force, and mineral salts are left in the soil by evaporation.

Soils of these dry environments are divided into two major classes. The _saline soils_ are rich in salts chemically classified as chlorides, sulfates, carbonates, and bicarbonates of sodium, calcium, magnesium, and potassium. These soils are light colored and show poorly developed horizons. Although there are many species of plants adapted to saline soils, the plant cover is at best sparse. Agriculture is not possible on saline soils unless they are flushed out with large quantities of irrigation water to remove the salts. This has been done to advantage to reclaim much good agricultural land in the southwestern United States.

The second major group of poorly drained desert soils are the _alkali soils,_ in which sodium salts predominate. The soil profile is characterized by the presence of a dark, hard layer with prismatic structure. Although salts of the alkali soils are of somewhat different chemical properties from salts of the saline soils, both soils occur in the same areas. Alkali soils occupy areas of somewhat better drainage than the saline soils. Grass and shrub vegetation is found on alkali soils, but the species are particularly alkali resistant.

Global soils in review

We have painted the picture of global soils in broad brush strokes, leaving out much detail that would be meaningful. However, some important basic concepts have been presented and illustrated with specific examples. The dominant role of climate in governing the soil-forming processes is a persistent theme of this chapter. It is a theme that allows us to make good use of the concepts of the soil-water budget. Soils associated with water-surplus budgets are strikingly different in structure, composition, and fertility from those associated with budgets featuring a large soil-water shortage. Surprisingly, the most fertile soils from the standpoint of plant nutrients are those of the dry climates. A great abundance of water is not conducive to producing rich agricultural soils since, under such conditions, the nutrient bases are leached from the soil and can only be maintained by repeated application of fertilizers.

Soil character is also shaped by organisms living in and on the soil. Plant tissues supply humus that profoundly influences the soil profile. Decomposers prepare this humus. Moreover, plants recycle nutrient bases and so prevent their escape from the soil. Climate remains the dominant role through the input of heat and water needed for all the soil-forming processes, whether organic or inorganic, physical or chemical.

The Flow of Energy and Materials in the Biosphere

Chapter 11

ALL LIVING ORGANISMS of the earth, together with the environments with which those organisms interact constitute the *biosphere*. Organisms, whether belonging to the plant kingdom or to the animal kingdom, also interact with each other. Study of these interactions — in the form of exchanges of matter, energy, and stimuli of various sorts — between life forms and the environment is the science of *ecology*, very broadly defined. The total assemblage of components entering into the interactions of a group of organisms is known as an ecological system, or more simply, an *ecosystem*. The root "eco" comes from a Greek word connoting a house in the sense of household, which implies that a family lives together and interacts within a functional physical structure.

Ecosystems have inputs of matter and energy used to build biological structures, to reproduce, and to maintain necessary internal energy levels. Matter and energy are also exported from an ecosystem. An ecosystem tends to achieve a balance of the various processes and activities within it. For the most part these balances are quite sensitive and can be easily upset or destroyed. Physical geography meshes closely with ecology. We have already stressed the importance of organisms and their life processes in shaping the characteristics of the soil layer.

To the geographer, ecosystems are a part of the physical composition of the life layer. A forest, for example, is not only a living community of organisms; it is also a collection of physical objects on the land surface. A forest is physically very different from an expanse of prairie grassland. Plant geographers are aware of these place-to-place differences in the physical aspect of vegetation. They attempt to categorize the varied types; they map the global distribution of vegetation forms.

Geographers also view ecosystems as natural resource systems. Food, fiber, fuel, and structural material are products of ecosystems — they represent organic compounds placed in storage by organisms through the expenditure of energy basically derived from the sun. Geographers are interested in the influence of climate on ecosystem productivity. Our basic understanding of global climates and the global range of soil-water budgets will prove most useful in explaining the global pattern of ecosystems of the lands.

The ecosystem and the food chain

As an example of an ecosystem, consider a salt marsh (Figure 11.1). A variety of organisms are present: algae and aquatic plants, microorganisms, insects, snails, and crayfish, as well as larger organisms such as fishes, birds, shrews, mice, and rats. Inorganic components will be found as well: water, air, clay particles and organic sediment, inorganic nutrients, trace elements, and light energy. Energy transformations in the ecosystem occur by means of a series of steps or levels, referred to as a *food chain.*

The plants and algae in this chain are the *primary producers.* They use light energy to convert carbon dioxide and water to carbohydrates (long chains of sugar molecules) and eventually to other biochemical molecules needed for the support of life. This process of energy conversion is called photosynthesis. Organisms engaged in photosynthesis form the base of the food chain.

At the next level are the *primary consumers* (the snails, insects, and fishes); they live by feeding on the producers. At a still higher level are the *secondary consumers* (the mammals and birds); they feed on the primary consumers. As in many ecosystems, still higher levels of feeding are evident (the marsh hawks and owls). Feeding on dead organisms from all

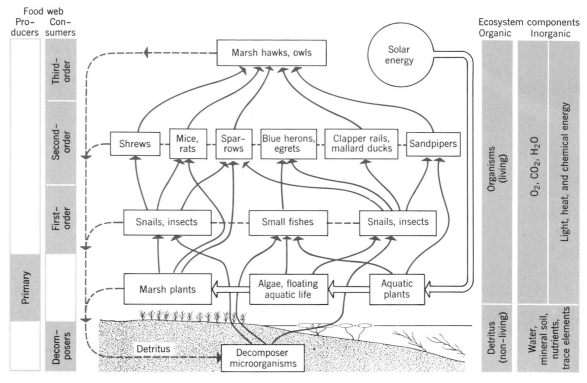

figure 11.1

Flow diagram of a salt marsh ecosystem in winter. Arrows show how energy flows from the sun to producers, consumers, and decomposers. (Food web after R. L. Smith, 1966, Ecology and Field Biology, Harper & Row, New York, p. 30, Figure 3.1.)

levels are the *decomposers;* they are mostly microorganisms and bacteria feeding on detritus (decaying organic matter).

The food chain is really an energy flow system, tracing the path of solar energy through the ecosystem. Solar energy is stored by one class of organisms in the chemical products of photosynthesis. As these organisms are consumed, chemical energy is released in metabolism. Metabolism is a general term for the sum of the biochemical processes within living cells; it includes both the building up and the breaking down of complex organic molecules. Chemical energy so released is used to power new biochemical reactions,

which again produce stored chemical energy in the bodies of the consumers.

At each level of energy transformation, energy is lost as waste heat. In addition, much of the energy input to each organism must be used in respiration. *Respiration* can be thought of as burning of fuel to keep the organism operating. Respiration is basically the oxidation of organic compounds, accompanied by release of heat energy. Energy expended in respiration is used for bodily maintenance and cannot be stored for use by other organisms higher up in the food web. This means that generally both the numbers of organisms and their total amount of living tissue must decrease drastically as one goes up the food chain.

Photosynthesis and respiration

Stated in the simplest possible terms, *photosynthesis* is the production of carbohydrate by the union of water (H_2O) and carbon dioxide (CO_2) while absorbing light energy. We can express carbohydrate by the formula CH_2O. The chemical reaction can then be written:

$$H_2O + CO_2 + \text{light energy} \rightarrow CH_2O + O_2$$

(handwritten: $C_6H_{12}O_6$)

Oxygen in the form of gas molecules (O_2) is a by-product of photosynthesis.

To state respiration in the simplest possible terms, we simply reverse the above reaction:

$$CH_2O + O_2 \rightarrow H_2O + CO_2 + \text{chemical energy}$$

(handwritten: $C_6H_{12}O_6$)

Neither photosynthesis nor respiration is as simple as we have made it out to be. The compounds involved in respiration include many that are far more complex than simple carbohydrate, but they all represent the products of photosynthesis and subsequent synthesis of more complex organic molecules.

At this point, it is helpful to link photosynthesis and respiration in a continuous cycle involving both the primary producer and the decomposer. Figure 11.2 shows one closed loop for hydrogen (H), one for carbon (C), and two loops for oxygen (O). We are not taking into account that there are two atoms of hydrogen in each molecule of water and carbohydrate, or that there are two atoms of oxygen in each molecule of carbon dioxide and oxygen gas. Only the flow pattern counts in this representation.

A good place to start is the soil, from which water is drawn up into the body of a living plant. In the green leaves of the plant photosynthesis takes place while light energy is absorbed by the leaf cells. Carbon dioxide is being brought in from the atmosphere at this point. Here oxygen is liberated and begins its atmospheric cycle. The plant tissue then dies and falls to the ground, where it is acted on by

figure 11.2

A simplified flow diagram of the essential components of photosynthesis and respiration.

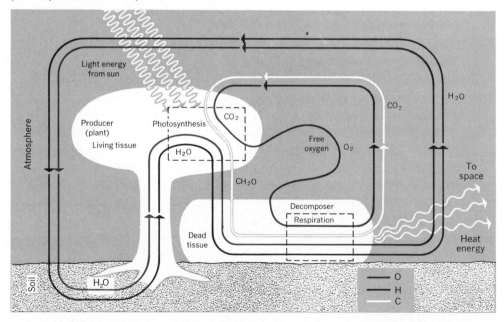

the decomposer. Through respiration oxygen is taken out of the atmosphere or soil air and combined with the hydrogen freed by breakdown of the carbohydrate. Energy is now liberated. Here both carbon dioxide and water enter the atmosphere as gases. The water vapor eventually condenses and falls to earth as precipitation.

An important concept emerges from this flow diagram. Energy passes through the system. It comes from the sun and returns eventually to outer space. On the other hand, the material components, hydrogen, oxygen, and carbon, are recycled within the total system. Of course, many other material components are recycled in the same way. Recall from Chapter 10 that plant nutrients — among them the base cations — are essential in the growth of plants and are constantly recycled. Because the earth as a planet is a closed system, the material components never leave the total system, but they can be stored in other ways and forms where they are unavailable for use by plants for prolonged periods of geologic time. We shall develop this concept later.

Net photosynthesis

Both photosynthesis and respiration go on simultaneously in a plant, so that the amount of new carbohydrate placed in storage is less than the total carbohydrate being synthesized. We therefore distinguish between gross photosynthesis and net photosynthesis. _Gross photosynthesis_ is the total amount of carbohydrate produced by photosynthesis; _net photosynthesis_ is the amount of carbohydrate remaining after respiration has broken down sufficient carbohydrate to power the plant. Stated as an equation,

Net photosynthesis =
        Gross photosynthesis − Respiration

Since both photosynthesis and respiration occur in the same cell, gross photosynthesis

cannot be readily measured. Instead, we shall deal with net photosynthesis. In most cases, respiration will be held constant, so that use of the net instead of the gross will show the same trends.

The rate of net photosynthesis is strongly dependent on the intensity of light energy available, up to a limit. Figure 11.3 shows this principle. On the vertical axis the rate of net photosynthesis is indicated by the rate at which a plant takes up carbon dioxide. On the horizontal axis light intensity increases from left to right. At first, net photosynthesis rises rapidly as light intensity increases, but the rate then slows and reaches a maximum value shown by the plateau in the curve. Above this maximum, the rate falls off because the incoming light is also causing heating. Increased temperature, in turn, increases the rate of respiration, which offsets gross production by photosynthesis.

Light intensity sufficient to allow maximum net photosynthesis in only 10 to 30 percent of full summer sunlight for most green plants. Additional light energy is simply ineffective. The factor of duration of daylight then becomes the important factor in the rate at which products of photosynthesis accumulate as plant tissues. On this subject, you can draw

figure 11.3
The curve of net photosynthesis shows a steep initial rise, then levels off as light intensity rises.

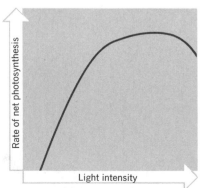

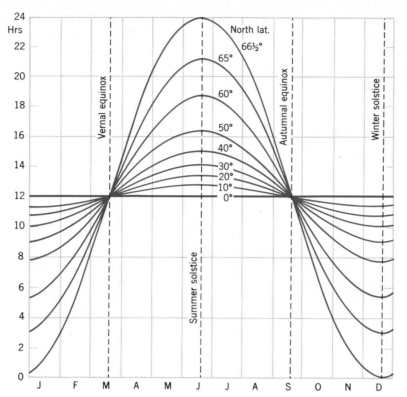

figure 11.4
Duration of the day at various latitudes throughout the year. Vertical scale gives number of hours the sun is above the horizon.

on your knowledge of the seasons and the changing angle of the sun's rays with latitude. Figure 11.4 shows the duration of the daylight period with changing seasons for a wide range of latitudes in the northern hemisphere. At low latitudes, days are not far from the average 12-hour length throughout the year, whereas at high latitudes, days are short in winter but long in summer. The seasonal contrast in day length increases with latitude. It is obvious that in subarctic latitudes photosynthesis can go on in summer during most of the 24 hours, and this factor can compensate partly for the shortness of the season.

Photosynthesis also increases in rate with increase in air temperature, up to a limit. Figure 11.5 shows the results of a laboratory experiment in which sphagnum moss was grown under constant illumination. Gross photosynthesis increased rapidly to a maximum at about 70° F (20° C), then leveled

figure 11.5
Respiration and gross and net photosynthesis vary with temperature. (Data of Stofelt, 1937, in A. C. Leopold, 1964, Plant Growth and Development, McGraw-Hill, New York, p. 31, Figure 2.26.)

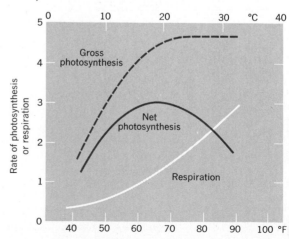

off. Respiration increased quite steadily to the limit of the experiment. Net photosynthesis, which is the difference between the values in the two curves, peaked at about 65° F (18° C), then fell off rapidly.

Net primary production

Plant ecologists measure the accumulated net production by photosynthesis in terms of the *biomass,* which is the dry weight of the organic matter. This quantity could, of course, be stated for a single plant or animal, but a more useful statement is made in terms of the biomass per unit of surface area within the ecosystem — the acre, square foot, or square meter. Of all ecosystems, forests have the greatest biomass; that of grasslands and croplands is very small in comparison. For freshwater bodies and the oceans the biomass is even smaller — on the order of one-hundredth that of the grasslands and croplands.

The annual rate of production of organic matter is the vital information we need, not the biomass itself. Granted that forests have a very large biomass and grasslands very little, what we want to know is: which ecosystem is the more productive? In other words, which ecosystem produces the greatest annual yield in terms of resources useful to Man? For the answer we turn to figures on the *net primary production* of various ecosystems. Table 11.1 gives this information in units of grams of dry organic matter produced annually from 1 sq m of surface. The figures are only rough estimates, but they are nevertheless highly meaningful. Since we wish only to make comparisons, English units are omitted. Note that the highest values are in two quite unlike environments: forests and estuaries. Agricultural land compares favorably with grassland, but the range is very large in agricultural land, reflecting many factors such as availability of soil water, soil fertility, and use of fertilizers and machinery.

	Grams per Square Meter per Year	
	Average	Typical Range
Lands		
Rainforest of the equatorial zone	2000	1000-5000
Freshwater swamps and marshes	2000	800-4000
Midlatitude forest	1300	600-2500
Midlatitude grassland	500	150-1500
Agricultural land	650	100-4000
Lakes and streams	500	100-1500
Extreme desert	3	0-10
Oceans		
Estuaries (tidal)	2000	500-4000
Continental shelf	350	200-600
Open ocean	125	1-400

table 11.1
Net primary production for various ecosystems

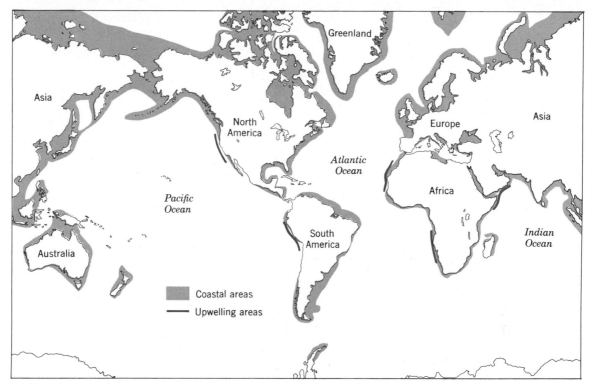

figure 11.6
Distribution of world fisheries. Coastal areas and upwelling areas together supply over 99 percent of world production. (Compiled by the National Science Board, National Science Foundation.)

Productivity of the oceans is generally low. The deep-water oceanic zone is the least productive of the marine ecosystems. At the same time, however, it comprises about 90 percent of the world ocean area. Continental shelf areas are a good deal more productive and, in fact, support much of the world's fishing industry at the present time (Figure 11.6). Except for the shallow-water estuarine and reef ecosystems, the upwelling zones, shown on the world map, are most productive.

Upwelling of cold water from great depths brings nutrients to the surface and greatly increases the growth of microscopic floating plants (phytoplankton). These, in turn, serve as food sources for marine animals in the food chain. Consequently, zones of upwelling are fisheries of great wealth. An example is the

Peru Current off the west coast of South America. Here, myriads of a species of small fish, the anchoveta, provide food for larger fish and for birds. The birds, in turn, excrete their wastes on the mainland coast of Peru. The accumulated deposit, called guano, is a rich source of fertilizer.

Net production and climate

The geographer is interested in the climatic factors controlling organic productivity. Besides light intensity and temperature, an obvious factor will be availability of water. Soil-water shortage or surplus might be the best climatic factor to examine, but data are not available. Ecologists have related net annual primary production to mean annual pre-

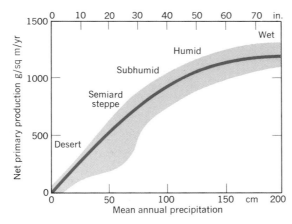

figure 11.7

Net primary production increases rapidly with increasing precipitation, but levels off in the higher values. Observed values fall mostly within the shaded zone. (Data of Whittaker, 1970).

cipitation, as shown in Figure 11.7. The production values are for plant structures above the ground surface. Although the productivity increases rapidly with precipitation in the lower range from desert through semiarid to subhumid climates, it seems to level off in the humid range. Apparently, a large soil-water surplus carries with it some compensating influence, such as removal of plant nutrients by leaching.

Combining the effects of light intensity, temperature, and precipitation, you would guess that the maximum net productivity year around would be in wet equatorial and tropical climates where soil water is adequate at all times. Productivity would be low in all desert climates and would decrease generally with increasing latitude. Productivity would be reduced to near zero in the arctic zone where the combination of a short growing season and low temperatures would act in concert to halt the growth process.

These deductions are nicely confirmed by a schematic diagram of net primary productivity on an imagined supercontinent (Figure 11.8).

Comparing this diagram with the world climate diagram (Figure 8.7), we can assign rough values of productivity to each of the climates:

Highest (over 800)	Wet equatorial (1)
Very high (600-800)	Trade-wind and monsoon coastal (2) Tropical wet-dry (3)
High (400-600)	Tropical wet-dry (3) (S.E. Asia) Humid subtropical (5) Marine west coast (8)
Moderate (200-400)	Mediterranean (7) Humid continental (6)
Low (100-200)	Tropical steppe (4b) Midlatitude steppe (9b) Continental subarctic (10)
Very low (0-100)	Tropical desert (4a) Midlatitude desert (9a) Continental subarctic (10) Tundra (11)

The tropical wet-dry climate (3) is in a narrow transition zone between very high productivity and low productivity so that some areas are in the "very high" rating and others are in the "high" rating. Transition zones such as this are highly sensitive to alternating cycles of drought and of excessive rainfall. They pose the most difficult famine problem of all global regions because they carry dense populations, which can be adequately fed only under the best of conditions.

Energy flow along the food chain

The primary producers trap the energy of the sun and make it available to support consuming organisms at higher levels in the food chain. But remember that energy is lost during each upward step in the chain, so that the number of steps is limited. In general, anywhere from 10 to 50 percent of the energy stored in organic matter at one level can be passed up the chain to the next level. Also, the higher values of energy loss are found at the higher levels of the chain, so that a limit is

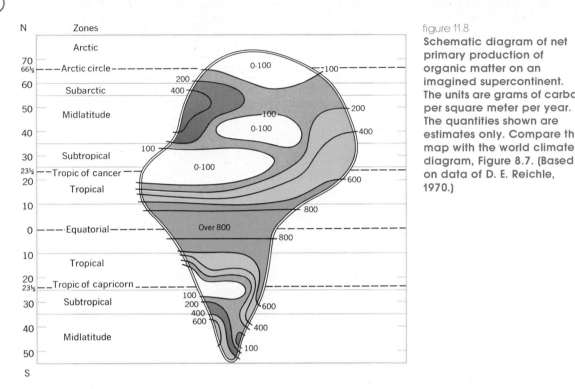

figure 11.8
Schematic diagram of net primary production of organic matter on an imagined supercontinent. The units are grams of carbon per square meter per year. The quantities shown are estimates only. Compare this map with the world climate diagram, Figure 8.7. (Based on data of D. E. Reichle, 1970.)

quickly reached. Four levels of consumers are about the normal limit.

Figure 11.9 is a bar graph showing the percent of energy passed up the chain when only 10 percent moves from one level to the next. The horizontal scale is in powers of ten. In ecosystems of the lands the biomass also decreases with each upward step in the chain; the number of individuals of the consuming animals also decreases with each upward step. In the food chain shown in Figure 11.1, there are only a few individuals of marsh hawks and owls in the third level of consumers, whereas there are countless individuals in the primary level. These facts serve as a clear message to mankind; they tell us that when we raise beef cattle, swine, sheep, and fowl to eat, we are extremely wasteful of the world's total food resource. These animals, in storing food for humans to consume have already wasted perhaps nine tenths of the energy they consumed as plant matter from the producing

level. If, instead, we subsisted on grains and other food plants such as legumes, grown on the available farm lands, the food resource could support a much greater population. This observation doesn't apply to sheep and cattle subsisting entirely on grazing lands unfit for cereal crops, but it is a potent concept nevertheless.

figure 11.9
Percentage of energy passed up the steps of the food chain, assuming 90 percent is lost energy at each step.

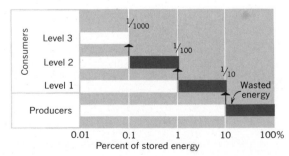

Agricultural ecosystems

The principles of energy use and flow in natural ecosystems also apply to agricultural ecosystems. Important differences exist, however, between natural ecosystems and those that are highly managed for agriculture. The first major difference is the reliance of agricultural ecosystems on inputs of energy that are ultimately derived from fossil fuels. The most obvious of these inputs is the fuel that runs the machinery used to plant, cultivate, and harvest the crops. Another is application of fertilizers and pesticides requiring large expenditures of fuel to extract or synthesize and to transport. Fossil fuels also power activities such as the breeding of plants, which have higher yields and are resistant to disease, and developing new chemicals to combat insect pests. Fuel is expended in transporting crops to distant sources of consumption, thus enabling large areas of similar climate and soils to be used for the same crop. In these and many other ways, the high yields obtained today are brought about only at the cost of a large energy subsidy derived from fossil fuels.

In terms of ecosystem structure and function, agricultural ecosystems are very simple. They usually consist of one genetic strain of one species. Such ecosystems are overly sensitive to attacks by one or two well-adapted insects that can multiply very rapidly to take advantage of an abundant food source. Thus pesticides are constantly needed to reduce insect populations. Weeds, too, are a problem, adapted as they are to rapid growth on disturbed soil in sunny environments. Weeds can divert much of the productivity to undesirable forms. Herbicides are often the immediate solution to these problems.

Application of agricultural chemicals is one of the ways in which Man uses energy inputs to increase net primary productivity. Large increases in productivity are achieved by application of nutrient elements and compounds — usually of nitrogen and phosphorus — which are in short supply in most soils. In natural ecosystems these elements are returned to the soil following the death of the plants that concentrate them. In agricultural ecosystems this recycling is interrupted by harvesting the crop for consumption at a distant location. Nutrients must be added each year in the form of fertilizers, and these are synthesized or mined only with considerable energy input from fossil fuels.

The inputs of energy that Man adds to managed ecosystems in the form of agricultural chemicals and fertilizers, as well as those in the form of the work of farm machinery, have acted to boost greatly the net primary productivity of the land. Table 11.2 gives some examples of how agricultural yields have responded to these inputs. (Units used in the table are the energy equivalents of the biomass.) The table shows that Man has acted to raise the net primary productivity of his agricultural ecosystems more than five times

table 11.2

Crop productivity with and without fossil fuel energy subsidy

Crop	Net Productivity, as Harvested (kcal/sq. m/day)
Without fossil fuel energy inputs	
Grain, Africa, 1936	0.71
All U.S. farms, 1880	1.3
With fossil fuel energy inputs	
Grain, North American average, 1960	5
Rice, United States, 1964	10

Source: Data of several sources, compiled in H. T. Odum, 1971, *Environment, Power and Society,* Wiley-Interscience, New York, p. 116, Table 4.1.

over through the use of energy contained in fossil fuels.

Diverting solar energy to human service through management of ecosystems is only one example of the great changes Man has produced in natural environments. Other changes are concerned with the movement of matter through ecosystems, influenced not only by the agricultural demands mentioned above, but also through many other types of impact.

Cycling of materials in ecosystems

From an investigation of energy flow in ecosystems, we now turn to the flow of materials used by organisms. The elements hydrogen, carbon, and oxygen are basic materials whose pathways we have already followed in the photosynthesis-respiration circuits (Figure 11.2). At this point, a broader view of the cycling of other essential elements will give you an increased understanding of the role of geological processes in storing and releasing these elements and of Man's role in upsetting the natural balance of these cycles.

We will now take the fifteen most abundant elements in living matter as a whole and designate their total mass as 100 percent. Percentages of each of the elements are given in Table 11.3. The three principal components of carbohydrate — hydrogen, carbon, and oxygen — account for almost all living matter and are called _macronutrients_. The remaining ½ percent is divided among twelve elements. Six of these are also macronutrients: nitrogen, calcium, potassium, magnesium, sulfur, and phosphorus. The macronutrients are all required in substantial quantities for organic life to thrive.

Nitrogen is much more abundant than the other five lesser macronutrients, since it is contained within each amino acid building block of protein. Almost all the nitrogen available to living organisms is obtained from

table 11.3

Elements comprising global living matter, taking 100 percent as the total of the 15 most abundant elements.

Basic carbohydrate	
Hydrogen (M)	49.74
Carbon (M)	24.90
Oxygen (M)	24.83
Subtotal	99.47
Other nutrients	
Nitrogen (M)	0.272
Calcium (M)	0.072
Potassium (M)	0.044
Silicon	0.033
Magnesium (M)	0.031
Sulfur (M)	0.017
Aluminum	0.016
Phosphorus (M)	0.013
Chlorine	0.011
Sodium	0.006
Iron	0.005
Manganese	0.003
	M - macronutrient

Data of E. S. Deevey, Jr., 1970, _Scientific American_, Vol. 223.

the atmosphere through the efforts of nitrogen-fixing bacteria. These bacteria have a biochemical system that can convert nitrogen from its inert atmospheric form, molecular nitrogen, to other forms, such as ammonia and nitrate, which can be used directly by plants. The plants assimilate these forms of nitrogen and convert them to organic forms, which are consumed by animals grazing on the plants. In this way nitrogen is passed up the food chain. Decomposers convert the nitrogen from organic forms in decaying plant and animal matter to simple, inorganic forms that the plants can absorb from soil or water. Certain bacteria can also return nitrogen to its inert atmospheric form. This function completes the nitrogen cycle, by which nitrogen moves from the atmosphere to organisms in the life layer and back again to the atmosphere.

Of the remaining macronutrients, you will recognize three — calcium, potassium, and magnesium — as the base cations so important in soils. They are all derived from rocks through mineral weathering, along with silicon, sodium, sulfur, phosphorus, iron, and manganese. Some of the sulfur gets into the soil as a component of rainwater, having originally been swept from breaking waves on the sea surface and carried aloft as minute suspended salt particles.

Quite a number of additional elements, not among the nine macronutrients, are also vital to life processes. Their presence is needed in mere traces and, for this reason, they are called *micronutrients*. The micronutrient list includes iron, copper, zinc, boron, molybdenum, manganese, and chlorine.

Figure 11.10 is a flow diagram showing the pathways taken by the more important mineral elements from the inorganic realms of the lithosphere, hydrosphere, and atmosphere, through the realm of living tissues, or biosphere, and returning to the inorganic realm. Within the large box representing the lithosphere are smaller compartments representing the parent matter of the soil and the soil itself. In the soil, nutrients are held as ions on the surfaces of soil colloids and are readily available to plants. These same elements are also held in enormous storage pools where they are unavailable to organisms. These storage pools include seawater (unavailable to land organisms), sediments on the sea floor, and enormous accumulations of sedimentary rock beneath both lands and oceans. Eventually, elements held in the geologic storage pools are released into the soil by weathering. Soil particles are lifted into the atmosphere by winds and fall back to earth or are washed down by precipitation. Chlorine

figure 11.10

A flow diagram of the mineral portion of materials cycle in and out of the biosphere and within the inorganic realm of the lithosphere, hydrosphere, and atmosphere. The four macronutrients of atmospheric origin — hydrogen, carbon, oxygen, and nitrogen — are omitted.

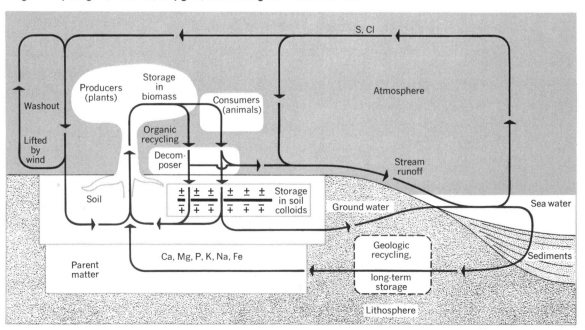

and sulfur are shown as passing from the ocean into the atmosphere and entering the soil by the same mechanisms of fallout and washout. The organic realm, or biosphere, is shown in three compartments: producers, consumers, and decomposers. Considerable element recycling occurs between these organisms and the soil. However, there is a continual process of escape of these elements to the sea as ions dissolved in stream runoff and ground water flow.

Ecosystems in review

This chapter has reviewed principles of ecology, a field within the biological sciences. Physical geography draws on many and diverse fields of science — meteorology, oceanography, and geology, among others — to explain the spatial patterns of Man's physical environment over the face of the globe. The flow of energy and materials through ecosystems helps the plant geographer and soils geographer to interpret the patterns of soil and vegetation through the controlling factor of climate. In the next chapter we shall move into the area of plant geography to investigate the global patterns of vegetation developed in terrestrial ecosystems that have come into a balance, or equilibrium, with the climate and soil types that support them.

Geographers are greatly concerned with Man's use of the land and the patterns of human settlement, agriculture, and urbanization. These patterns have been superimposed over many centuries on natural ecosystems in which Man was once a very insignificant agent of change.

Environments of Natural Vegetation

Chapter 12

IN THIS CHAPTER we continue to study the biosphere, but with our focus on vegetation of the lands. Thinking of plants as stationary objects on the landscape places them in much the same frame as other physical elements of that landscape such as landforms, soils, streams, and lakes. Plants are also consumable and renewable sources of food, medicinals, fuel, clothing, shelter, and a host of other life essentials. The ways in which humans have used this plant resource to their advantage — or have been hindered by plants in their progress — have been persistent themes in the writings of geographers.

Two concepts of the geography of world vegetation stand opposed but inseparable, like the two sides of a coin. One is the concept of *natural vegetation* (or native vegetation), a plant cover that attains its development without appreciable interference by Man and that is subject to natural forces of modification and destruction, such as storms or fires. The other concept is that of vegetation sustained in a modified state by Man's activities. Extremes of both cases are found over the world. Natural vegetation can still be seen over vast areas of the wet equatorial climate where rainforests are as yet scarcely touched by Man. Much of the arctic tundra and the forest of the subarctic zones is in a natural state. In contrast, much of the continental surface in midlatitudes is almost totally under Man's control through intensive agriculture, grazing, or urbanization. You can drive across an entire state, such as Ohio or Iowa, without seeing a vestige of the plant cover it bore before the coming of the white man. Only if you know where to look for them can you find a few small plots of virgin prairie or virgin forest.

Some areas of natural vegetation appear to be untouched but are dominated by Man in a subtle manner. Certain of our national parks and national forests have been protected from fire for many decades, setting up an unnatural condition in terms of what might be expected of the natural ecosystem. When lightning starts a forest fire, the smoke-eaters parachute down and put out the flames as fast as possible. But we realize that periodic burning of forests and grasslands must be a natural phenomenon, and it must perform some vital function in the ecosystem. One vital function is to release nutrients from storage in the biomass so that the soil can be revitalized. Already those who manage our parks and forests are experimenting with a "hands-off" policy in fire control.

Man has influenced vegetation in yet another way — by moving plant species from their indigenous habitats to foreign lands and foreign environments. The eucalyptus tree is a striking example. From Australia the various species of eucalypts have been transplanted to such far-off lands as California, North Africa, and India. Sometimes these exported plants thrive like weeds, forcing out natural species and becoming a major nuisance. It is said that scarcely one of the grasses seen clothing the coast ranges of California is a native species, yet the casual observer might think that these represent the native vegetation. Even so, all plants have limits of tolerance to the environmental conditions of soil water, heat and cold, and soil nutrients. Consequently, the structure and outward appearance of the plant cover conforms to the basic environmental controls, and each vegetation type stays within a characteristic geographical region, whether this be forest, grassland, or desert.

Structure and life form of plants

The plant geographer classifies plants in terms of the *life form*, which is the physical structure, size, and shape of the plant. Botanical associations are not necessarily related to life form. Although the life forms go by common names and are well understood by almost everyone, we shall review them to establish a uniform set of meanings.

Both trees and shrubs are erect, woody plants (Figure 12.1). They are perennial, meaning that the woody tissues endure from year to year, and most have life spans of many years. _Trees_ are large, woody perennial plants having a single upright main trunk, often with few branches in the lower part, but branching in the upper part to form a crown. _Shrubs_ are woody perennial plants having several stems branching from a base near the soil surface, so as to place the mass of foliage close to ground level.

Lianas are also woody plants, but they take the form of vines supported on trees and shrubs.

figure 12.1

This forest of mature beech and hemlock trees has a lower layer of scattered shrubs and seedling trees and a basal layer of herbs. Allegheny National Forest, Pennsylvania. (U.S. Forest Service.)

Lianas include not only the tall heavy vines of the wet equatorial and tropical rainforests (see Figure 12.7), but also some woody vines of midlatitude forests.

Herbs comprise a major class of plant life forms. _Herbs_ are tender plants, lacking woody stems, and usually small. They occur in a wide range of shapes and leaf types. Some are annuals, others are perennials. Some are broad leaved, others are grasses. Herbs as a class share few characteristics in common except that they form a low layer as compared with shrubs and trees.

Forest is a vegetation structure in which trees grow close together with crowns in contact, so that the foliage largely shades the ground. In most forests of the humid climates the life forms are arranged in distinct layers, and the vegetation is said to be stratified (Figure 12.2). Tree crowns form the uppermost layer, shrubs an intermediate layer, and herbs a lower layer. The lowermost layer consists of mosses and related small plants. In _woodland,_ crowns of trees are mostly separated by open areas, usually having a low herb or shrub layer.

Lichens represent yet another life form seen in a layer close to the ground. Lichens are plant forms in which algae and fungi live together to form a single plant structure. In some alpine and arctic environments lichens grow in profusion and dominate the vegetation (see Figure 12.13).

Plant habitats

As we travel through a hilly wooded area it becomes obvious that the vegetation is strongly influenced by landform and soil. Landform refers to the configuration of the land surface, including features such as hills, valleys, ridges, or cliffs. Vegetation on an upland — relatively high ground with thick soil and good drainage — is quite different from that on an adjacent valley floor, where water lies near the surface

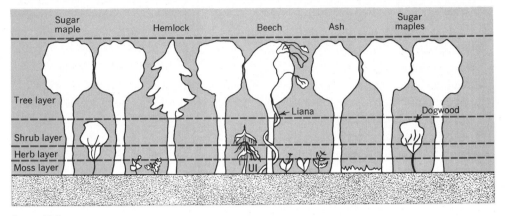

figure 12.2
This schematic diagram shows the layers of a beech-maple-hemlock forest, similar to that pictured in Figure 12.1. The vertical dimensions of the lower layers are greatly exaggerated. (Modified from Pierre Dansereau, 1951, _Ecology_, Vol. 32.)

much of the time. Vegetation is also strikingly different in form on rocky ridges and on steep cliffs, where water drains away rapidly and soil is thin or largely absent.

The total vegetation cover is actually a mosaic of small units reflecting the inequalities in conditions of slope, drainage, and soil type. Such subdivisions of the plant environment are described as _habitats_. In the example shown in Figure 12.3 the Canadian forest actually comprises at least six habitats: upland, bog, bottomland, ridge, cliff, and active dune. Just where each habitat is located and how large an area it occupies depends largely on the geologic history of the region and on processes of erosion and deposition that have acted in the past to shape the landforms.

Closely tied in with the effect of landform on habitat is the distribution of water in the soil and rock. Finally, each habitat has its own set of soil properties, determined not only by the conditions of slope and water but, in part, by the plants themselves.

In establishing the larger units of vegetation for generalized maps, the plant geographer usually bases his classes upon the life-form assemblages of the upland habitat, because it is here that a middle range of environmental conditions prevails.

Awareness that varied habitats present a wide spectrum of controls over plants leads us to look further into the role of soil-water availability in favoring one form of plant over another.

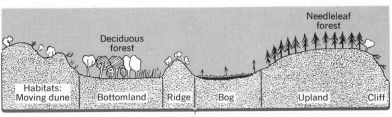

Canadian forest

figure 12.3
Habitats within the Canadian forest. (Modified from Pierre Dansereau, 1951, _Ecology_, Vol. 32.)

Plants and water need

Because the soil-water budget has been a dominant theme of earlier chapters, we can start with an investigation of the response of plants to the degree of availability of water. The green plants, or leaf-bearing plants, give off large quantities of water to the atmosphere through transpiration. This water loss takes the form of evaporation from water films on exposed surfaces of certain leaf cells. (Only a very small proportion of the water consumed by plants is used in photosynthesis.) Most plants derive their water through roots in the soil zone. These plants must have mechanisms by which to control the excessive loss of water by transpiration at times when stored soil water is depleted and a shortage becomes severe.

The adaptation of plant structures to water budgets with large water shortages is of particular interest to the plant geographer. Transpiration occurs largely from specialized leaf pores, which provide openings in the outer cell layer. These pores allow water vapor and other gases to pass into and out of the leaf. Surrounding these openings are cells that can open and close the openings and thus, to some extent, regulate the flow of water vapor and other gases. Water vapor may also pass through the cuticle, or outermost protective layer of the leaf. This form of water loss is reduced in some plants by thickening of the outer layers of cells or by the deposition of wax or waxlike material on the leaf surface. Many desert plants have thickened cuticle or wax-coated leaves, stems, or branches. A plant may also adapt to a desert environment by greatly reducing the leaf area or by bearing no leaves at all. Needle-leaves and spines representing leaves greatly reduce loss from transpiration. In cactus plants the foliage leaf is not present, and transpiration is limited to fleshy stems.

In addition to developing leaf structures that reduce water loss by transpiration, plants in a water-scarce environment improve their means of obtaining water and of storing it. Roots become greatly extended to reach soil moisture at increased depth. In cases where the roots reach to the ground water table, a steady supply of water is assured. Plants drawing from such a source may be found along dry stream channels and valley floors in desert regions. Other desert plants produce a widespread but shallow root system enabling them to absorb the maximum quantity of water from sporadic desert downpours that saturate only the uppermost soil layer. Stems of desert plants are commonly greatly thickened by a spongy tissue in which much water can be stored.

A quite different adaptation to extreme aridity is seen in many small desert plants that complete a very short cycle of germination, leafing, flowering, fruiting, and seed dispersal immediately following a desert downpour.

Plants may be classified according to their water requirements. Terms associated with the water factor are built on three simple prefixes of Greek roots: *xero-,* dry; *hygro-,* wet; and *meso-,* intermediate. Plants that grow in dry habitats are _xerophytes;_ those that grow in water or in wet habitats are _hygrophytes;_ those of habitats of an intermediate degree of wetness and relatively uniform soil-water availability are _mesophytes_.

The xerophytes are highly tolerant of drought and can survive in habitats that dry quickly following rapid drainage of precipitation (e.g., on sand dunes, beaches, and bare rock surfaces). The plants typical of dry climates are also xerophytes; cactus is an example. The hygrophytes are tolerant of excessive water and may be found in shallow streams, lakes, marshes, swamps, and bogs; an example is the water lily. The mesophytes are found in upland habitats in regions of ample rainfall. Here the drainage of soil water is good, and moisture penetrates deeply where it can later be used by the plants.

Certain climates, such as the tropical wet-dry climate and the humid continental climate, have a yearly cycle with one season in which water is unavailable to plants because of lack of precipitation or because the soil water is frozen. This season alternates with one in which there is abundant water. Plants adapted to such regimes are called *tropophytes,* from the Greek word *trophos,* meaning change, or turn. Tropophytes meet the impact of the season of unavailable water by dropping their leaves and becoming dormant. When water is again available, they leaf out and grow at a rapid rate. Trees and shrubs that seasonally shed their leaves are *deciduous plants;* in distinction, *evergreen plants* retain most of their leaves in a green state through the year.

The Mediterranean climate also has a strong seasonal wet-dry alternation, the summers being dry and the winters wet. Plants in this climate adopt the habit of xerophytic plants and characteristically have hard, thick leathery leaves. An example is the live oak, which holds most of its leaves through the dry season. Such hard-leaved evergreen trees and woody shrubs are called *sclerophylls.* The prefix *scler* is from the Greek word for "hard." Plants that hold their leaves through a dry or cold season have the advantage of being able to resume photosynthesis immediately when growing conditions become favorable, whereas the deciduous plants must grow a new set of leaves.

Plants and temperature

Temperature, another of the important climatic factors in plant ecology, acts directly on plants through its influence on the rates at which the physiological processes take place. In general, we can say that each plant species has an optimum temperature associated with each of its functions, such as photosynthesis, flowering, fruiting, or seed germination. Some overall optimum yearly temperature conditions exist for its growth in terms of size and numbers of individuals. There are also limiting lower and

upper temperatures for the individual functions of the plant and for its total survival. Temperature acts as an indirect factor in many other ways. Higher air temperatures increase the water-vapor capacity of the air, inducing greater transpiration as well as faster rates of direct evaporation of soil water.

In general, the colder the climate, the fewer are the species that are capable of surviving. A large number of tropical plant species cannot survive below-freezing temperatures. In the severely cold arctic and alpine environments of high latitudes and high altitudes, only a few species can survive. This principle explains why a forest in the equatorial zone has many species of trees, whereas a forest of the subarctic zone may be dominantly of one, two, or three tree species. Tolerance to cold is closely tied up with the ability of the plant to withstand the physical disruption that accompanies the freezing of water. If the plant has no means of disposing of the excess water in its tissues, the freezing of that water will damage the cell tissue.

Plant geographers recognize that there is a critical level of climatic stress beyond which a plant species cannot survive, and that a geographical boundary will exist marking the limits of its distribution. Such a boundary is sometimes referred to as a frontier. Although the frontier is determined by a complex of climatic elements, it is sometimes possible to single out one climate element related to soil water or temperature that coincides with the plant frontier.

Plant succession and Man's impact on vegetation

The phenomenon of change in ecosystems through time is familiar to us all. A drive in the country reveals patches of vegetation in many stages of development — from open, cultivated fields to grassy shrublands to forests. Clear lakes gradually fill in with sediment from the rivers that drain into them and become marshes. These kinds of changes, in which plant and animal communities succeed one another on the way to a stable endpoint, are referred to as *ecological succession*. In general, succession leads to formation of the most complex community of organisms possible in an area, given its physical controlling factors of climate, soil, and water. The stable community, which is the endpoint of succession, is the *climax*. Succession may begin on a newly constructed mineral deposit such as a sand dune or river bar. It may also occur on a previously vegetated area that has been recently disturbed by agents such as fire, flood, windstorms, and Man.

An important case of succession is that on farmlands, intensively cultivated for decades, then abandoned to the natural sequence of events. An example of this "old-field" succession comes from the southeastern United States, where the humid subtropical climate provides ample soil water much of the year and the growing season is long. Figure 12.4 shows the succession, starting at the left with the bare field. The first plants to take hold in the hostile environment are called pioneers. They are annual herbs most people would call weeds. Crabgrass, horseweed, and aster thrive during the first two years. These plants supply organic matter to the soil; they also begin to shade the soil and reduce the extremes of soil temperature. Next, grasses and shrubs move in, and these occupy the old field during the first two decades or so. Pine seedlings now enter the habitat and, as these grow, they develop a shade cover, greatly changing the climate near the ground. In the next half century, as the pine forest matures, broad-leaved deciduous trees begin to come in, displacing the pines. By the middle of the second century, a mature forest of oak and hickory has dominated the habitat. This climax forest has a lower shrub layer, a basal herb layer, and a substantial accumulation of organic matter in various stages of decomposition. This is an environment in which soil water tends to be conserved and the thermal environment is protected from extremes of heat or cold. Although generally stabilized in its composition and structure, serious natural upsets of equilibrium may occur through fires and severe storms. Insect hordes and disease epidemics may also radically alter the climax forest. Fire will be followed by an appropriate succession, with stages leading to the return of the oak-hickory forest.

figure 12.4

Old-field succession in the Piedmont region of the southeastern United States, following abandonment of corn fields and cotton fields. This is a pictorial graph of continuously changing plant composition spanning about 150 years. (After E. P. Odum, 1973, _Fundamentals of Ecology_, 3rd ed., W. B. Saunders Co., Philadelphia.)

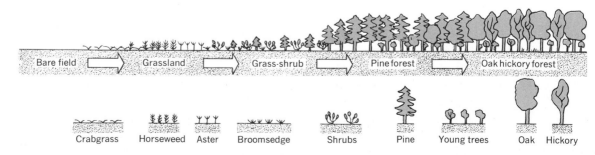

Bare field → Grassland → Grass-shrub → Pine forest → Oak hickory forest

Crabgrass   Horseweed   Aster   Broomsedge   Shrubs   Pine   Young trees   Oak   Hickory

Cutting and clearing of forest trees brings a man-made demise to the forest climax. Introduction of a plant disease from a foreign continent can cause the extinction of a particular plant species. An example is the chestnut blight, which eliminated the American chestnut from the forests of the northeastern United States. Imported insects may also wipe out most of the mature individuals of a plant species when there is no native predator available to combat the invasion. These are just a few of the ways in which Man interferes with natural vegetation.

Terrestrial ecosystems — the biomes

Ecosystems fall into two major groups: aquatic and terrestrial. The *aquatic ecosystems* include life forms of the marine environments and the freshwater environments of the lands. Marine ecosystems include the open ocean, coastal estuaries, and coral reefs. Freshwater ecosystems include lakes, ponds, streams, marshes, and bogs. Our survey of physical geography will not include the ecology of these aquatic environments. Instead, we shall focus on the *terrestrial ecosystems;* they comprise the assemblages of land plants spread widely over the upland surfaces of the continents. The terrestrial ecosystems are directly impacted by climate and interact with the soil and, in this way, are closely woven into the fabric of physical geography.

Within terrestrial ecosystems the largest recognizable subdivision is the *biome*. Although the biome includes the total assemblage of plant and animal life interacting within the life layer, the green plants dominate the biome physically because of their enormous biomass, as compared with that of other organisms. The plant geographer concentrates on the characteristic life form of the green plants within the biome. As we have already seen, these life forms are principally trees, shrubs, lianas, and herbs, but other life forms are important in certain biomes. Biomes are recognized and mapped on the concept of the climax vegetation. Areas intensively modified by Man through agriculture and urbanization are assigned to a biome according to the climax vegetation assumed to have once been present.

The following are the principal biomes, listed in order of availability of soil water and heat.

✓ *Forest*	(ample soil water and heat)
Savanna	(transitional between forest and grassland)
Grassland	(moderate shortage of soil water; adequate heat)
Desert	(extreme shortage of soil water; adequate heat)
Tundra	(insufficient heat)

For the plant geographer the biomes are broken down into smaller vegetation units, called *formation classes,* concentrating on the life form of the plants. For example, at least four and perhaps as many as six kinds of forests are easily recognizable within the forest biome. At least three kinds of grasslands are easily recognizable. Deserts, too, span a wide range in terms of the abundance and life form of plants.

The formation classes we shall introduce are major widespread types clearly associated with eleven climate types and their soil-water budgets. Association with major soil types will also be important. In this way we survey the global scope of plant geography as a synthesis of climate with organic and physical processes of the life layer. Table 12.1 lists the formation classes and their approximate associations with climate types and soil types.

Figure 12.5 is a schematic diagram to show how the vegetation formation classes would be arranged on an idealized continent, combining

table 12.1
The major formation classes

Formation Class	Associated Climate Type		Associated Soil Type
Forest biome			
Equatorial rainforest	(1)	Wet equatorial	Latosols
tropical rainforest	(2)	Trade-wind and monsoon coastal	
Monsoon forest	(3)	Tropical wet-dry (Asia)	Latosols Vertisols
Temperate rainforest (laurel forest)	(5) (8)	Humid subtropical Marine west coast	Latosolic to podzolic
Midlatitude deciduous forest	(5) (6)	Humid subtropical Humid continental (long summers)	Podzolic forest
Needleleaf forest	(6) (10)	Humid continental (short summers) Continental subarctic	Podzolic
Sclerophyll forest	(7)	Mediterranean	Reddish and reddish-brown
Savanna biome			
Tropical savanna (incl. woodland, thornbush, scrub, grassland)	(3)	Tropical wet-dry	Latosols Vertisols
Grassland biome			
Prairie (tall-grass)	(6)	Humid subtropical (subhumid parts)	Prairie Chernozem
Steppe (short-grass)	(4b) (9b)	Tropical steppe Midlatitude steppe	Chestnut, brown, reddish-brown, reddish-chestnut
Desert biome			
Semidesert	(4) (9)	Tropical desert and steppe Midlatitude desert and steppe	Brown soils Desert soils
Dry desert	(4a) (9a)	Tropical desert Midlatitude desert	Red and gray desert
Tundra biome			
Arctic grassy tundra	(11)	Tundra	Tundra
Alpine tundra		Highland climate (alpine zone)	Tundra

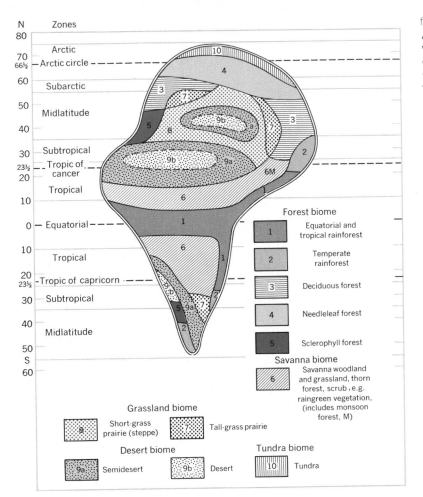

figure 12.5

A schematic diagram of the vegetation formation classes on an idealized continent. Compare with the world vegetation map, Figure 12.6.

Eurasia and Africa into a single supercontinent. Figure 12.6 is a highly generalized world map of the formation classes. It simplifies the very complex patterns of natural vegetation to create large uniform regions in which a given formation class might be expected to occur.

Forest biome

Equatorial rainforest and *tropical rainforest* consist of tall, closely set trees; their crowns form a continuous canopy of foliage and provide dense shade for the ground and lower layers (Figure 12.7). The trees are characteristically smooth barked and unbranched in the lower two thirds. Trunks are sometimes buttressed at the base by radiating, wall-like roots (Figure 12.8). Tree leaves are large and evergreen; from this characteristic the equatorial rainforest is often described as "broadleaf evergreen forest." Crowns of the trees tend to form into two or three layers, or strata, of which the highest layer consists of scattered emergent crowns rising to 130 ft (40 m) and protruding conspicuously above a second layer, 50 to 100 ft (15 to 30 m), which is continuous (Figure 12.9). A third, lower layer consists of small, slender trees 15 to 50 ft (5 to 15 m) high with narrow crowns.

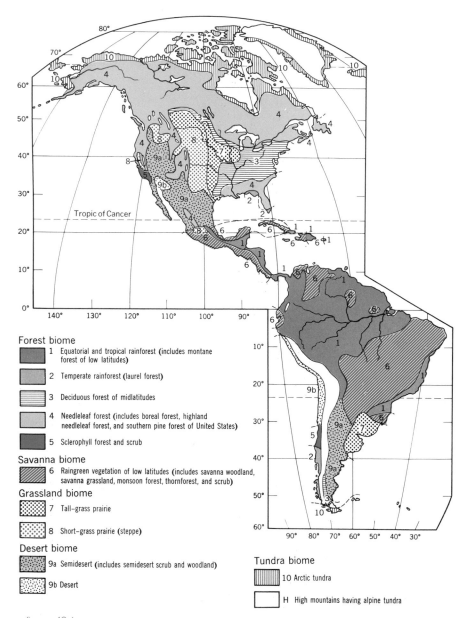

Forest biome

■ 1 Equatorial and tropical rainforest (includes montane forest of low latitudes)

□ 2 Temperate rainforest (laurel forest)

▤ 3 Deciduous forest of midlatitudes

□ 4 Needleleaf forest (includes boreal forest, highland needleleaf forest, and southern pine forest of United States)

■ 5 Sclerophyll forest and scrub

Savanna biome

▨ 6 Raingreen vegetation of low latitudes (includes savanna woodland, savanna grassland, monsoon forest, thornforest, and scrub)

Grassland biome

▨ 7 Tall-grass prairie

▨ 8 Short-grass prairie (steppe)

Desert biome

▨ 9a Semidesert (includes semidesert scrub and woodland)

□ 9b Desert

Tundra biome

▥ 10 Arctic tundra

□ H High mountains having alpine tundra

figure 12.6

World map of the major vegetation formation classes. This map is very generalized and is intended only to delineate large regions where the formation class might be expected to occur. (Data of S. R. Eyre and other sources. Based on Goode Base Map.)

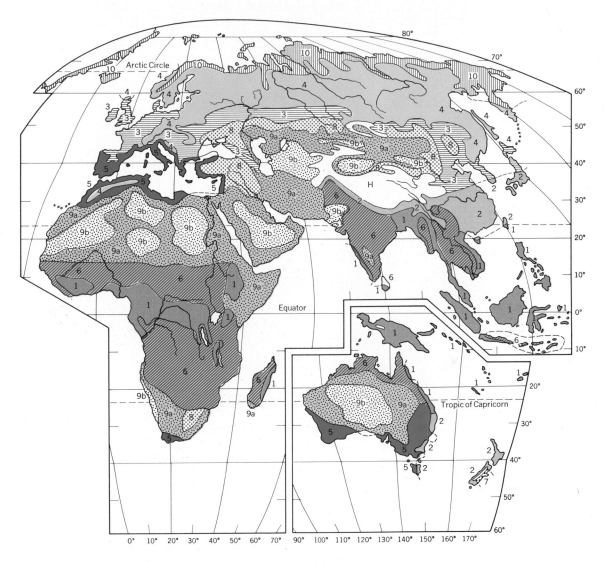

Typical of the equatorial rainforest are lianas, thick woody vines supported by the trunks and branches of trees. Some are slender, like ropes; others reach thicknesses of 8 in. (20 cm). They rise to heights of the upper tree levels where light is available and may have profusely branched crowns. Lianas may depend on a growing tree to be carried upward, where the liana has no devices with which to climb by itself. Other woody climbers rise by winding about the tree trunk. *Epiphytes* are numerous in the equatorial rainforest. These are plants attached to the trunk, branches, and foliage of trees and lianas, using the "host" solely as a means of physical support. Epiphytes are of many plant classes and include ferns, orchids, mosses, and lichens.

A particularly important botanical characteristic of the equatorial rainforest is the large number of species of trees that coexist. It is said that as many as 3000 species may be found in a square mile. Individuals of a given species are often widely separated. Consequently, if a particular tree species is to be extracted from the forest for commerical

figure 12.7
Equatorial rainforest of the Amazon lowland in Brazil. This forest of tall broad-leaved evergreen trees and numerous lianas is on the relatively high ground. (Otto Penner, Instituto Agronómico do Norte.)

Equatorial rainforest is a response to a climate that is continuously warm, frost-free, and has abundant precipitation in all months of the year (or, at most, only one or two dry months). A large water surplus characterizes the annual water budget so that soil water is adequate at all times. In the absence of a cold or dry season, plant growth goes on continuously throughout the year. Individual species have their own seasons of leaf shedding, possibly caused by slight changes in the light period.

World distribution of equatorial rainforest is shown on the vegetation map (Figure 12.6). Rainforest lies astride the equator. The principal world areas are: the Amazon lowland of South America; the Congo lowland of Africa and a coastal zone extending westward from Nigeria to Guinea; and the East Indian region, from Sumatra on the west to the islands of the western Pacific on the east.

Tropical rainforest is quite similar in structure to the equatorial variety, but extends through the tropical zone (lat. 10°-25° N and S) along coasts windward to the trades. The trade-wind coastal climate in which the tropical rainforest

uses, considerable labor is involved in seeking out the trees and transporting them from their isolated positions.

The floor of the equatorial rainforest is usually so densely shaded that plant foliage is sparse close to the ground and gives the forest an open aspect, making it easy to traverse. The ground surface is covered only by a thin litter of leaves. Rapid consumption of dead plant matter by bacterial action results in the absence of humus on the soil surface and within the soil profile. These conditions are typical of the laterization process with which the rainforest is identified.

figure 12.8
Buttress roots at the base of a large rainforest tree in the Canal Zone. The man standing at the right of center gives a sense of scale. Lianas can be seen near the center of the picture. (American Museum of National History.)

figure 12.9
**This diagram shows the typical structure of tropical rainforest. (After
J. S. Beard, 1946.)**

thrives has a short dry season, but not intense enough to deplete the soil water.

In the southern hemisphere belts of tropical rainforest extend down the eastern Brazilian coast, the Madagascar coast, and the coast of northeastern Australia. Tropical rainforest is important along east coasts in Central America and the West Indies. These highlands receive abundant orographic rainfall in the belt of trades. A good example seen by many Americans is the rainforest of the eastern mountains of Puerto Rico.

In Southeast Asia tropical rainforest is extensive in coastal zones and highlands, which have heavy monsoon rainfall and a very short dry season. The western coasts of India and Burma have tropical rainforest supported by orographic rains of the southwest monsoon.

Monsoon forest presents a more open tree growth than the equatorial and tropical rainforests (Figure 12.10). Consequently, there is less competition among trees for light but a

greater development of vegetation in the lower layers. Tree heights are less than in the equatorial rainforest. Many tree species are present and may number 30 to 40 species in a small tract. Tree trunks are massive; the bark is often thick and rough. Branching starts at a comparatively low level and produces large, round crowns.

Perhaps the most important feature of the monsoon forest is the deciduous habit of most of the tree species present. The shedding of leaves results from the stress of a long dry season that occurs at time of low sun and cooler temperatures. In the dry season the forest has somewhat the dormant winter aspect of deciduous forests of midlatitudes. A representative example of a monsoon forest tree is the teakwood tree.

Monsoon forest is a response to a wet-dry tropical climate in which a long rainy season with a water surplus alternates with a dry, rather cool season having a soil-water deficit. These conditions are most strongly developed

figure 12.10
Monsoon forest in Chieng Mai Province, northern Thailand. Scale is indicated by a line of people crossing the clearing. (Robert L. Pendleton, American Geographical Society.)

in the Asiatic monsoon climate, but are not limited to that area. Perhaps the type regions of monsoon forest are in Burma, Thailand, and Cambodia. Large areas of deciduous tropical forest occur in west Africa and in Central and South America, bordering the equatorial and tropical rainforests.

Temperate rainforest is an evergreen forest differing widely in composition from one global region to another. In some northern hemisphere areas this forest consists of broad-leaved trees such as evergreen oaks, and trees of the laurel and magnolia families. The name "laurel forest" is applied to these forests, which are associated with the humid subtropical climate in the southeastern United States, southern China, and southern Japan. However, these lands are under intense crop cultivation and have been largely deforested for several centuries, so that little natural forest remains.

Another climatic association of temperate rainforest is with the marine west coast climate in the southern hemisphere, notably in New Zealand and Chile. Here the kinds of trees are quite different from those of the northern hemisphere.

Temperate rainforests tend to have a well-developed lower layer of vegetation that, in different places, may include tree ferns, small palms, bamboos, shrubs, and herbaceous plants. Lianas and epiphytes are abundant. Particularly striking at higher elevations, where fog and cloud are persistent, is the sheathing of tree trunks and branches by mosses.

Midlatitude deciduous forest is familiar to inhabitants of eastern North America and western Europe as a native forest type. It is dominated by tall, broadleaf trees that provide a continuous and dense canopy in summer but shed their leaves completely in the winter (Figure 12.11). Lower layers of small trees and shrubs are weakly developed. In the spring a luxuriant low layer of herbs quickly develops, but this is greatly reduced after the trees have reached full foliage and shaded the ground.

The deciduous forest is almost entirely limited to the midlatitude landmasses of the northern hemisphere. Common trees of the deciduous forests of eastern North America, southeastern Europe, and eastern Asia are oak, beech, birch, hickory, walnut, maple, basswood, elm, ash, tulip, sweet chestnut, and hornbeam. In western and central Europe, under a marine west coast climate, dominant trees are mostly oak and ash, with beech in cooler and moister areas. Where the deciduous forests have been cleared in lumbering, pines readily develop as second-growth forest.

The midlatitude deciduous forest represents a response to a moist continental climate that receives adequate precipitation in all months. There is a strong annual temperature cycle with a cold winter season and a warm summer. Precipitation is markedly greater in the summer

figure 12.11
A stand of mature sugar maples, many over 200 years old, in the Allegheny National Forest, Pennsylvania. (U.S. Forest Service.)

months (especially in eastern Asia) and increases at the time of year when water need is greatest. Only a small water shortage is incurred in the summer, while a large surplus normally develops in spring.

Needleleaf forest is composed of straight-trunked, conical trees with relatively short branches and small, narrow, needlelike leaves. These trees are conifers. The needleleaf forest provides deep shade to the ground so that lower layers of vegetation are sparse or absent, except for a thick carpet of mosses in many places. Species are few, and large tracts of forest consist almost entirely of only one or two species (Figure 12.12).

Needleleaf forest forms two great continental belts, one in North America and one in Eurasia. These belts span the landmasses from west to east in lat. 45° to 75° N. The needleleaf forest of North America, Europe, and western Siberia is composed of evergreen conifers such as spruce, fir, and pine. The forest of north-central and eastern Siberia is dominantly of larch, which sheds its needles in winter and is a deciduous forest.

Needleleaf evergreen forest extends into lower latitudes wherever there are mountain ranges and high plateaus. In western North America this forest reaches southward in long fingers into the Cascade, Sierra Nevada, and Rocky

figure 12.12
Spruce trees make up almost the entire needleleaf forest in this locality in Quebec, Canada. (Pierre Dansereau.)

Mountain ranges and over parts of the higher plateaus of the southwest. In Europe needleleaf evergreen forests flourish on all of the higher mountain ranges and well into Scandinavia.

The needleleaf forest of the subarctic continental climate is an expression of a regime of long, severely cold winters and a short, cool summer. Podzols are the dominant soil type. Soil water is adequate in the growing season, but it is locked up in the soil as ice much of the year. Toward its northern limits this forest gives way to cold woodland in which trees are smaller and more widely spaced apart. A cover of lichens on the open ground between trees gives the woodland a parklike appearance (see Figure 15.3). In places, the northern limit of growth, or tree line, is quite distinct. Beyond lies the treeless tundra.

Sclerophyll forest consists of low trees with small, hard, leathery leaves. Typically, the trees are low branched and gnarled, with thick bark. The formation class includes much woodland, an open forest in which the canopy coverage is 25 to 60 percent (Figure 12.13). Included

are extensive areas of scrub, a formation type consisting of shrubs having a canopy coverage of perhaps 50 percent. The trees and shrubs are evergreen. Their thickened leaves are retained despite a severe summer drought. There is little stratification in the sclerophyll forest and scrub, although there may be a spring herb layer.

Sclerophyll forest and woodland is closely associated with the Mediterranean climate and is quite narrowly limited to west coasts between lat. 30° or 45° N and S. In the Mediterranean lands sclerophyll woodland forms a narrow peripheral coastal belt. What may have once been luxuriant forests were greatly disturbed by Man over the centuries and reduced to open woodland or entirely destroyed.

The other northern hemisphere region of evergreen hardwood forest is that of the California coast ranges. Much of the vegetation is a scrub known as "chaparral" (Figures 12.14 and 12.15). Important areas of sclerophyll forest, woodland, and scrub are also found in South Africa and Australia.

figure 12.13
This eucalyptus woodland was natural sclerophyll vegetation of the Darling Range, Western Australia. Since 1930, when the photo was taken, this woodland has been largely cut away. (Agent General for Western Australia, courtesy of the American Geographical Society.)

figure 12.14
Chaparral on steep mountain slopes of the San Dimas Experimental Forest, near Glendora, California. This vegetation had been protected from burning for many years. (A. N. Strahler.)

figure 12.15
Detail of chaparral vegetation, San Dimas Experimental Forest (see Figure 12.14). A tilted rain gauge, 8 in. (20 cm) in diameter, rests on bare ground at the left. (A. N. Strahler.)

The Mediterranean climate of the sclerophyll forest is one of great environmental stress, because the severe drought season coincides with high air temperatures. A large soil-water shortage accumulates in the summer and lasts far into autumn. The wet, mild winter is, by contrast, highly favorable to rapid plant growth.

Savanna biome

Savanna woodland consists of trees spaced rather widely apart. The intervening ground has a dense, low layer that may consist of grasses or low shrubs. This formation class is sometimes called "parkland" because of the open, parklike appearance of the vegetation.

Although this vegetation formation can appear in a wide range of latitudes, many geographers associate savanna woodland closely with the tropical wet-dry climate. In the tropical savanna woodland, the trees are of medium height, the crowns are flattened or umbrella shaped, and the trunks have thick, rough bark (see Figure 12.17). Some species of trees are xerophytic forms with small leaves and thorns. Others are deciduous, shedding their leaves in the dry season. In this respect, tropical savanna woodland is closely akin to the monsoon forest into which it grades. The trees are capable of withstanding the fires that sweep through the lower layer in dry season.

Tropical savanna woodland is found widely throughout Africa, South America, Southeast Asia, northern Australia, and in Central America and the Caribbean islands.

Thornbush is made up of xerophytic trees and shrubs responding to a climate with a very long dry season and only a short, but intense, rainy season. This vegetation formation consists of tall, closely spaced woody shrubs commonly bearing thorns and largely deciduous (Figure 12.16). Cactus plants may also be present. The lower layer of herbs may consist of annuals,

figure 12.16
Thornbush in central Africa during the cool, dry season. The plants have shed their leaves and the grasses have turned to straw. (Akeley Expedition, American Museum of Natural History.)

which largely disappear in the dry season, or of grasses.

Tropical scrub is a dense growth of low woody shrubs occurring in patches or clumps separated by barren ground. Scrub may develop in stony, sandy, or gravelly sites in areas of thornbush.

Savanna grassland has widely scattered trees rising from broad expanses of grasses (Figure 12.17). This formation class might equally well be placed in the grassland biome.

As with the savanna woodland, the savanna grassland supports trees and shrubs that are xerophytic or deciduous. In some localities the trees are palms, giving a "palm savanna." Grasses in the tropical savanna are characteristically tall, with stiff coarse blades, commonly higher than the height of a man, and even up to 12 ft (4 m) high. In the dry season these grasses form a yellowish straw mat that is highly flammable and subject to periodic burning. Many plant geographers hold the view that periodic burning of the savanna

grasses is responsible for the maintenance of the grassland against the invasion of forest. Fire does not kill the underground parts of grass plants, but limits tree growth to a few individuals of fire-resistant species. The browsing of animals, which kills many young trees, is also a factor in maintaining grassland at the expense of forest.

The African savanna is perhaps the most celebrated of the tropical grasslands, spanning the continent from west to east in two great belts, centered about on lat. 10° N and S, and connected in eastern Africa by a broad north-south belt crossing the equatorial zone.

All formation classes of the savanna biome within the tropical zone feel the strong control of the tropical wet-dry climate and its feast-or-famine soil-water budget. In all these formations plants quickly turn green and grow vigorously at the onset of the rainy season. On the world vegetation map (Figure 12.6), the savanna biome formation classes are grouped as one, along with the monsoon forest, which shares the same climate controls. This single map unit is called "raingreen vegetation." Soils of the savanna biome include latosols, reddish-brown soils, and vertisols.

figure 12.17
African tall-grass savanna in Kenya. (Richard U. Light, American Geographical Society.)

Grassland biome

Prairie consists largely of tall grasses. *Forbs,* which are broad-leaved herbs, are also present, but they are secondary in abundance. Trees and shrubs are almost totally absent but may occur in the same region as narrow patches of forest in valleys. The grasses are deeply rooted and form a continuous and dense sward.

Prairie grasslands are best developed in the midlatitude and subtropical zones where winter and summer seasons are well developed. The grasses flower in spring and early summer, the forbs in late summer. In Iowa a representative region of tall-grass prairie, typical grasses are the bluestems; a typical forb is black-eyed Susan (Figure 12.18).

The tall-grass prairies are closely associated with the humid continental climate. They are found in the subhumid sections of that climate in which water need exceeds precipitation during the summer months. The North American prairies are found in a broad belt extending from Illinois northwestward to southern Alberta and Saskatchewan. Areas of forest are mixed with areas of prairies in a transitional belt between forest and prairie regions.

figure 12.18
Kalsow Prairie, Iowa, has been set aside as an example of virgin prairie. (State Conservation Commission of Iowa.)

The tall-grass prairies grade into short-grass prairies and then into steppe grasslands in the direction of increasing aridity. In Europe, a typical region of tall-grass prairie is the puszta of Hungary. The Argentine pampa is another region of prairie vegetation. The tall-grass prairies are formed on highly fertile prairie soils and chernozem soils. For this reason they have been intensively cultivated, and few vestiges of natural prairie remain.

Steppe grassland, or *short-grass prairie,* is a formation class consisting of short grasses tending to be bunched and sparsely distributed (see Figure 14.25). Scattered shrubs and low trees may also be found in the steppe grasslands. Ground coverage is small, and much bare soil is exposed. Many species of grasses and other herbs occur. A typical grass of the American steppe is buffalo grass; other typical plants are the sunflower and loco weed.

Steppe grasslands range from lat. 25° to 50° N and S. Steppes of midlatitudes are associated with the midlatitude steppe climate in which there is a large soil-water shortage in summer and no water surplus in spring. Winters in the midlatitude steppes are cold; the summers are warm to hot. The dominant soil-forming process is calcification. Soils contain a large excess of precipitated calcium carbonate and are very rich in nutrient bases. Brown soils are typical.

Desert biome

Semidesert is a xerophytic shrub vegetation with a very sparse lower herb layer. Semidesert shrub vegetation is well developed in subtropical and midlatitude dry climates having a small annual total rainfall and high summer temperatures. An example is the sagebrush vegetation of the middle and southern Rocky Mountain region and Colorado Plateau (Figure 12.19). Creosote bush is a dominant shrub in the semidesert plains of the Mexican

figure 12.19
Sage-brush semidesert, Vermilion Cliffs, near Kanab, Utah, 1906. (Douglas Johnson.)

border region of Texas, New Mexico, and Arizona. Semidesert shrub vegetation has spread widely into areas of the western United States that were formerly steppe grasslands as a result of overgrazing and trampling by livestock. The world map of vegetation (Figure 12.6) shows that semidesert vegetation occurs along the less arid margins of the deserts and in favorable highland locations within the desert.

In midlatitudes soils of the semidesert are typically the brown soils, with an excess of calcium carbonate. Although there is a precipitation maximum in the summer, the soil-water shortage is severe.

Dry desert is a formation class of xerophytic plants widely dispersed and providing almost negligible ground cover. The visible vegetation normally consists of small hard-leaved or spiny shrubs, cactus, or hard grasses. Many species of small annuals may be present, but these appear only after a rare but heavy desert downpour.

Desert floras differ greatly from one part of the world to another. In the Mohave and Sonoran deserts of the southwestern United States, plants are often large and, in places, give a near-woodland appearance (Figure 12.20). Well-known desert plants are the treelike saguaro cactus, prickly-pear cactus, ocotillo, creosote bush, and smoke tree.

In the Sahara Desert (most of it very much drier than in the American desert) a typical plant is a hard grass; another, found along the dry beds of watercourses, is the tamarisk, a shrub or tree.

Global distribution of dry desert vegetation is shown on the world vegetation map (Figure 12.8). Much of the area assigned to desert vegetation has no plants of visible dimensions, because the surface consists of shifting dune sands or sterile salt flats.

The soil-water budget of the dry-desert vegetation usually shows a greater quantity of water need than of precipitation in every one of the twelve months. The annual total soil-water shortage is therefore very large. When rain falls in rare, torrential rainstorms, the soil water is used to maximum advantage by the plants and is rapidly depleted. During long periods when soil water storage is near zero, the plants must maintain a nearly dormant state to survive.

Tundra biome

The *arctic grassy tundra* is a formation limited to very cold climates having ample soil water and often saturated soils. Grassy tundra flourishes under an arctic climate of long summer days during which time the ground ice melts only in a shallow surface layer. The frozen ground beneath (permafrost) remains impermeable, and melt water cannot readily escape. Consequently, in summer, a marshy condition prevails for at least a short time over wide areas. Humus accumulates in well-developed layers in the soil.

figure 12.20
Vegetation of the Sonoran Desert, southwestern Arizona. Ocotillo plant in the left foreground; saguaro cactus plants in left distance. (A. N. Strahler.)

Plants of the arctic grassy tundra are small. Sedges, grasses, mosses, and lichens dominate the tundra in a low layer (Figure 12.21). There are also many species of forbs, which flower brightly in the summer. Considerable variations in composition of the tundra are seen in the range from wet to well-drained habitats. One form of tundra consists of sturdy hummocks of plants with low, water-covered ground between. In the regions of grassy tundra areas of arctic scrub vegetation composed of stunted willows and birches will also be found.

Size of tundra plants is partly limited by the mechanical rupture of roots during freeze and thaw of the surface layer of soil, producing shallow-rooted plants. In the winter drying winds and mechanical abrasion by wind-driven snow tend to reduce any portions of a plant that project above the snow.

In all latitudes, where altitude is sufficiently high, an *alpine tundra* is developed above the limit of tree growth and below the vegetation-free zone of barren rock and perpetual snow. Alpine tundra resembles arctic tundra in many physical respects.

figure 12.21
Cotton-grass meadows of the arctic tundra on the coastal plain of Alaska. (William R. Farrand.)

S — Equatorial rainforest — Savanna woodland — Savanna grassland — Tropical scrub — Tropical desert — N

Profile from equator to tropic of cancer, Africa

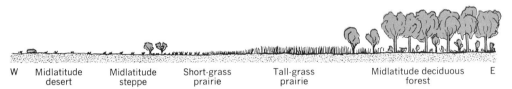

S — Tropical desert — Subtropical steppe — Sclerophyll forest — Midlatitude deciduous forest — Subarctic needleleaf forest — Subarctic woodland (taiga) — Arctic tundra — N

Profile from tropic of cancer to Arctic circle, Africa-Eurasia

W — Midlatitude desert — Midlatitude steppe — Short-grass prairie — Tall-grass prairie — Midlatitude deciduous forest — E

Profile across United States, 40°N, Nevada to Ohio

figure 12.22
Three schematic profiles showing the succession of plant formation classes across climatic gradients.

Global vegetation in review

The leading theme of this chapter has been that the structure of the natural plant cover is a response to environments dictated by climate. Each of the life forms favored by green plants has a capability for survival and a limit to survival as well. Forests exist where the environment is most favorable to net productivity of the biomass because of the abundance of heat and of soil water during a long growth season. Where soil water is in short supply but heat remains adequate, forest gives way to the savanna structure in which shrubs and herbs have the upper hand over trees (Figure 12.22). The savanna woodland grades into grassland and this, in turn, grades

into desert, where only those plants capable of living through long drought periods can survive. Traced into colder, higher latitudes, where heat supply becomes reduced below the optimum level, the forest gives way to arctic tundra. Here plants contend with the effects of prolonged freezing of soil water and have only a short cold annual period of growth.

We have found a close correspondence of vegetation structure with soil type, since soil-forming processes are also strongly dominated by climate. But organic processes also shape soil character, so that the close associations of vegetation formation classes and soil types are partly a result of interaction.

Low-Latitude Environments

Chapter 13

NOW THAT WE have assembled the climate-soils-vegetation complex, it will be meaningful to survey the climate regions of the world in a new perspective, viewing them as a global system of _environmental regions_. Each environmental region has its own set of opportunities for and restraints on expanded utilization by Man. Especially crucial are the opportunities that Man has to modify the natural terrestrial ecosystems into food-producing ecosystems. This transformation has been remarkably successful in some parts of the world, but highly unsuccessful in others. Can the production of food be further extended into regions now clothed in equatorial rainforests or savanna grasslands? Can more deserts be made to bloom by massive applications of irrigation water? Partial answers to these questions lie in the environmental factors relating to physical geography. Other parts of the answers lie in the application of technology and the availability of supplies of energy. Ethical issues and political policies are also factors to be weighed in seeking these answers.

In this chapter we investigate the environments of climates belonging to Group I: climates controlled largely by equatorial and tropical air masses. Climates of Group I cover an enormous part of the earth's land surface, most of it under the control of the developing nations. Most of the enormous population within the scope of these environments is subsisting under an agricultural economy barely able to support its burden of human lives.

In past decades great colonial empires were carved out of the low-latitude regions by the Western powers. Today, nearly all of the former colonies are politically independent. Yet the physical environment of each has not changed. The same complex of climate, soils, and vegetation confronts the free peoples of these nations. Years of colonialism brought

figure 13.1

World map of the wet low-latitude environments, combining the wet equatorial climate (1) and the trade-wind and monsoon coastal climate (2). (Based on Goode Base Map.)

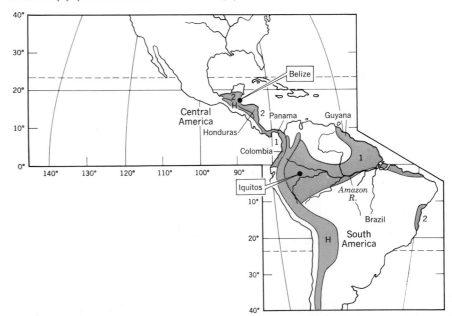

major changes to the natural ecosystems of the low latitudes; many of these changes were dictated by the desires of the ruling European nations to exploit natural resources for gain. As the developing nations plan their futures, they must cope with the inheritance of the past as well as with the constraints of the natural environmental controls.

(1) Wet equatorial climate (Köppen: Af)

(2) Trade-wind and monsoon coastal climate (Köppen: Af, Am)

To unify and simplify our investigation, we will group into one environmental region the wet equatorial climate (1) and the trade-wind littoral and monsoon coastal climate (2). We can refer to this unified region as the wet low-latitude environment (Figure 13.1). The rainforest mentioned here includes the equatorial and tropical varieties, which are very similar in structure. As a unit, the wet low-latitude environment has a common denominator through its large water surplus, its latosolic soils, and its rainforest.

The wet equatorial climate (1) is found in a belt from about lat. 10° N to 10° S. The major regions in which it occurs are the Amazon lowland of South America, the Congo basin of equatorial Africa, and the East Indies, from Sumatra to New Guinea.

In the wet equatorial climate average monthly and annual temperatures are always close to 80° F (27° C). The seasonal range of temperature is so slight as to be imperceptible because the sun is not far from the zenith throughout the year. Rainfall is heavy during the entire year, but with considerable differences in monthly averages because of the seasonal shifting of the intertropical convergence zone (ITC). Usually the rainfall is heaviest when the ITC is close by on its annual migration northward or southward.

Figure 13.2 is a climograph for Iquitos, Peru, a typical wet equatorial station located close to the equator in the broad, low basin of the Amazon River. Note that the annual range in temperature is only 4 F° (2.2 C°) and that the

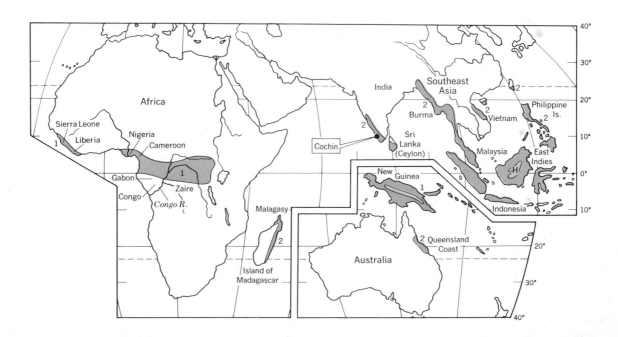

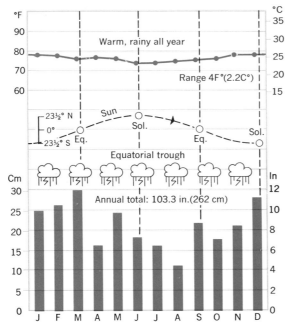

figure 13.2

Iquitos, Peru, is a rainforest station near the equator. Temperatures differ very little from month to month, and there is copious rainfall throughout the year.

monthly mean temperatures. In other words, daily variations far exceed seasonal variations in the wet equatorial climate.

The soil-water budget of the wet equatorial climate typically shows a large water surplus in every month (see Figure 9.10a). The storage capacity of the soil remains at its maximum much of the time, and the water need of plants is met almost all of the time. As a result, plant photosynthesis can be sustained at its maximum value.

The trade-wind and monsoon coastal climate (2) is found between lat. 10° and 25° N and S along narrow coastal zones windward to the trades. The climate map (Figure 13.1) shows this climate along the east coasts of Central and South America, the Caribbean islands, Madagascar (Malagasy), Indochina, the Philippines, and northeastern Australia. Figure 13.4 is a climograph for Belize in Central America, a coastal city on the Caribbean Sea. Moist maritime tropical (mT) air masses reach this coast freely, and rainfall is copious from June through November. The rainfall occurs in periods when weak easterly air waves move onshore or when a tropical storm strikes. Then, after winter solstice, a short period of reduced precipitation occurs, with the minimum in March and April. Soil water remains ample in all months for rainforest, and there is only a brief period when soil water in storage is drawn to fulfill the water need. Compared with the equatorial zone, air temperatures show a marked annual cycle with a maximum in the

annual rainfall total is more than 100 in. (250 cm). In all but one month the mean monthly rainfall is more than 6 in. (15 cm).

You can get some idea of the extreme monotony of the daily temperature cycle in this climate from Figure 13.3. This graph shows minimum and maximum daily temperatures for 2 months at Panama, lat. 9° N. The daily range is normally from 15 to 20 F° (8 to 11 C°), a vastly greater range than the annual range of

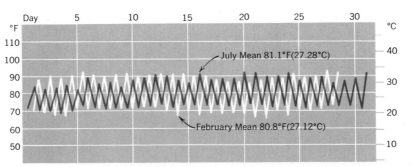

figure 13.3
July and February temperatures at Panama, lat. 9° N. The sawtooth graph shows daily maximum and minimum readings for each day of the month. (After Mark Jefferson, Geographical Review.)

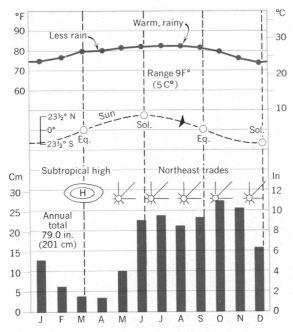

figure 13.4

This climograph for Belize, a Central American east-coast city, at lat. 17° N, shows a marked season of low rainfall following the period of low sun.

monsoon coastal climate also extends along the entire coast of Burma, reaching to a latitude of 25° N.

Vegetation and animal life Throughout all three kinds of wet low-latitude climates natural vegetation is of the equatorial and tropical rainforest formation classes described in Chapter 12. _Selva_ is another name widely used for the equatorial rainforest; it is a Portuguese word originally applied to the rainforest of the Amazon lowland (Figure 13.6).

Animal life of the rainforest is most abundant in the upper layers of the vegetation. Above the canopy, birds and bats are important carnivores, feeding largely on insects above and within the topmost canopy. Below this level live a wide variety of birds, mammals, reptiles, and invertebrates; they feed on the leaves, fruit, and nectar abundantly available in the main part of the canopy. Ranging

high-sun months. The annual range, 9 F° (5 C°), is about double that in the equatorial zone.

We have also included in this climate the hilly and mountainous coasts of Southeast Asia, exposed to maritime equatorial (mE) air masses. Here, a strong annual precipitation cycle is dominant. There is a wet season with very high monthly rainfall in 2 or 3 successive months, starting in June, and a pronounced season of low rainfall at time of low sun. The climograph for Cochin, India, illustrates this cycle (Figure 13.5). Cochin is situated on the southwest coast of India and receives onshore movements of moist air from the Indian Ocean. Both June and July receive over 25 in. (64 cm) of rainfall per month, whereas in January and February the monthly amount is only about 1 in. (2.5 cm). Air temperatures show only a very weak annual cycle, dropping a bit during the rains, but the annual range is small. This

figure 13.5

Cochin, India, on a windward coast at lat. 10° N, shows an extreme peak of rainfall during the rainy monsoon, contrasting with a short dry season at time of low sun.

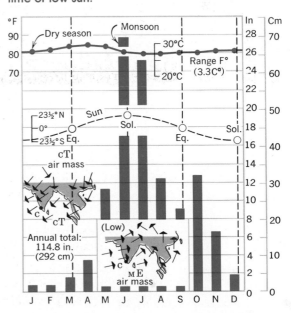

figure 13.6
This air view over the rainforest, or selva, of the Amazon Basin shows the Rio Negro, a tributary of the Amazon River. The locality is close to the equator. (Colonel Richmond, American Geographical Society.)

between the canopy and the ground are the climbing mammals, which forage in both layers. At the surface are the larger ground mammals, including herbivores, which graze the low leaves and fallen fruits, and carnivores, which prey on the abundant vertebrates and invertebrates found at the surface.

A major difference between the low-latitude rainforests and forests of higher latitudes is the great diversity of species that it possesses. The rainforest may have as many as 3000 different tree species in a square mile area — midlatitude forests would be considered very diverse indeed with only one tenth that number! The fauna of the rainforest is also very rich. A 6-sq mi (16-sq km) area in the Canal Zone, for example, contains about 20,000

species of insects, whereas there are only a few hundred in all of France. This large number of species occurs because the rainforest environment is so uniform and free of physical stress.

In the wet low-latitude environment the large water surplus and prevailing warm soil temperatures promote decay and decomposition of rock to great depths, so that a thick regolith is typical of most areas. Streams flow copiously throughout most of the year and river channels are lined along the banks with dense forest vegetation. Natives of the rainforest travel the rivers in dugout canoes, and these travelers now enjoy self-propulsion by attaching an outboard motor. Larger shallow-draft river craft have turned the major waterways into the main arteries of trade; towns are situated on the river banks. Aircraft have added a new dimension in mobility, crisscrossing the almost trackless green sea of forest to find landings in clearings or on reaches of the broader rivers.

In many parts of the wet low-latitude environment the terrain is mountainous, with steep slopes. Slides and avalanches of soil are common and can strip away the forest and soil to expose regolith and bedrock.

Economic products and food resources

Several forest products are of economic value. Rainforest lumber, such as mahogany, ebony, or balsawood, is an important export. Quinine, cocaine, and other drugs come from the bark and leaves of tropical plants; cocoa comes from the seed kernel of the cacao plant. Natural rubber is made from the sap of the rubber tree. The tree comes from South America, where it was first exploited. Rubber trees also are widely distributed through the rainforest of Africa. Today, the principal production is from plantations in Indonesia, Malaya, Thailand, Vietnam, and Sri Lanka (Ceylon).

An important class of food plants native to the wet low-latitude environment are starchy staples; some are root structures, others are fruits. Manioc, also known as cassava, is one of these staples. The plant has a tuberous root — something like a sweet potato — that reaches lengths over a foot and may weigh several pounds (Figure 13.7). Prepared so as to remove a poisonous cyanide compound, the roots yield a starchy food with very little (1 percent or less) protein. Manioc was used as a food in the Amazon basin of Brazil; it was taken to Africa in the sixteenth century by the Portuguese. In the present century, its use in Africa has increased widely. Manioc is now also important in Indonesia. The food unfortunately contributes to malnutrition because it contains too little protein. Manioc is

cultivated in small plantations placed in forest clearings.

Yams are another starchy staple of the wet equatorial regions. Like the manioc, the yam is a large underground tuber. Yams are a major source of food in West Africa. The plant was introduced into the Caribbean region during the era of slavery and is an important food in the region today. The yam has a higher protein content than the manioc.

Taro is another starchy staple of humid low-latitude climates. The taro plant has large, elephantlike ears, which are edible, but the food value lies mostly in an enlarged underground portion of the plant, called a

figure 13.7
Tuberous manioc roots, brought in dugout canoes from villages in the Brazilian rainforest, are being loaded on a river boat for transport to market. At one time manioc, in the form of tapioca, was popular in the United States as a dessert pudding. (Charles Perry Weimer)

corm. Visitors to Hawaii know the taro plant through its transformation into poi, a fermented paste. Few mainland tourists eat poi a second time, but it has long been a favored food of the native Hawaiians. Taro corms consist of about 30 percent starch, 3 percent sugar, and only about 1 percent protein. Taro was imported into Africa from Southeast Asia and eventually reached the Caribbean region.

An important starchy staple in the form of a fruit is the breadfruit. A single breadfruit can attain a weight of 10 lb and is rich in carbohydrate. A native of the Pacific islands, the breadfruit has long been a staple of the diet of the Polynesians. In 1789 Captain Bligh in his ship, H.M.S. Bounty, attempted to bring a cargo of small breadfruit trees to the West Indies to be cultivated as a food supply for African slaves. As everyone knows, something went wrong with his plans.

The banana and plantain are starchy staples familiar to everyone. The banana plant lacks woody tissue and is, in fact, a perennial herb; the fruit is classed as a berry by botanists. The banana was first cultivated for food in Southeast Asia, then spread to Africa. It was imported into the Americas in the sixteenth century and quickly became well established. The plantain is a coarse variety of the banana that is starchy, has little sugar, and requires cooking.

Perhaps the plant most important to Man in the low latitudes is the coconut palm. Besides being a staple food, it provides a multitude of useful products in the form of fiber and structural materials. The coconut palm flourishes on islands and coastal fringes of the wet equatorial and tropical climates (Figure 13.8). Copra, the dried meat of the coconut, is a valuable source of vegetable oil. Copra and coconut oil are major products of Indonesia, the Philippines, and New Guinea. Palm oil and palm kernels of other palm species are an important product of the equatorial zone of West Africa and the Congo River basin.

figure 13.8
A grove of coconut palms along a sandy beach in the Solomon Islands. Wave action is undermining the trees. (American Museum of Natural History.)

The rainforest ecosystem In the past, Man has farmed the low-latitude rainforest by the *slash-and-burn* method — cutting down all the vegetation in a small area, then burning it. Most of the nutrients in a rainforest ecosystem are tied up in the biomass rather than in the soil. Burning the slash on the site releases the trapped nutrients, returning them to the soil. The supply of nutrients derived from the original biomass is small, however, and the harvesting of crops rapidly depletes the nutrients. After a few seasons of cultivation, the soil loses much of its productivity. A new field is then cleared in another area, and the old field is abandoned. Plants of the rainforest are able to reinvade the abandoned area because fruiting species and animal seed carriers are close by, and so the rainforest soon returns to its original state. In this way the primitive slash-and-burn agriculture is compatible with the maintenance of the rainforest ecosystem.

On the other hand, modern intensive agriculture uses large areas of land and is not compatible with the rainforest ecosystem. When such lands are abandoned, seed sources are so far away that the forest species cannot take hold. Instead, secondary species dominate, often accompanied by species from other vegetation types. The dominance of these secondary species is permanent, at least on the human time scale. Thus the tropical rainforest ecosystem is, in this sense, a nonrenewable genetic resource of many, many species of plants and animals that, once displaced by large-scale cultivation, can never return to reoccupy the area. Ecologists warn that the disappearance of thousands of species of organisms from the rainforest environment would mean the loss of millions of years of evolution, together with the destruction of the most complex ecosystem on the globe.

Future prospects for the wet low-latitude environment   Now we can take up a question raised in the opening paragraphs of this chapter: can the wet low-latitude environment be exploited as a new source of food for mankind? Some agricultural specialists are highly optimistic; but others, specializing in ecology, are pessimistic, or at best very dubious. What we can do here is to point out restraints on such expansion, where it relates to raising staples such as rice, soy beans, corn, or sugar cane, which are field crops harvested seasonally.

First, the low content of nutrients in the latosolic soils will require massive and repeated applications of fertilizers. These applied nutrients are not held in storage in substantial quantities in the soil and are quickly exported from the land in runoff because of the large water surplus. Second, there is no dry season in which crops can reach maturity and be harvested under dry conditions, as is the case in the tropical wet-dry climates and the subhumid and semiarid continental climates of midlatitudes. Third, the latosolic soils are capable, in a few areas, of becoming lithified — literally turning to rock — when denuded of plant cover and exposed to the atmosphere. This has actually happened on agricultural test farms in the selva of Brazil.

It is easy to say that agricultural technology can overcome these and other difficulties. Even if this conclusion is valid, no one questions the fact that the cost will be enormous in terms of energy input through all parts of the agricultural system. Fertilizers, pesticides, machinery, and fuel are only a part of that energy input. Large amounts of capital are needed to set up efficient management and marketing systems. In view of the pending destruction of the rainforest ecosystem as a consequence of agricultural expansion, we can justifiably raise the question as to whether benefits will outweigh losses. Would not the human effort and energy expenditure be more effectively applied in those environments where many natural factors favor the increased cultivation of protein-rich food staples?

(3) The tropical wet-dry climate (Köppen: <u>Aw</u>, <u>Cwa</u>)

Lying between the water-rich equatorial zone and the water-poor tropical deserts is an environmental region that shares the qualities of each of those zones in an appointed season of the year. We have already referred to the soil-water budget of this region as "feast-or-famine," in terms of reliability of the wet season. The dry season is absolutely guaranteed, but the wet season is a gamble of nature on which the lives of hundreds of millions of humans are wagered each year. In 1974 the human race was on the losing end of the wager in a number of places — from the Sahel of West Africa to the highlands of Ethiopia, and in parts of northwest India and Pakistan.

Even a feast year is no picnic because, with the best of harvests, there is not enough surplus to place in reserve for famine years. Moreover, the same rains that can bring good crops can bring devastating floods that raise havoc with crop yields. Despite these recurring misfortunes, the tropical wet-dry environment

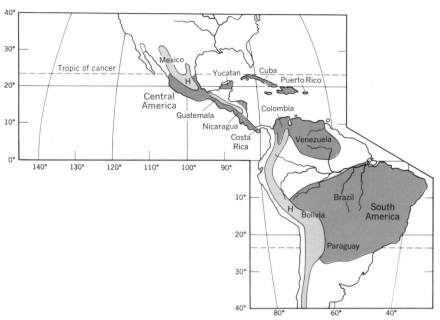

figure 13.9
World map of the tropical wet-dry climate. (Based on Goode Base Map.)

holds some of the most densely populated regions of the world.

The tropical wet-dry climate (3) has a wet season controlled by moist, warm maritime equatorial (mE) air masses at time of high sun and a dry season controlled by the continental tropical (cT) air masses at time of low sun.

The latitude belts in which tropical wet-dry climate are found lie roughly between lat. 5° and 25° N and S, throughout Central and South America, Africa, and Australia (Figure 13.9). In Southeast Asia this zone is pushed northward to lat. 10° to 30° N because the Asiatic landmass generates a powerful monsoon effect to reinforce the migration of the ITC.

Climate features of the tropical wet-dry climate can be judged from Figure 13.10, a climograph for Timbo, Guinea, a representative west African station. Rains begin just after vernal equinox and build to a maximum in the 2 months following summer solstice, when the ITC migrates to its most northerly position. Rainfall then declines as the low-sun season

arrives. Three months — December through February — are practically rainless. At this season the subtropical high-pressure cell dominates the climate. Its subsiding air mass is stable and dry.

The temperature cycle is particularly interesting, since it is linked to precipitation. In February and March air temperature rises sharply, bringing on the hot season. As soon as the rains set in, the effect of cloud cover and evaporation of rain is to cause the temperatures to decline. By July, temperatures resume an even level.

Let us now move a few degrees farther north, toward the Sahara Desert, to see how the climate changes. We are now in the Sahelian zone, or Sahel, which is a belt transitional from the tropical wet-dry climate to the tropical steppe. Most of the Sahel is regarded as tropical steppe, but it follows the rhythm of the wet-dry climate. Figure 13.11 is a climograph for Kayes, Mali, a tropical steppe station at lat. 14½° N. It shows that the wet season is

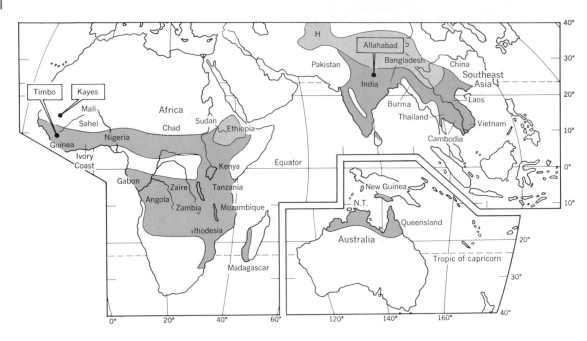

figure 13.10

Timbo, Guinea, at lat. 10½° N, is in the wet-dry tropical climate of West Africa. A long wet season at time of high sun alternates with an almost rainless dry season at time of low sun.

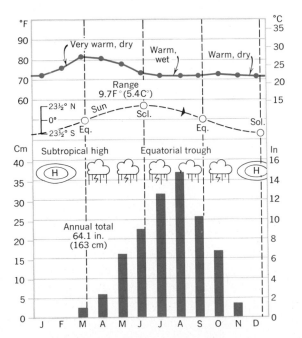

figure 13.11

Kayes, Mali, at lat. 14½° N, lies in the drought-stricken Sahel of West Africa. In normal years there is a short wet season with adequate rainfall. The dry season is long, with a succession of rainless months.

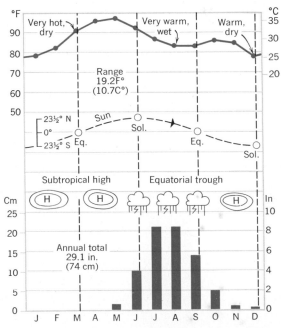

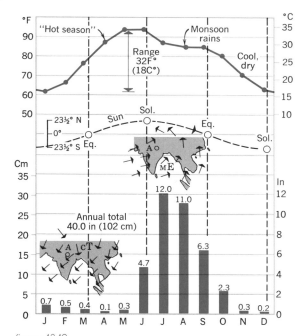

figure 13.12

Allahabad, India, lat. 25° N, lies in the rich agricultural lowland of the Ganges River. Here the monsoon rains are usually ample, but there is a long dry season. Note the very high temperatures in May and June, just before the onset of rains.

reduced in length and the monthly totals are decreased, while the dry season is lengthened to nearly 7 months. The annual total rainfall is only about half that at Timbo, located only about 250 mi (400 km) farther south. Air temperatures run a full 10 F° (5.5 C°) higher in all months than at Timbo. What we observe here is a steep climate gradient. Drought in the Sahel began in 1968 with failure of the rains and has continued for several years. In the news media, this change is described as a southward advance of the Sahara Desert. The change in rainfall is probably cyclic and should reverse in time, because this is the nature of climatic boundary zones.

The Asiatic wet-dry climate has some special features induced by the powerful monsoon controls. Figure 13.12 is a climograph for Allahabad, India. This station is located at lat.

25° N, about 900 mi farther north than Timbo. The main difference in the climate of the two stations is in the temperature cycle. Allahabad has a much greater annual range than Timbo. The hot season is just as hot as at Kayes, but the cool season at Allahabad is much cooler. The rainy season of Allahabad peaks more sharply than at Timbo, and there is a small amount of rain (on the average) in most of the drier months.

The soil-water budget of a tropical wet-dry station is illustrated by Calcutta, India (Figure 9.10c). The wet season is longer and wetter than at Allahabad, but the pattern is the same. The rains bring a substantial surplus, but there is a long period of storage withdrawal during the dry season, and the accumulated water need is large by the time the rains again arrive.

Soils, vegetation, and animal life Large areas of the wet-dry tropical climate are underlain by latosols or soils of latosolic affiliation. In the drier areas hard crusts of laterite lie near the ground surface or are exposed by erosion. Other areas have very dark-colored tropical soils, such as the vertisols described in Chapter 10. A third important class of soils are the alluvial soils, brought by flooding rivers, that spread silt over wide floodplains. Soils of the lower Ganges-Brahmaputra floodplain and deltaic plain are of this origin and are extremely fertile. Other zones of densely inhabited productive alluvial soils border such major rivers as the Mekong River in Indochina and the Irrawaddy River in Burma.

The vegetation formation classes of the tropical wet-dry environment have been covered in some detail in Chapter 12. The common denominator is the raingreen character of the plants, which grow vigorously after the rains arrive, but remain dormant for many months in the dry season. Savanna woodland and grassland, thornbush, and scrub form a mosaic throughout the wet-dry environment. Monsoon forest occupies some areas in Southeast Asia, although it has been extensively destroyed or modified by Man.

Closely geared in to vegetation and climate is the natural animal life of the savanna grasslands and woodlands. These are the regions of the carnivorous game animals and a vast multitude of grazing animals on which they feed (Figure 13.13). The savannas of Africa are the natural home of herbivores such as wildebeest, gazelle, deer, antelope, buffalo, rhinoceros, zebra, giraffe, and elephant. On them feed the lion, leopard, hyena, and jackal. Some of the herbivores depend on fleetness of foot to escape the predators. Others, such as the rhinoceros, buffalo, and elephant, defend themselves by their size, strength, or armor-thick hide. The giraffe is a peculiar adaptation to savanna woodlands; his long neck permits browsing on the higher foliage of scattered trees.

The dry season brings a severe struggle for existence to animals of the African savanna. As streams and hollows dry up, the few muddy waterholes must supply all drinking water. Danger of attack by carnivores is greatly increased.

The savanna ecosystem in Africa faces the prospect of widespread destruction through the impact of Man. Parks set aside to preserve the ecosystem confine the grazing animals to a narrow range and prevent their seasonal migrations in search of food. In some instances the growth in animal populations has been phenomenal because they have been protected from hunting. In confined preserves they rapidly consume all available vegetation in futile attempts to survive. Rapidly growing human populations are bringing increasing pressure to allow encroachment on game preserves in order to expand cattle grazing and agriculture.

Food crops Of the staple food crops grown widely in the tropical wet-dry environment, rice is perhaps the most important and the most closely linked to the wet-dry cycle. About one third of the human race subsists on rice, and most of this rice-eating mass of humanity is crowded into the arable lands of Southeast Asia. Most of the rice cultivation is within the monsoon variety of the wet-dry tropical climate and in the humid subtropical climate

figure 13.13
Giraffes and zebras roam free on the highland savanna of Kenya in East Africa. (Marc & Evelyne Bernheim, Woodfin Camp)

figure 13.14
In this scene in the northern Philippines rice seedlings are being transplanted into a paddy field flooded by monsoon rains. By the time this rice crop has been harvested, a single acre may have consumed a thousand person-hours of hand labor. (Department of Tourism, Philippines Board of Travel & Tourist Industry)

that lies at higher latitudes in China and Japan. Rice requires flooding of the ground at the time the seedling plants are cultivated, and this activity has been traditionally timed to coincide with the peak of the rainy monsoon season (Figure 13.14). The crop matures and is harvested in the dry season. Sugar cane is another important crop that grows rapidly during the rainy season and is harvested in the dry season.

Sorghum, which goes by other names such as kaffir corn, or guinea corn, is an important food crop of the tropical wet-dry environment in Africa and India. It is a grain capable of survival under conditions of a short wet season and a long, hot dry season.

Wheat is intensively cultivated in northern India and Pakistan, within the more arid part of the tropical wet-dry climate. This agricultural region includes the Punjab of northern Pakistan and northwestern India and the Indian state of Uttar Pradesh, lying mostly within the broad lowland of the Ganges River (Figure 13.15). Much of the region is irrigated to supplement the monsoon rains, and some areas produce two crops per year.

Peanuts are another major food crop of the tropical wet-dry environment in India, and they have been introduced into a corresponding climate zone of West Africa.

The green revolution   Prospects for exploiting the tropical wet-dry environment for increased food production rest today on concerted application of techniques embodied in the term *green revolution*. A major ingredient in that revolution has been the development of new genetic strains of rice and wheat capable of greatly increased production per unit of area of land now under cultivation. To achieve the potential of these new plant strains requires substantial applications of fertilizers. Changes in agricultural practices and increased use of machines are also a part of the total picture. While striking increases of rice and wheat yields were achieved in Southeast Asia by the green revolution, the recent sharp increases in price of fertilizers and fuel have dealt a staggering blow to the new agriculture. Because of the green revolution, wheat production in the Punjab of India was doubled in the five years between 1966 and 1971. It was hoped that this phenomenal increase would sustain the growing needs of India's population, which increases by some 17 million persons per year. However, wheat production leveled off and, in 1974, was forced into sharp decline by a number of negative factors. These included partial failure of the rains and the enormous increase in cost of fertilizers and fuels. Bore wells, drilled into the sand and gravel to reach the ground water, require electricity for pumps, and this energy source was also severely cut.

One hazard of the green revolution lies in the requirement that the genetic strains bred for high yields must be used to the exclusion of a variety of native strains. Should the high-yield strain prove vulnerable to an epidemic plant disease, the entire crop of a whole nation could be wiped out in one season. A second hazard lies in the need to increase the size of fields, merging many small plots into large ones to allow mechanized agriculture to work most efficiently. In so doing, a variety of food crops is no longer grown, while dependence for survival comes to rest on the single crop. Under traditional practices, the Asiatic farmer planted several food crops to ensure that if some failed, others would yield enough food to prevent starvation. The technique of the green revolution is being viewed more and more as a gamble with human life, imposing the technology of the industrial midlatitude nations on an ancient agricultural system that had developed important safeguards to insure its survival without outside aid.

Soil scientists have pointed out that large areas of vertisols remain to be placed under cultivation. The vertisols are rich in nutrient bases, but will require machine cultivation to overcome poor tillage. Most of the arable land of the wet-dry tropical environment in Southeast Asia has already been intensively developed by existing standards. To hope for major expansion of agriculture into poor latosolic soils and steppe zones marginal to the tropical deserts is, at best, unrealistic.

(4) The tropical dry climates (Köppen: <u>BWh</u>, <u>BSh</u>)

The tropical dry environment includes tropical deserts (4a) and steppes (4b) of the continental interiors and the narrow west-coast deserts (4c). Although there are important climatic differences between the interior and coastal deserts, we shall combine them for simplicity.

The tropical desert climate (4a) and tropical steppe (4b) are sustained by subsiding air of the continental high-pressure cells. These cells dominate much of the earth's land areas at lat. 15° to 35° N and S, a belt roughly centered on the tropics of cancer and capricorn (Figure 13.16). Here lie the source regions of the continental tropical (cT) air masses. The vast

figure 13.15

Here in the Punjab province of north India, wheat sheaves are being loaded on a wagon for transport to the threshing site. The tractor, a European import, symbolizes the technology and input of fossil fuels required to gain full benefits of the green revolution. (Jehangir Gazdar © Woodfin Camp)

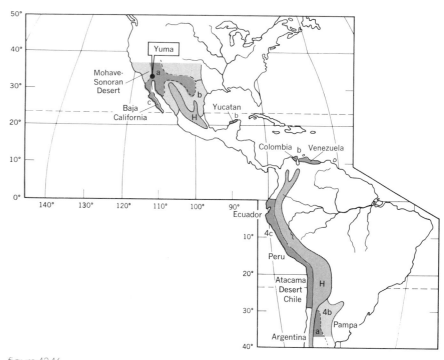

figure 13.16
World map of the tropical dry climates. (Based on Goode Base Map.)

deserts of North Africa, Arabia, Iran, and Pakistan exemplify this climate type. Other important deserts of this belt are the Sonoran Desert of the southwestern United States and northern Mexico, the interior Kalahari Desert of South Africa, and the Australian Desert.

Within this belt of general aridity the truly arid zones, or deserts, have an annual rainfall less than 10 in. (4 cm). In the semiarid zones, or steppes, rainfall is from 10 to 30 in. (25 to 76 cm) annually. In the continental interiors, far removed from oceanic sources of moisture, extreme aridity prevails. Average annual rainfall is often less than 5 in. (2 cm); at some localities, several years may pass without measurable rainfall. As an example, at In Salah, Algeria, the average annual rainfall for a 15-year period of observation was 0.6 in. (1.5 cm). Other Algerian stations receive even less.

The capacity of tropical desert air for evaporation of exposed water surfaces is enormous.

In the Sonoran Desert, annual evaporation from a free water surface exceeds 80 in. (200 cm) annually, or about twenty times as much as falls in rain. It is obvious that the full amount of evaporation that is possible does not take place in the tropical deserts. Once the stream channels and soil have become dry after a rain, further evaporation is limited to a small amount of moisture slowly brought to the surface locally by capillary movement from moist soil or rock at depth. Figure 9.10*d* shows the soil-water balance for Alice Springs, Australia. There is a soil-water shortage in every month. Very little water is stored in the soil at any time.

Although dryness is the dominant characteristic of the continental tropical air-mass source regions, sporadic heavy rainfall does occur from intense convectional storms. Penetration of maritime tropical or equatorial air may be responsible for such storms. During a single

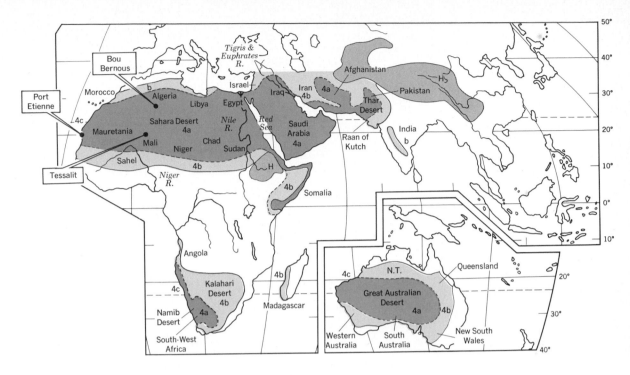

cloudburst confined to an area of a few square miles, the major portion of rainfall of one or more year's total may fall, producing raging floods in the stream channels. Figure 13.17 shows the yearly cycle of monthly mean temperatures at Yuma, Arizona, a representative North American station of the tropical desert. Two points are noteworthy: (1) temperatures are very high during the period of high sun; and (2) annual range is moderately strong. Normally the annual range in these climates is 30 to 40 F° (17 to 22 C°) and is directly related to the height of the sun in the sky.

For those who consider southern Arizona to be a hot desert, the data of Bou-Bernous, Algeria, may be enlightening (Figure 13.18). Here the mean daily temperature averages over 100° F (38° C) in July — a full 10 F° (5.5 C°) higher than Yuma, Arizona. Notice that the mean value of the highest temperature observed in July is 120° F (49° C).

The west-coast desert climate (4c) is a special variation of the interior desert at the same

figure 13.17

Yuma, Arizona, represents the tropical desert climate in North America, although its latitude (33° N) is somewhat higher than the major areas of tropical desert in Africa, Asia, and Australia. There is little rainfall in any month, on the average.

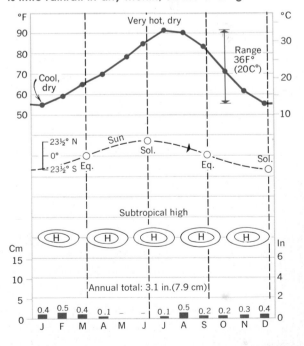

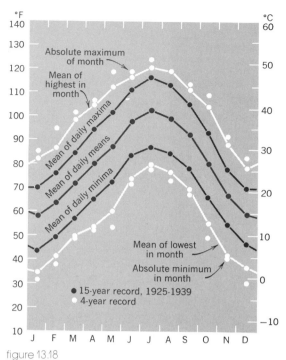

figure 13.18

Monthly air temperature data for Bou-Bernous, Algeria, at lat. 27½° N in the heart of the Sahara Desert of North Africa.

latitude. The world rainfall map (Figure 8.3) shows that all west coasts at lat. 15° to 30° N and S are extremely dry, generally with less than 10 in. (25 cm) of rainfall annually. The Atacama Desert of Chile and the Namib Desert of coastal southwest Africa are perhaps the most celebrated of these deserts, but they exist also in Lower (Baja) California, the Atlantic coast of North Africa, and the west coast of Australia. The arid belt extends inland to join the continental tropical deserts.

It may seem strange that extreme dryness exists immediately along the shores of the oceans, close to possible sources of moist maritime air masses. The key to these coastal deserts lies in the concept that the oceanic subtropical high-pressure cells are inherently dry on their eastern sides. The circulation in these cells is such that the air on the east sides is subsiding as it moves outward. The air is adiabatically

heated and its humidity reduced. The result is dry, stable air masses that bring an arid zone not only to the coast but extending far seaward as well (Figure 13.19). A cold ocean current with upwelling from great depth lies offshore. The cold ocean surface absorbs heat from the overlying air layer. Fog is a persistent feature of this cool air layer and often spreads inland a short distance, as shown in Figure 13.20.

Air temperatures in the west-coast deserts are remarkably cool at the time of high sun, when inland deserts at the same latitude are extremely hot. The yearly air temperature cycle is very weak and has a remarkably small annual range. Figure 13.21 shows the monthly mean temperatures for a desert station on the Atlantic coast of Africa compared with a station in the heart of the Sahara Desert. Both are at the same latitude. Both have about the same minimum month temperature, but the summer maximum is much higher in the interior desert.

Soils, vegetation, and animal life Desert soils and vegetation were described in Chapters 10 and 12. Reddish desert soils are typical of the interior continental deserts, and these grade into reddish-brown and brown soils in the bordering steppes. Over large expanses of the most arid desert land there is little or no plant cover in view. Much of the land surface consists of barren rock, coarse stream gravels, or drifting dune sands.

Poorly drained, shallow basins accumulate thick salt deposits close to the surface. These form white salt flats, entirely sterile and perfectly smooth. On rare occasions these flats are covered by a shallow layer of water, brought by flooding streams heading in adjacent highlands. A number of desert salts are of economic value and have been profitably extracted. An example well known to most persons is borax (sodium borate), widely used as a water-softening agent. In shallow coastal estuaries in the desert climate sea salt is

figure 13.19
In this air view of the barren coast of Peru, a fog bank can be seen in the distance, lying over the cold Peru current. (Ministerio de Fomento, Peru.)

commercially harvested by allowing it to evaporate in shallow basins. One well-known salt source of this kind is the Rann of Kutch, a coastal lowland in the tropical desert of westernmost India, close to Pakistan. Here the evaporation of shallow water of the Arabian Sea has long provided a major source of salt, the only product of value in the region.

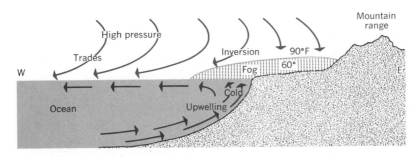

figure 13.20
Subsiding air over a western continental coast produces a persistent upper-air temperature inversion, trapping cool air and fog in a surface layer near the shore.

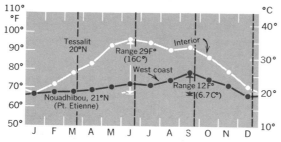

figure 13.21

These two temperature graphs for stations in the Republic of Mauretania, North Africa, show the low annual range of the west-coast desert (Port Etienne) in contrast with the large annual range within the interior desert (Tessalit).

In Chapter 12 we described the adaptation of desert plants to prolonged water shortages of the desert environment. In addition to these xerophytic plants, the desert ecosystem has a full complement of animals also adapted to survival with little or no free water. The mammals are, by nature, poorly adapted to desert environments, yet many survive there because they use a variety of mechanisms to avoid water loss. Just as plants reduce transpiration to conserve water, so many desert mammals do not sweat through skin glands; they rely instead on other methods of cooling. For example, the huge ears of the jackrabbit serve as efficient radiators of heat to the sky. Calculations have shown that a jackrabbit sitting in a shaded depression could dispose of about one third of his metabolic heat load through longwave radiation from his un-insulated ears to the clear desert sky above. Many of the desert mammals conserve water by excreting highly concentrated urine and relatively dry feces. The desert mammals also evade the heat by nocturnal activity. In this respect, they are joined by most of the rest of the desert fauna, spending their days in cool burrows in the soil and their nights foraging for food.

Desert irrigation Man's interaction with the tropical desert environment is as old as civilization itself. Two of the earliest sites of civilization — Egypt and Mesopotamia — lie in the tropical deserts. The key to Man's successful occupation of the deserts lies in availability of large supplies of water from nondesert sources. This is a concept so familiar to all that it scarcely needs to be stated. For Egypt and Mesopotamia, the water sources of ancient times were _exotic rivers_ deriving their flow from regions having a water surplus and flowing across the desert region because of geologic events and controls having nothing to do with climate.

A look at a global population map shows that population density is less than two persons per square mile over nearly all of the area of the tropical deserts. Only where exotic streams cross the desert does the population density rise sharply. Valleys of the Nile, the Tigris and Euphrates, and the Indus are striking examples in the Old World. But we also find in the coastal desert of Peru a substantial population long dependent on exotic streams fed from the Andes range and crossing the coastal desert to reach the Pacific Ocean.

Can Man increase production of food by expanding agriculture into the deserts, both in tropical and midlatitude regions? Making the desert bloom is a romantic concept fostered on the American scene for generations by bureaucrats, politicians, and land developers. Were these promoters of vast irrigation schemes working in the long-term public interest? Only in recent years have the undesirable environmental impacts of desert irrigation come to the forefront. Yet we could have read the modern scenario in the history of rise and fall of the Mesopotamian civilization.

Irrigation systems in arid lands divert the discharge of a large river, such as the Nile, Indus, Jordan, and Colorado, into a distributary system that allows the water to infiltrate the soil of areas under crop cultivation. Ultimately, such irrigation projects suffer from two undesirable side effects — salinization and waterlogging of the soil.

The irrigated area is subject to very heavy soil-water losses through evapotranspiration. Salts contained in the irrigation water remain in the soil and increase in concentration. This process is called _salinization_. Ultimately, when salinity of the soil reaches the limit of tolerance of the plants, the land must be abandoned. Prevention or cure of salinization may be possible by flushing the soil salts downward to lower levels by use of more water. This remedy requires greater water use than for crop growth alone.

Infiltration of large volumes of water causes a rise in the water table and may, in time, bring the zone of saturation close to the surface. This phenomenon is called _waterlogging_. Crops cannot grow in perpetually saturated soils. Furthermore, when the water table rises to the point that upward movement under capillary action can bring water to the surface, evaporation is increased and salinization is intensified.

One of the largest of the modern irrigation projects affected adversely by salinization and waterlogging lies within the basin of the lower Indus River in Pakistan. Here the annual rate of rise of the water table has averaged about 1 ft (0.3 m), while the annual increase in land area adversely affected is on the order of 50,000 acres (20,000 hectares).

The question of expanding irrigation agriculture in the southwestern United States was recently studied by the Committee on Arid Lands of the American Association for the Advancement of Science. In their 1972 report this body recommended that additional large-scale importation of irrigation water from distant sources should be made only where there are compelling reasons to do so. One such reason is to augment rapidly failing ground water supplies in districts already under irrigation. A second is to arrest the progress of salinization in areas already under irrigation. In short, they agreed that only to prevent social and economic disruption in established irrigated areas should additional water be imported. The lesson is that the search for new regions in which to expand agriculture should be directed to other, more favorable environments where the scales are not so heavily weighted by enormous evaporative water losses.

The low-latitude environments in review

No generalization is meaningful for the low-latitude zones considered together, because they include great extremes of the climatic spectrum. Anyone who makes a sweeping statement about the ''tropics'' is very poorly informed. How can the extreme aridity of the low-latitude deserts be equated with the extreme wetness of the equatorial zone in reference to the thermal environment, the soil-water balance, or the food resource? If diversity is the salient quality of the low-latitude environments, shall we look to the midlatitudes for some sweeping simplicity of character? The midlatitude zone has for centuries been called the ''temperate zone.'' How ''temperate'' is the midlatitude zone? Keep this question in mind as we turn next to the environmental regions controlled by climates of Group II.

The Midlatitude Environments

Chapter 14

THE MIDLATITUDE ENVIRONMENTS are identified with the climates of Group II, occupying the polar front zone in which both tropical and polar air masses play an important climatic role. The midlatitudes are a belt subject to wave cyclones; much of the precipitation of this belt comes from lifting of moist air masses along fronts within those cyclones. Strong climatic seasons are a characteristic of the midlatitude climates — seasons in which temperatures as well as precipitation show strong annual cycles. The midlatitude climates have a distinct winter season, which the low-latitude climates do not.

Those who haven't studied physical geography refer to the midlatitude region of the globe as the "temperate zone." Nothing could be more misleading. "Temperate" means a regime marked by moderation — not extreme or excessive. Intemperance is the rule of most of the midlatitude zone. Not only do seasons swing from one extreme to the other in heat and cold or in dryness and wetness, but the day-to-day swings in weather as cyclones pass by provide a pattern of intemperance often more dramatic than that of the seasons. Intemperance in the midlatitudes is partly due to the occurrence of huge continents at that location in the northern hemisphere. Continentality is synonymous with intemperance of climate. Only along the west coasts do we find narrow strips of land where a temperate climate actually prevails. These are concepts we must develop fully.

(5) The humid subtropical climate (Köppen: Cfa)

The humid subtropical climate, by its very name, borders on the tropical zone and is really transitional between the low-latitude and midlatitude environments. This climate has a winter season, but toward its low-latitude borders the winter season weakens in intensity.

Here it is not easy to decide where the change to the wet trade-wind coastal climate sets in. This transitional character of climate also is reflected in soils and vegetation. The red-yellow podzolic soils of the humid subtropical climate are transitional between podzols of the cold climates and true latosols of the warm climates. In the warmer part of the humid subtropical climate the natural vegetation is temperate rainforest, but this gives way in the colder part to midlatitude deciduous forest. While being aware of the transitional nature of the environment, we can try to make some broad generalizations about the more typical parts of the humid subtropical environment.

The humid subtropical climate is found on the eastern sides of landmasses at lat. 25° to 30° N and S (Figure 14.1). Looking back to Figure 5.12, you will see that the oceanic high-pressure cells are pumping air into the eastern sides of the continents at lat. 25° to 35°. This effect is strongest in the summer, when the highs are strong and persistent. The maritime tropical (mT) air mass involved in this motion has a high moisture content and produces ample rainfall in the summer. Recall, too, that tropical cyclones arrive on these shores after their long westward paths through the trades. As these storms turn inland, they bring very heavy rainfall and flooding conditions.

The humid subtropical climate of the northern hemisphere is well developed in the southeastern United States, southern China, and the southern part of Japan (Figure 14.1). In the southern hemisphere it is best developed over a large area of South America in eastern Argentina, southern Brazil, and Uruguay.

The climograph for Charleston, South Carolina, illustrates the summer rainfall maximum of the humid subtropical climate at a coastal location (Figure 14.2). This rainfall is mostly of the convectional type. Showers and thunderstorms

figure 14.1
World map of the humid subtropical climate (5). (Based on Goode Base Map.)

break out in the moist, tropical air mass. When a weak cold front moves through the area, lines of intense thunderstorms are formed, often with severe squall winds. In winter the precipitation is reduced, but remains substantial in all months. This is the season when wave cyclones dominate the weather. Rain and occasionally snow are produced on warm fronts in the cyclones. (Some inland places in this climate have nearly uniform precipitation throughout the year.)

Temperatures show a moderately strong annual cycle in the humid subtropical climate. As the graph for Charleston shows, winters are mild, with the January average well above freezing, while the summers are warm. The summers are also humid, and the combination of heat and a high relative humidity make the summer

climate very much like that of the wet equatorial climate for long spells. Mild as the winters might seem, judged from a mean monthly temperature of 50°, occasional strong bursts of cold polar air sweep over this zone, bringing killing frosts. No one who lives in this region would describe the winter climate as "temperate."

In Southeast Asia the humid subtropical climate is somewhat modified by intensive monsoon development. Winter air masses from interior Asia are very dry, and a winter scarcity of precipitation develops. In summer the strong inflow of maritime air masses, together with occasional typhoons, cause a strongly accentuated maximum in the summer. This shows clearly in the precipitation cycle of Shanghai, China (Figure 14.3), which has a

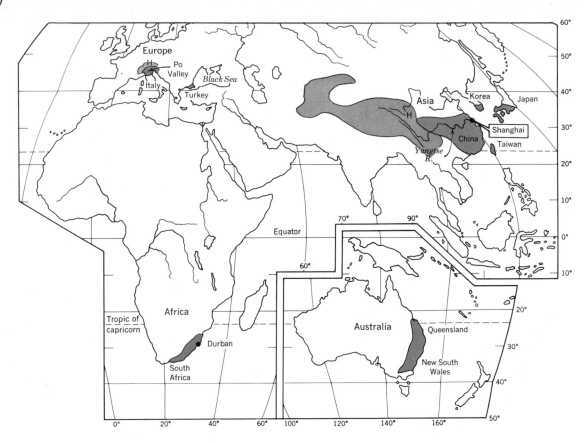

winter both drier and cooler than Charleston.

A large winter water surplus characterizes the soil-water balance of the humid subtropical climate while, by comparison, the summer shortage is typically very small. Refer back to Figure 9.10*E*, the soil-water budget for Baton Rouge, to verify these points.

Soils and vegetation   Soils of the moister, warmer parts of the humid subtropical regions are strongly leached red-yellow soils related to the latosols of the humid tropical and equatorial climates. Rich in iron and aluminum oxides, these soils are deficient in plant nutrients essential for successful agricultural production. Fertilizers are needed in substantial amounts to give good crop yields.

Forest is the natural vegetation of most of the areas having the humid subtropical climate. Temperate rainforest was the dominant type

in the warmer, moister parts. Some areas of tropical rainforest occupied the coastal belts in lower latitudes. Little of this forest remains, because of intensive land use. Much of the sandy coastal region of the southeastern United States today has a second-growth forest of longleaf, loblolly, and slash pine (Figure 14.4). The colder, more northerly inland region has summer-green deciduous forest. Toward higher latitudes forest gives way to tall-grass prairie such as the Pampa of Argentina and Uruguay and the prairies of Oklahoma and Missouri. Here the soils are of dark prairie and chernozem groups, typical of a less humid continental climate. These grasslands occupy regions that are subhumid. They are best considered transitional to the semiarid steppes, because they have less than 40 in. (100 cm) annual precipitation and a marked dryness of the winter season.

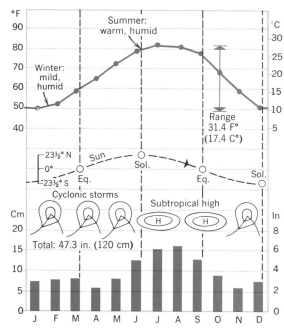

figure 14.2

Charleston, South Carolina, lat. 33° N, has a mild winter and a warm summer. There is ample precipitation in all months, but a definite summer maximum.

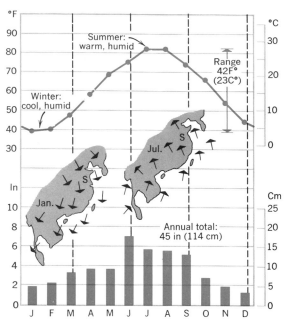

figure 14.3

At Shanghai, China, lat. 31° N, winters are colder than at Charleston, and the summer maximum of rainfall is strongly developed because of the monsoon effect.

Agricultural potential Because the humid subtropical climate includes so much area that is transitional to adjacent climates, it is difficult to equate it to a single distinctive agricultural region. One common denominator is ample soil water and heat for production of field crops on a big scale. In looking over the northern hemisphere areas, some major

figure 14.4

This plantation of slash pine grows on sandy soil of the Georgia coastal plain. A cup is attached to the base of each tree to catch sap exuding from a cut into the sapwood. Turpentine is distilled from the sap. (Grant Heilman)

figure 14.5
Tea leaves are carefully picked from new growth of a cultivated evergreen shrub (Camellia) on this tea plantation near Kyoto, Honshu Island, Japan. (Rene Burri, © Magnum Photos)

differences in food crops show up between the southern United States and those parts of China and Japan with similar climate. The differences are partly historical and cultural, but partly reflect the dominance of the monsoon regime in Asia. The human population is vastly denser in this part of Asia than in the New World. It subsists largely on rice, the dominant staple food crop, while legumes are an important food source. Except in the Mississippi delta region, little rice is produced in the southern United States. Both American and Asiatic regions produce sugar cane, peanuts, tobacco, and cotton, although not on an equal intensity in both regions. One striking difference is that tea is widely cultivated in Southeast Asia, but not at all in the southern United States (Figure 14.5). Corn is a major crop in both the southern United States and Argentina, but not in the corresponding climate of Asia.

The potential of the humid subtropical climate to produce more food rests in more intensive land use instead of in expansion of the area now under cultivation. The best land is already in use and, in Southeast Asia, elaborate terrace systems have been in use for centuries to allow farming of steep hill slopes (Figure 14.6). New genetic strains of rice and corn offer promise of greatly increased yields when the necessary fertilizers are used. Referring now to the southern United States, cattle production is another direction of increased food production, making use of soils too sandy for field crops.

figure 14.6
Terraced mountainsides on the Island of Shikoku, Japan. Without terraces, very little land would be available for agriculture in this region. (Burt Glinn, © Magnum Photos)

figure 14.7
World map of the humid continental climate (6). (Based on Goode Base Map.)

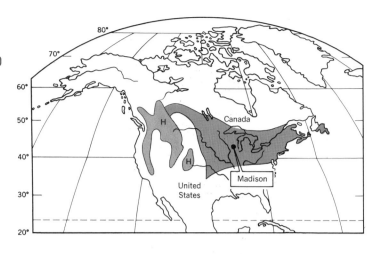

With soil water adequate through the long, warm summer, pasture and range land is continuously productive. Tree farming is also an important use of sandy soils. Pines are well adapted to rapid growth on sandy soils and thrive where nutrient bases are in short supply.

The large water surplus of the humid subtropical climate has important implications in terms of economic development. The large flows of rivers can furnish abundant freshwater resources for urbanization and industry without competition from irrigation demands. Evaporative losses from reservoirs are much less important than in arid lands. The maintenance of copious stream flows tends to reduce the dangers of severe water pollution and its adverse effects on ecosystems of streams and estuaries.

(6) The humid continental climate (Köppen: <u>Dfa</u>, <u>Dfb</u>, <u>Dwa</u>, <u>Dwb</u>)

Exclusively a northern hemisphere region, the humid continental climate covers vast interior expanses of the landmasses of North America and Eurasia between lat. 40° and 55° N (Figure 14.7). The humid continental climate occupies a position intermediate between the source region of polar continental air masses on the north and maritime or continental tropical air masses on the south and southeast. In this polar-front zone, maximum interaction between polar and tropical air masses can be expected along warm and cold fronts associated with east-moving cyclones. In winter the continental polar air masses dominate, and much cold weather prevails; in summer tropical air masses dominate, and high temperatures prevail. Strong seasonal temperature contrasts must be expected in this region. Precipitation is distributed throughout the year because the region lies in a frontal zone. However, a summer maximum of precipitation is the rule because warmer air has greater moisture-holding capacity. Strong contrasts in air masses result in strong frontal activity and highly changeable weather.

The climate picture we have sketched applies well to the north-central and northeastern United States and southeastern Canada as well as to northern China (including Manchuria), Korea, and northern Japan. Very similar climatic conditions also hold for much of central and eastern Europe, the Balkan countries, and Russia. This European region differs from the first two in that it is influenced by a great source region of continental-tropical air masses lying to the south and southeast.

We can get a good grasp of the typical features

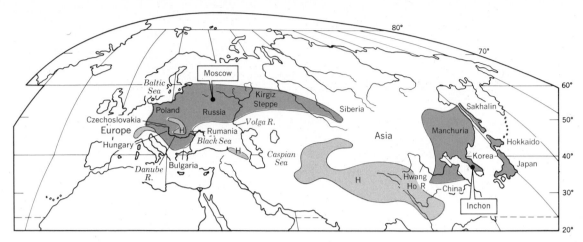

of the humid continental climate by comparing the climographs of three widely separated stations: Madison, Wisconsin, Moscow, Russia, and Inchon, Korea (Figures 14.8, 14.9, and 14.10). All three places have very strong annual temperature cycles and similar ranges — somewhat over 50 F° (28 C°). In all three places the coldest-month average is well below freezing. Moscow, located over 1000 mi (600 km) farther north than the other two, is colder in winter and not as warm in summer. All three have a snowy winter when soil water is solidly frozen. All three stations have a summer maximum of precipitation, but that at Inchon makes a remarkable summer peak in response to the strong Asiatic monsoon control. In July

figure 14.8

Madison, Wisconsin, lat. 43° N, has cold winters and warm summers, making the annual temperature range very large.

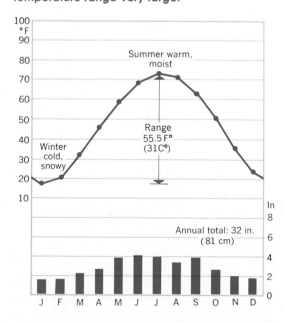

figure 14.9

Moscow, Russia, lat. 56° N, has an annual range about the same as at Madison, Wisconsin, but summers in Moscow are not as warm.

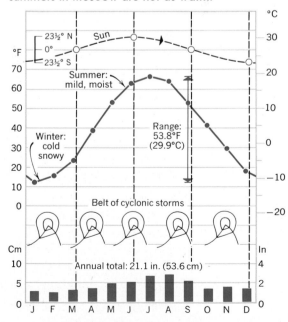

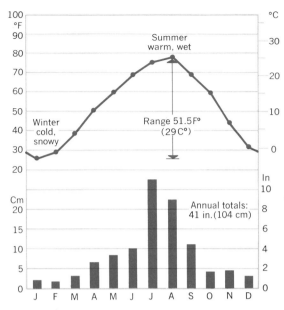

figure 14.10

Inchon, Korea, lat. 37½° N, has two very rainy months during the summer monsoon, but very little precipitation in winter.

and August moist maritime air masses invade Korea freely, whereas the other two stations are far from oceanic air mass sources.

The soil-water budget for the humid continental climate is well illustrated by the graph for Pittsburgh (Figure 9.10F). The water shortage of summer is very small, while a substantial water surplus develops in late winter and early spring. Inchon's water budget would show no shortage at all but, instead, a large surplus during the summer rainy season. Moscow's summer shortage would be substantial.

Soils and vegetation Soils of the humid continental climate are dominantly podzolic. Podzols occur in the colder northern portions, and these support a needleleaf forest to which they are always closely tied. The central and southerly portions have brown forest soils that are moderately podzolized, but richer in humus and not as acid. Here the forest is of the mixed needleleaf-deciduous type, grading

southward into deciduous forest. In the United States in the western part of the humid continental environment a transitional zone exists, grading into the bordering semiarid climate. This transitional zone includes the tall-grass prairie and its thick brown prairie soils. Here forest occurs in fingerlike extensions in the major stream valleys. Most of this tall-grass prairie has disappeared, replaced by some of the richest farmland in the world (Figure 14.11). Throughout the more humid easterly areas forest is widespread in mountainous terrain of the Appalachians and in woodlots throughout the farmed belt. Much of this forest consists of second- or third-growth tree stands, but the forest structure is preserved. In China and Korea the effects of prolonged deforestation are everywhere evident. Throughout central and western Europe, large areas have been under field crops, pastures, and vineyards for centuries, while at the same time forests have been carefully cultivated over large areas. In this environment the potentials for both crop agriculture and forest culture (silviculture) have reached a near-optimum adjustment in terms of the soils and terrain.

Agriculture Because of the availability of soil water through a warm summer growing season, the humid continental climate has an enormous potential for food production. The cooler, more northerly sections in North America and Europe support dairy farming on a large scale. Landforms in these areas were shaped by ice sheets that disappeared only about 10,000 to 15,000 years ago. Unfavorable glacial terrain in the form of bogs and lakes, rocky hills, and stony soils has deterred crop farming in many parts. Farther south, plains formed on former lake floors and undulating uplands, bearing a cover of wind-deposited silt (loess), are ideally suited to crop farming. Cereals grown extensively in North America and Europe include corn, wheat, rye (especially in Europe), oats, and

figure 14.11
Rich farmlands of the corn belt in Iowa are laid out on a grid pattern. There are about four farms per square mile in this area. (Aero Service, Western Geophysical Co., Inc.)

barley. The famed corn belt of the Midwest can be singled out for special mention, but corn is also a dominant crop in Hungary and Rumania. Beet sugar is an important product of this environmental region in Europe, but not in North America. On the other hand, soybeans are intensively cultivated in the midwestern United States and, in China, on the plains of Manchuria, but little in Europe. Rice is a dominant crop in both South Korea and Japan, much farther poleward than elsewhere in Asia. The rice seedlings can be planted in paddies flooded during the brief but copious rains of midsummer, then harvested in the dry autumn. Among geographers, this northern rice area is included in the region called Monsoon Asia.

figure 14.12
Combines harvesting winter wheat on the High Plains of western Kansas. (Grant Heilman)

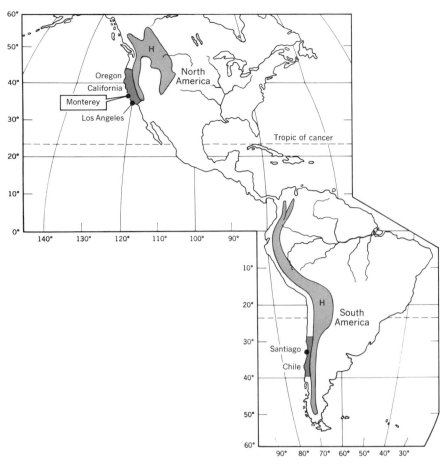

figure 14.13
World map of the Mediterranean climate (7). (Based on Goode Base Map.)

Wheat and, to a lesser but important degree, oats, rye, and barley, are dominant crops in the transition zone in which the humid continental climate grades into the midlatitude steppe climate. North American Great Plains, the Ukraine, and parts of north China are all within this transition zone (Figure 14.12). Wheat farming extends into a semiarid steppe climate in which there is a substantial soil-water shortage in summer and no water surplus during the year.

The humid continental climate and its transition zone into adjacent steppes is a region of huge grain surpluses and represents a very advanced agricultural system in terms of inputs of energy from fossil fuels and the application of technology. During the current world food crisis, arable land in the United States previously withdrawn from cultivation has been returned to use, and there is little good farm land left to exploit. To produce greatly increased American grain for export to the famine lands would require greatly curtailed meat production. This change involves a diversion of production from animal feeds to cereals and soybeans capable of being directly consumed by humans instead of being inefficiently passed up the food chain to animal consumers.

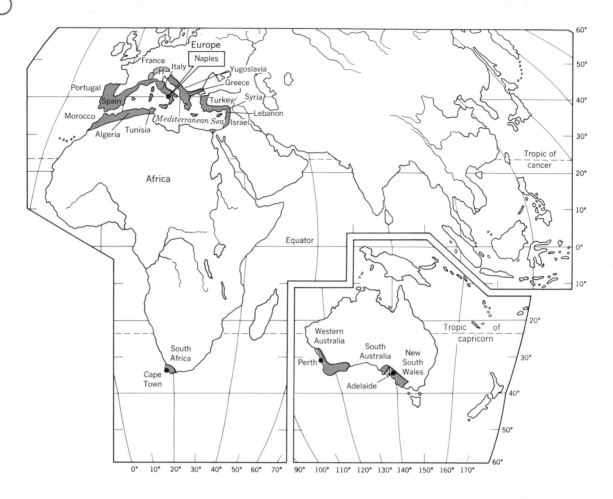

(7) The Mediterranean climate (Köppen: <u>Csa</u>, <u>Csb</u>)

We now jump to the west sides of continents to investigate a distinctive climate located in subtropical and lower midlatitude zones. The Mediterranean climate, as its name implies, is found in lands bordering the Mediterranean Sea (Figure 14.13). An alternative name, the dry-summer subtropical climate, is in some respects better because it gives the key to the climatic pattern. Dry summers alternate with moist winters, a cycle just the opposite from that of the humid subtropical and humid continental climates.

The Mediterranean climate has an important occurrence in central and southern California, extending from the Pacific shores inland into the Great Valley and other valleys within the Coast Ranges. The climograph for Monterey illustrates the features of the Mediterranean climate (Figure 14.14).

In summer, when the oceanic subtropical high is most powerfully developed and farthest north, the same desert conditions that prevail permanently at lower latitudes take over control of the climate and bring a severe drought. However, the proximity of the ocean with its cool current keeps summer temperatures to a mild 60° F (16° C) average. In winter the polar front zone with its procession of wave cyclones moves over this coastal area. Moist maritime polar (mP) air masses furnish the moisture for ample precipitation, which is in the form of rain at low altitudes.

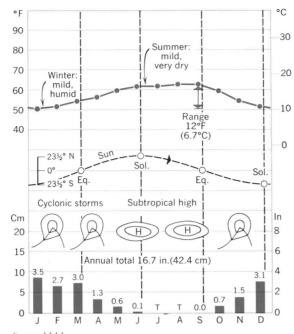

figure 14.14

Monterey, California, lat. 36½° N, has a very weak annual temperature cycle because of its closeness to the Pacific Ocean. The summer is very dry.

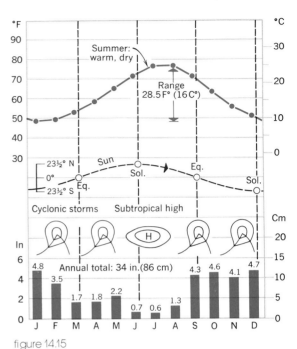

figure 14.15

Naples, Italy, lat. 40½° N, has a much warmer summer than Monterey, but winter temperatures are much the same.

The soil-water budget of the Mediterranean climate is well illustrated by Los Angeles (Figure 9.10G). The nearly rainless summer coincides with a strong peak in water need, so that a large shortage develops. Soil-water recharge goes on through the rainy winter, but there is no surplus.

In lands surrounding the Mediterranean Sea, this climate shows a much stronger annual temperature cycle, illustrated by the climograph for Naples, Italy (Figure 14.15). Total rainfall is also greater here, and there is a small amount of rainfall, on the average, in each of the summer months.

Other occurrences of the Mediterranean climate are on the coast of Chile, the tip of South Africa, and the southwestern and southern coastal belts of Australia.

Soils and vegetation   Soils of the Mediterranean climate are not easy to describe and classify in simple terms. Most are described as brown, reddish-brown, reddish chestnut, and red in terms of color. The climate is effectively semiarid, and the soils are closely affiliated with the grassland soils. Severe and prolonged soil erosion following deforestation and overgrazing has left the Mediterranean lands with a poor heritage in terms of mature soils. Much regolith and bedrock is now exposed in hillsides and upland surfaces.

Vegetation of the Mediterranean climate is adapted to a long summer drought during which soil water is severely depleted. Sclerophyll forest, woodland, and scrub is the typical vegetation. Trees common in the evergreen forest and woodland of the Mediterranean lands are the cork oak (Figure

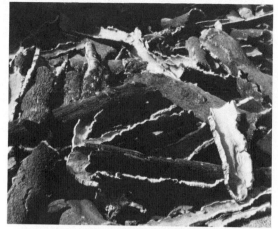

figure 14.16
This bark, stripped from the cork oak (Quercus suber), will be ground up and cemented into cork board and other structural products. Thick bark of choice quality is used for wine corks. Algeria, North Africa. (A. N. Strahler)

14.16), live oak, pine, and olive. Large areas consist of dense scrub, called maquis; it is composed of many species, some of them very spiny. In coastal California the sclerophyll vegetation includes a woodland of live oak and white oak, with grassland (Figure 14.17). A distinctive "dwarf forest," called chaparral,

covers large expanses of hill and mountain slopes. Among the woody shrubs of the chaparral are the wild lilac, manzanita, and mountain mahogany (see Figures 12.14 and 12.15). Extreme flammability characterizes the chaparral during the long, dry summer. Brush fires are an everpresent threat to Californians wherever suburban housing has penetrated vulnerable canyons and expanded over chaparral-covered hillsides.

Agriculture   Lands bordering the Mediterranean Sea produce cereals — wheat, oats, and barley — where arable soils are extensive enough to be cultivated. However, we usually think of the region as an important source of citrus fruits, grapes, and olives. Cork from the bark of the cork oak is also a product of economic importance. In central and southern California citrus, grapes, avocadoes, nuts (almond, walnut), and deciduous fruits are extensively grown (Figure 14.18). Irrigated alluvial soils are also highly productive of vegetable crops such as carrots, lettuce, artichokes, strawberries, and sugar beets, and also of forage crops (alfalfa). Cattle ranching and sheep grazing is of major importance on

figure 14.17
This heavily grazed woodland of oak and open grassland is in Monterey County, California. (U.S. Forest Service.)

figure 14.18
Lemon orchards in Ventura County, California, show two geometrical patterns. Narrow terraces on steep hillsides (left) follow curving contours. The grid pattern (right) characterizes older orchards on gently sloping lowland surfaces. (U.S. Department of Agriculture)

grassy hillslopes, unsuited to field crops and orchards, and on irrigated lowland pastures.

Because the Mediterranean environment is limited in extent to comparatively small land areas, it offers little prospect for important additions to the world's food supply. Irrigation is essential for high productivity, and we have already stressed the hazards associated with heavy irrigation of lowland soils: salinization and waterlogging. Urbanization and industrial development also face major problems of obtaining water supplies through importation over long distances by aqueduct. Nevertheless, the mild, sunny climate of southern California has proved a powerful population magnet, and water importation has been developed on a mammoth scale.

(8) Marine west coast climate (Köppen: Cfb, Cfc)

A logical geographical step is to move poleward along the west coasts of the continents to arrive in a midlatitude zone where the succession of wave cyclones within the prevailing westerlies dominates the climate throughout most of the year. The marine west coast climate lies poleward of lat. 40°, too far removed from the oceanic high-pressure cells to experience summer aridity (Figure 14.19). However, in summer, there is the same diminished precipitation seen so strongly in the Mediterranean climate.

The climograph for Brest, France, illustrates the marine west coast climate in Europe (Figure 14.20). The annual temperature cycle is weak for a station this far north. In winter the great heat reservoir of the Atlantic Ocean prevents monthly mean temperatures from falling to the below-freezing levels found inland. Summers escape warming normal for this latitude, and the annual range is only 20 F° (11 C°). Precipitation is not large in any month, but runs markedly higher in winter than in summer.

The soil-water budget for the marine west coast climate is illustrated by Cork, Ireland (Figure 9.10H). Here precipitation is quite a bit greater than at Brest, but the cycle is similar. Only a

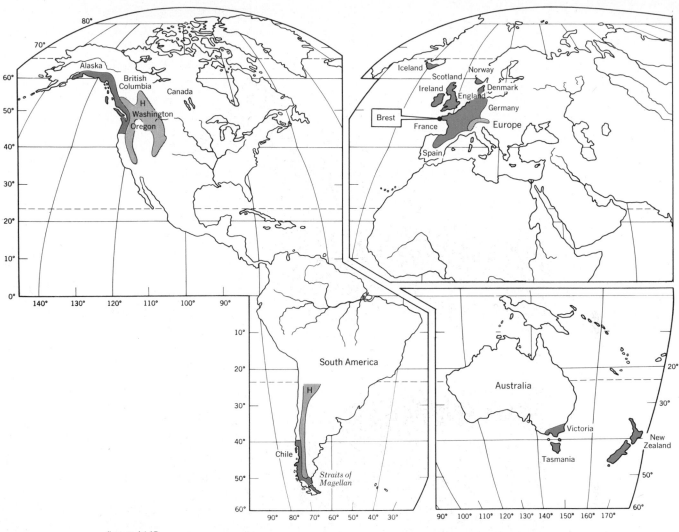

figure 14.19

World map of the marine west-coast climate (8). (Based on Goode Base Map.)

small shortage occurs in summer, while the total water surplus is substantial.

The marine west coast climate is found in both hemispheres (Figure 14.19). The entire Pacific coast from northern California to Alaska is one occurrence; another is in western Europe and includes much of France, the low countries, Ireland and England, and the coast of Norway. In the southern hemisphere an important belt lies along the coast of Chile; another takes in the southeastern tip of Australia, the island of

Tasmania, and western sides of the two islands of New Zealand.

The influence of coastal ranges on rainfall is strong in the midlatitudes. Low-lying coasts, such as in northern France or southern England, receive only 30 to 40 in. (75 to 100 cm) of precipitation annually. In contrast, mountainous coasts of British Columbia, Alaska, Norway, and Chile get 60 to 80 in. (150 to 200 cm) and over. This precipitation has been a major factor in the development of

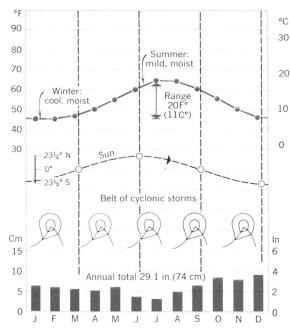

figure 14.20

Brest, France, lat. 49° N, has a mild summer and a small annual temperature range. Precipitation shows a marked decline in the summer months.

Agricultural and water resources   The marine west coast environment of western Europe and the British Isles has been intensively developed for centuries for such diverse uses as crop farming, dairying, orchards, and forests. It is an environment in most respects similar in character and productivity to the humid continental environment with which it merges toward the east.

In North America the mountainous terrain of the coastal belt offers only limited valley floors for agriculture, which is generally diversified farming. Forests are the primary plant resource, and here they constitute perhaps the greatest structural and pulpwood timber resource on earth. Douglas fir, western cedar, and western hemlock are the principal lumber trees of the Pacific Northwest (Figure 14.21). Redwood is

figure 14.21

A forest of Douglas fir in Mount Hood National Forest, Oregon. (U.S. Forest Service.)

fiords along the seacoasts. Heavy snows in the glacial period nourished vigorous valley glaciers that descended to the sea, scouring deep troughs below sea level (see Chapter 22).

Natural vegetation of the marine west coast climate is forest, but of widely different classes from one world locality to another. Much of western Europe has a summer-green deciduous forest, whereas that of west-coast North America is of the needleleaf class. On coast ranges of the Pacific Northwest, Douglas fir, red cedar, and spruce grow in magnificent forests (Figure 14.21). Forests of the marine west coast climate belts of Chile and New Zealand are of the temperate rainforest class.

Soils of the marine west coast climate regions bearing needleleaf forests are podzols. Soils of western Europe are of the gray-brown podzolic group, typically associated with deciduous forests in midlatitudes.

an important lumber tree in coastal forests of northern California.

The same mountainous terrain that limits agriculture in the Pacific Northwest is a producer of enormous water surpluses that run to the sea in rivers. Including now the ranges of the northern Rocky Mountains with the Pacific coastal ranges, the potential for long-distance transfer of this excess water to dry regions of the western United States has not passed unnoticed.

(9) Midlatitude dry climates (Köppen: <u>BWk</u>, <u>BSk</u>)

In a general way, the dry midlatitude climates of the northern hemisphere are centrally located on the continents between the humid climates that occupy both the eastern and western sides of the continents (Figure 14.22). In the southern hemisphere, only South America projects far enough into midlatitudes to have a dry midlatitude climate, and it is only a narrow zone. The dry midlatitude climates are contiguous with the dry tropical climates in North America, Asia, and South America. In this contact zone there is no clear dividing line, since average temperatures decline steadily into higher latitudes.

We recognize a midlatitude desert climate (9a) and a midlatitude steppe climate (9b) on the basis of the degree of aridity but, here again, the dividing line is an arbitrary one. As a working figure, we can take the 8-in. (20-cm) isohyet of annual precipitation as a boundary between desert and steppe. Using this boundary, the midlatitude steppes are far more extensive than the true deserts.

The midlatitude dry climates have a somewhat complex set of causes. Three basic air mass controls operate. (1) In summer, when pressure and wind belts are shifted poleward, these regions temporarily become the source regions for dry continental air masses, developed because of intense heating of the large continental interiors. (2) In winter the strong Siberian and Canadian highs, which are source areas for continental polar air masses, furnish frequent invasions of dry air. (3) Mountain ranges separate these deserts from moist maritime polar and maritime tropical air masses, which supply abundant rainfall to the west and east coasts. Through forced ascent over these ranges, followed by adiabatic heating upon descending the lee slopes (see Chapter 6), the maritime air masses are deprived of their moisture and are raised in temperature as well. Altogether, these continental interiors are poorly situated for obtaining precipitation.

Of the small proportion of the dry area covered by true desert, most occurs in the Turkestan and Gobi regions of central Asia and in parts of the Great Basin in Nevada and Utah. A small strip of Patagonia in South America is also desert. These midlatitude deserts differ from the tropical deserts of lower latitudes in that their annual temperature range is much greater and the winter temperatues are much lower. The climograph for Kazalinsk, U.S.S.R., illustrates this feature nicely (Figure 14.23). Here, on a low desert plain near the Aral Sea, the enormous annual range of 67 F° (37 C°) is almost double that of Yuma, Arizona (Figure 13.17). The January mean temperature is a severe 10° F (−12° C).

The soil-water balance for a midlatitude desert station shows a large soil-water shortage in every summer month and no surplus water. Irrigation to meet fully the soil-water shortage would need to supply the equivalent of about 24 in. (60 cm) of rainfall during the 7 months of above-freezing temperatures. This quantity is about half the annual shortage of a tropical desert station.

The midlatitude steppe climate (9b) has an annual total precipitation in the range from 8 to 20 in. (20 to 50 cm). The larger

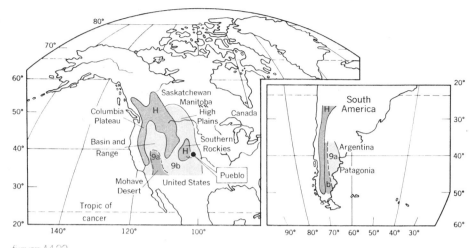

figure 14.22

World map in the midlatitude dry climates (9). (Based on Goode Base Map.)

figure 14.23

At Kazalinsk, Russia, lat. 46° N, winters are bitterly cold and summers are warm. Little precipitation falls in any month in this continental desert locality.

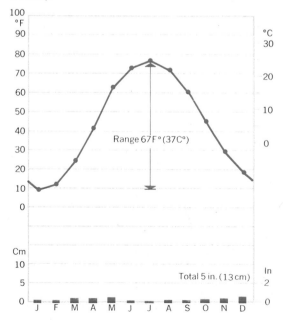

figure 14.24

Pueblo, Colorado, lat. 38° N, lies in the continental steppe climate, just east of the Rockies. Rainfall peaks strongly in the summer months.

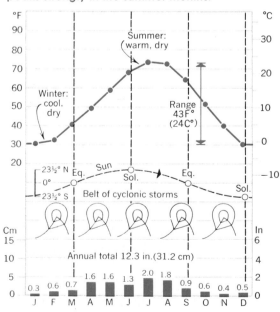

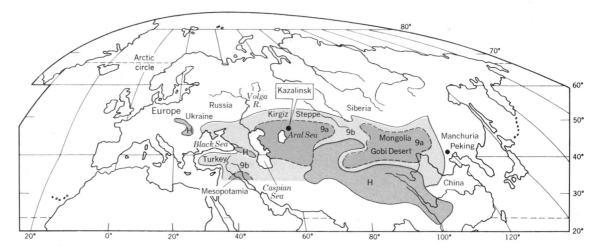

precipitation, as compared with the deserts, is partly because the land surface is at higher altitudes, and partly because the steppes occupy transitional belts with the bordering humid climates.

A typical steppe climate in North America is illustrated by Pueblo, Colorado (Figure 14.24). Compared with the Mediterranean climate of Monterey, California (Figure 14.14), which lies at about the same latitude, the annual temperature range at Pueblo is very much greater; the precipitation cycle is just reversed, with the summer maximum clearly marked.

The soil-water budget for a continental steppe climate is illustrated by the graph for Medicine Hat, Alberta (Figure 9.10*l*). The summer water shortage persists through 7 months. Recharge in the winter months is not sufficient to generate a surplus as runoff.

Soils, vegetation, and agriculture   Soils of the midlatitude desert are of the gray desert type, low in humus and containing deposits of excess calcium carbonate. Poorly drained areas develop saline soils, while dry lake beds are floored with salt deposits. Soils of the midlatitude steppes are dominantly the brown soils with substantial amounts of humus; they

figure 14.25
Cattle are grazing short-crested wheatgrass in this short-grass prairie landscape in Montana. The wheatgrass was introduced here from its native region in Russia. (U.S. Forest Service.)

are rich in nutrient bases. The *B* horizon of the brown soils has heavy accumulations of calcium carbonate. In some areas the zone of carbonate accumulation is actually a rocklike layer resembling a white limestone.

Midlatitude steppes are characterized by short-grass prairie and by local occurrences of woodland and semidesert shrubs (Figure 14.25). These steppes now constitute the great sheep and cattle ranges of the world. The steppes of central Asia have for centuries supported a nomadic population, with their sheep and goats finding subsistence on the scanty grassland. On the vast expanses of the High Plains, the American bison lived in great numbers until almost exterminated by hunters. The short-grass veldt of South Africa supported much game at one time. Steppe grasses do not form a complete sod cover; loose, bare soil is exposed between grass clumps. For this reason, overgrazing during a series of dry years can easily reduce the hold of grasses enough to permit destructive soil erosion and gullying.

We noted earlier that wheat is grown extensively in those areas of midlatitude steppe climate bordering the subhumid zone of the humid midlatitude continental climate. An important wheat-producing region largely within the steppe climate, as we have defined it, lies in southern Alberta and Saskatchewan and in the northern border region of Montana. Here the crop is spring wheat, planted in the spring of the year. Using the soil water that has been recharged in early spring and the precipitation that falls in late spring and early summer, the crop is able to reach maturity for midsummer harvesting.

The rich wheat region of the Ukraine is continued in a narrow zone far eastward across the steppes of Kazakhstan. In northern China wheat is grown within a steppe region bordering the humid midlatitude continental climate. The important wheat-producing region

of the western Argentine Pampa is partly within steppe climate, but this is in a subtropical zone (between lat. 30° and 40° S) and represents a transitional climate between the tropical steppe and the midlatitude steppe. The same can be said of the wheat-producing steppes of southeastern Australia.

Wheat production of the midlatitude steppes is very much at the mercy of variations in seasonal rainfall. Good years and poor years follow cyclic variations. Soil water, not soil fertility, is the key to wheat production over these vast midlatitude steppe lands lying beyond the practical limits of irrigation systems.

The midlatitude environments in review

The midlatitude environments run the gamut from very wet to very dry climates and from mild marine coastal climates to strongly seasonal continental climates. Soils and natural vegetation cover an equally great range of types befitting to the spectrum of climate. No useful generalization is possible for such a group of environments.

Production of food resources in the midlatitudes spans as wide a range of intensities as the climates themselves. Here we have the richest food-producing regions of the world, fully developed by the Western nations through massive inputs of fertilizers, pesticides, and fuels and guided by the most advanced technology. But here there are also deserts at the same latitudes. These midlatitude environments, taken as a whole, are neither temperate in climate nor uniform in plant resources. Perhaps this heterogeneous quality of the midlatitudes is just what we should expect of a global zone where polar and tropical air masses wage war incessantly over vast land areas that cut across the latitude zones and their prevailing westerly air flow.

High-Latitude and Highland Environments

Chapter 15

THE HIGH-LATITUDE environments are identified with the climates of Group III, controlled by polar and arctic air masses. As a group, these climates have low air temperatures, a severe winter season, and only small amounts of precipitation. A condition of frozen soil water for several consecutive months of winter is also typical, so that there is usually a long annual period with near-zero water use and near-zero water need.

Climates of highlands, the high-altitude zones of lofty mountain chains and plateaus, represent yet another dimension of the climate environment. Where mountains rise sharply from lowlands, the succession of highland climates occurs in very narrow bands and is geographically complex because of orographic effects on precipitation.

(10) Continental subarctic climate (Köppen: Dfc, Dfd, Dwc, Dwd)

In the two great landmasses of North America and Eurasia a vast expanse of continental interior lies between lat. 50° and 70°N. Here we find the continental subarctic climate; it occupies the source regions for the continental polar air masses (Figure 15.1).

In winter, when heat loss by radiation forms the prevailing Siberian and Canadian highs, severely cold air temperatures develop over snow-covered surfaces. At low levels a cold, dense air mass is formed. Typically, a severe temperature inversion prevails in winter. Low in moisture content, this air mass is generally stable.

In summer the source region is shifted farther north; air mass temperatures rise to moderately high levels, but the moisture content, although much greater in summer, is still small by comparison with that of maritime tropical air masses. We expect, therefore, a climate showing very great seasonal temperature range, with extremely severe winters and a small annual total precipitation concentrated in the warm months.

The continental subarctic climate is well illustrated by the climograph of Ft. Vermilion, Alberta, lat. 58° N (Figure 15.2). The annual range of 74 F° (41 C°) is remarkable enough, but it is greatly exceeded by that of Yakutsk, U.S.S.R. (Monthly mean temperatures at Yakutsk are shown on Figure 15.2 by a dashed line.) The annual temperature range of the continental subarctic climate is the greatest of any on earth, reaching 110 F° (61 C°) in Siberia. Even in central Antarctica, which has the lowest temperatures on earth, the annual range is not over 65° (36 C°).

Winter is the dominant season of the continental subarctic climate. Absolute minimum temperatures of −70° to −80° F (−57° to −62° C) are recorded in northwestern Canada. Perhaps the coldest place in the northern hemisphere is Verkhoyansk, U.S.S.R., which lies in this climate zone and has a January mean of −59° F (−51° C) and an

figure 15.1

World map of continental subarctic climate (10). (Based on Goode Base Map.)

absolute minimum recorded temperature of −92° F (−69° C).

Summer in the subarctic climate regions is very short. Referring back to the insolation graph (Figure 3.7), you will find that the insolation in June at lat. 60° N is greater than at the equator and at lat. 20° N. Figure 4.8 shows the radiation surplus at Yakutsk to be sharply peaked in June, whereas there is a deficit for 6 consecutive winter months. The warmest month average may not greatly exceed 50° F (10° C), and frosts can occur at any time during the summer. Daily maximum temperatures, however, commonly reach 70° F (21° C). At these high latitudes, the sun is above the horizon for 16 to 18 hours from May to August. (See graph on Figure 15.2.) The long hours of sunshine accelerate the growth of plants and make possible a short growing season.

Precipitation is largely from wave cyclones and shows a very definite maximum in the summer months. Snowfall, although conspicuous in winter because it remains on the ground, accounts for only a fraction of an inch of precipitation per month in the coldest months. Cyclones and fronts crossing the area bring little precipitation at this season. In summer cyclonic rains are frequent, although thunderstorms are few.

The soil-water balance of the continental subarctic climate is illustrated by the budget for Fort George, Quebec (Figure 9.10 J). Water need and water use are effectively zero from November to April, when the ground is frozen.

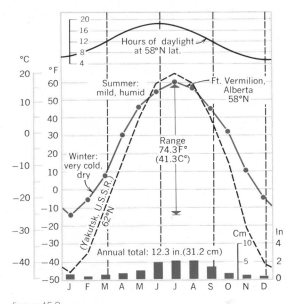

figure 15.2
Climographs for Ft. Vermilion, Alberta, and Yakutsk, Russia.

In summer the water need peaks sharply, and a very small shortage develops. Recharge accumulates as snow during the winter and is released to the soil during the spring thaw. Across the subarctic region of Canada, numerous lakes and flowing streams are a product of this water surplus. Other parts of the continental subarctic climate show no water surplus and represent a dry subtype. Much of eastern Siberia falls into this dry subarctic subtype.

Vegetation and soils The subarctic climate zone coincides with a great belt of needleleaf forest, often referred to as *boreal forest* (Figure

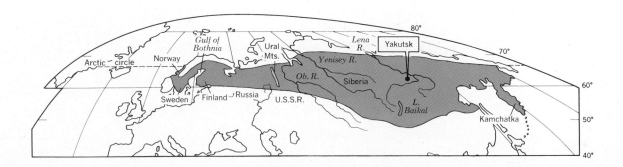

figure 15.3
Lichen woodland in Labrador. Black spruce and a few small white birch are widely spaced. The floor is covered by a lichen, called "reindeer moss." (F. Kenneth Hare, American Geographical Society.)

12.12), and open lichen woodland, the *taiga* (Figure 15.3). Trees are small, so that they are of less value for lumber than for pulpwood. In Siberia the dominant woodland tree is the larch, which sheds its leaves in winter (Figure 15.4).

Soils of the podzol group are associated with the cold needleleaf forest. These soils are strongly leached and acid. Podzols are extremely poor from the standpoint of agriculture.

Added to the natural inadequacy of soils of this region is a great prevalence of bogs and lakes left by the departed ice sheets. Some rock surfaces were scoured by ice, which stripped off the soil entirely. Elsewhere, rock basins were formed, or previous stream courses dammed, making countless lakes (see Figure 22.19).

Crop farming in the continental subarctic environment is largely limited to lands surrounding the Baltic Sea, in bordering Finland and Sweden. Cereals grown in this area include barley, oats, rye, and wheat. Along with dairying, these crops primarily provide food for subsistence. The principal nonmineral economic product throughout the subarctic lands of eastern Canada is pulpwood from the needleleaf forests. Logs are carried down the principal rivers to pulp mills and lumber mills (Figure 15.5). Forests of pine and fir in Finland and European Russia are the primary plant resource; the products are exported in the form of paper, pulp, cellulose, and construction lumber.

(11) Tundra Climate (Köppen: ET)

The tundra climate occupies the northern continental fringes of North America and Eurasia from the arctic circle to about the 75th parallel (Figure 15.6). These areas lie within the outer zone of control of arctic-type air masses, whose source region covers the Arctic Ocean and Greenland. To the south lie the continental polar (cP) and maritime polar (mP) air-mass source regions. In this respect, the arctic land fringes are in a frontal zone, which

figure 15.4
This open woodland of larch is located about lat. 64° N, in Russia. The larch tree is here near the northern limit of its growth. (Komarov Botanical Institute, Leningrad.)

figure 15.5
This saw mill at Hudson, Ontario processes logs floated across the surface of a sprawling lake in the Canadian Shield. (Ronny Jaques Studio, Toronto, from Black Star)

has been called the "arctic front." This frontal zone is more simply identified as belonging to a widely shifting polar front. Frontal activity produces many intense east-moving cyclonic storms with much severe weather.

The tundra climate is illustrated by the climograph for Upernivik at lat. 74° N on the west Greenland coast (Figure 15.7). Note that the temperature range is large, but not nearly as great as in the continental subarctic climate.

The warmest month averages are just over 40° F (4° C). Many climatologists consider this season too cool to be designated as "summer." The coldest month average is well below 0° F (−18° C). Total annual precipitation is less than 10 in. (25 cm). Nearness to the Arctic Ocean explains the somewhat more moderate temperature range and minimum, compared to the continental centers. Coolness of the summer is explained by the nearness to the large ocean body, keeping air temperatures

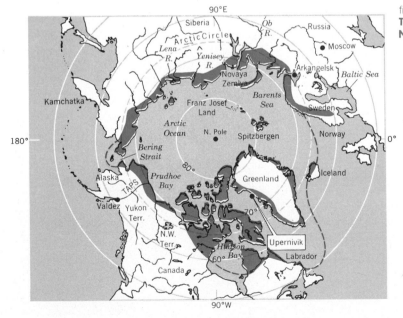

figure 15.6
Tundra climate in the Northern Hemisphere (11).

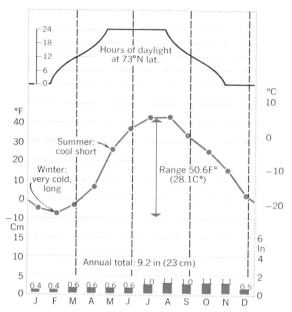

figure 15.7

Upernivik, Greenland, lat. 73° N, is a typical arctic tundra station.

down despite intense insolation at this latitude near summer solstice.

Vegetation, soils, and ecosystem   In tundra soils the mineral particles are produced almost entirely by mechanical breakup of the parent rock and have undergone little chemical alteration. Grayish loam and blue-gray clay layers are present, along with much peat.

Vegetation of the treeless tundra consists of grasses, sedges, and lichens, along with shrubs of willow. (The tundra biome was described in Chapter 12.) Traced southward the vegetation changes into birch-lichen woodland, then into the needleleaf forest. In some places a distinct tree line separates the forest and tundra. Coinciding approximately with the 50° isotherm of the warmest month, this line has been used by geographers as a boundary between subarctic and tundra climates.

Vegetation is scarce on dry, exposed slopes and summits — the rocky pavement of these areas gives them the name of *fell-field* (Figure 15.8).

The number of species in the tundra ecosystem is small, but the abundance of individuals is high. Among the animals, vast herds of caribou in North America or reindeer (their Eurasian relatives) roam the tundra, lightly grazing the lichens and plants and moving constantly. A smaller number of musk-oxen are also primary consumers of the tundra vegetation. Wolves and wolverines, arctic foxes, and polar bears are predators. Among the smaller mammals, snowshoe rabbits and lemmings are important herbivores. Invertebrates are scarce in the tundra, except for a small number of insect species. Black flies, deerflies, mosquitoes, and "no-see-ums" (tiny biting midges) are all abundant and can make July on the tundra most uncomfortable for both Man and beast. Reptiles and amphibians are also rare. The boggy tundra, however, offers an ideal summer environment for many migratory birds such as waterfowl, sandpipers, and plovers.

The food chain of the tundra ecosystem is simple and direct. The important producer is "reindeer moss," a lichen (Figure 15.9). In addition to the caribou and reindeer, lemmings, ptarmigan (arctic grouse), and snowshoe rabbits are important lichen grazers. The important predators are the fox, wolf, lynx, and Man, although all these animals may feed directly on plants as well. During the summer,

figure 15.8

Vegetation of the arctic fell-field. The plants grow in a small patch of soil surrounded by a ring of cobbles. (American Museum of Natural History.)

figure 15.9
Reindeer moss, a large lichen of the arctic woodland and tundra. (Larry West)

the abundant insects help support the migratory waterfowl populations. The directness of the tundra food chain makes it particularly vulnerable to fluctuations in the populations of a few species.

Arctic permafrost In the tundra subfreezing monthly average temperatures occur for 7 or more consecutive months. As a result, all moisture in the soil and subsoil is solidly frozen to depths of many feet. Summer warmth is insufficient to thaw more than the soil layer. This condition of perennially frozen ground, or *permafrost*, prevails over the tundra climate and a wide bordering area of continental subarctic climate. Seasonal thaw penetrates from 2 to 14 ft (0.6 to 4 m), depending on latitude and the nature of the ground. This shallow zone of alternate freeze and thaw is the *active zone*.

The distribution of permafrost in the northern hemisphere is shown in Figure 15.10. Three zones are recognized. Continuous permafrost, which extends without gaps or interruptions under all surface features, coincides largely with the tundra climate, but also includes a large part of the continental subarctic climate in Siberia. Discontinuous permafrost, which occurs in patches separated by frost-free zones under lakes and rivers, occupies much of the continental subarctic climate zone of North America and Eurasia. Sporadic occurence of permafrost in small patches extends into the southern limits of the continental subarctic climate.

Depth of permafrost reaches 1000 to 1500 ft (300 to 450 m) in the continuous zone near lat. 70°. Much of this permanent frost is an inheritance from more severe conditions of the last ice age, but some permafrost bodies may be growing under existing climate conditions. Bodies of ice have grown into large masses beneath the surface. A particularly distinctive form is the *ice wedge*, shown in Figure 15.11.

Environmental degradation of permafrost regions arises from man-made surface changes. The undesirable consequences are usually related to the destruction or removal of an insulating surface cover, which may consist of a moss or peat layer in combination with living plants of the tundra or arctic forest. When this layer is scraped off, the summer thaw is extended to a greater depth, with the result that ice wedges and other ice bodies melt in the summer and waste downward. This activity is called *thermal erosion*. Meltwater mixes with silt to form mud, which is then eroded and transported by water streams, with destructive effects that are evident in Figure 15.12.

The consequences of disturbance of permafrost terrain became evident in World War II, when military bases, airfields, and highways were hurriedly constructed without regard for maintenance of the natural protective surface insulation. In extreme cases, scraped areas turned into mud-filled depressions and even into small lakes that expanded in area with successive seasons of thaw, engulfing nearby buildings. Engineering practices now call for placing buildings on piles with an insulating air space below or for the deposition of a thick insulating pad of coarse gravel over the surface prior to construction. Steam and hot-water

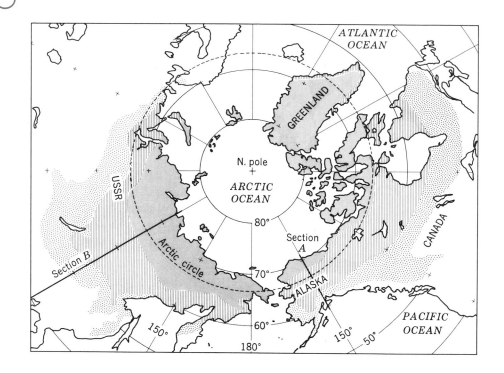

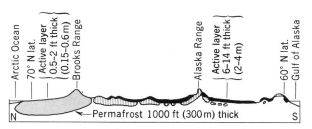

Section *A*: Alaska, on long. 150° W

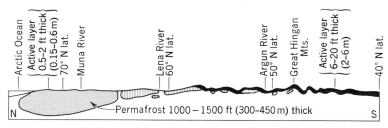

Section *B*: Asia, on long. 120° E
(Modified from I. V. Poiré)

Diagrammatic cross sections of permafrost

- ■ Active layer
- ▦ Discontinuous permafrost
- ▨ Continuous permafrost
- ░ Sporadic permafrost

figure 15.10
**Distribution of permafrost in the northern hemisphere, and
representative cross sections in Alaska and Asia. (From Robert F. Black,
"Permafrost," Chapter 14 of P. D. Trask's** Applied Sedimentation, **John
Wiley & Sons, New York, 1950.)**

figure 15.11
This vertical river-bank exposure near Livengood, Alaska, in the subarctic climate region reveals a V-shaped ice wedge surrounded by layered silt of alluvial origin. (T. L. Péwé, U.S. Geological Survey.)

lines are placed above ground to prevent thaw of the permafrost layer.

Another serious engineering problem of arctic regions is in the behavior of streams in winter. As the surfaces of streams and springs freeze over, the water beneath bursts out from place to place, freezing into huge accumulations of ice. Highways are thus made impassable.

The lessons of superimposing Man's technological ways on a highly sensitive natural environment were learned the hard way — by encountering unpleasant and costly effects that were not anticipated. The threat of environmental destruction persists as construction forges ahead on the Trans-Alaska Pipeline. This facility will carry hot oil from the northern shores of Alaska, across a permafrost landscape, to the port of Valdez on the south coast. Effects of this pipeline on permafrost and other elements of the environment were heavily debated and were the subject of intensive investigation. In addition to the

prospects of thaw of the permafrost layer by heat of the pipeline, there is the possibility of damage to the ecosystem from spills caused by pipe breakage. The possible effect of the pipeline as a barrier to animal migration remains to be seen.

(12) Ice sheet climate (Köppen: EF)

Three vast regions of ice exist on the earth. These are the ice sheets of Greenland and Antarctica and the large area of floating sea ice in the Arctic Ocean (Figure 15.13). (The ice sheets of Greenland and Antarctica are described in Chapter 22.)

Ice sheet climate has by far the lowest average annual temperature of all global climates. No month has an average above freezing (32° F, 0° C). The annual temperature graph for

figure 15.12
After one season of thaw this winter truck trail through the Alaskan arctic forest had suffered severe thermal and water erosion. (R. K. Haugen, U.S. Army Cold Regions Research & Engineering Laboratory.)

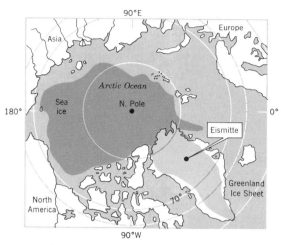

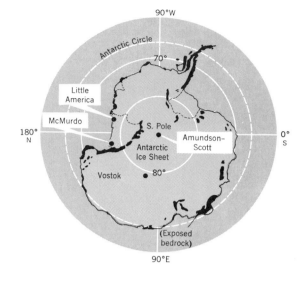

figure 15.13
Map of ice sheets and the area of arctic sea ice.

an interior Greenland ice sheet station, "Eismitte," is shown in Figure 15.14. This information was gathered by a German expedition from 1929 through 1931. Note that only in 3 months of the year did the mean monthly temperatures rise above 0° F (−18° C), and then only to a July maximum of 12° F (−11° C). The coldest month was February, with −53° F (−47° C), making an annual range of 65 F° (36 C°).

Temperatures in the interior of Antarctica have proved to be far lower than any place on earth. The Russian meteorological station Vostok, located about 800 mi (1300 km) from the south pole at an altitude of 11,440 ft (3488 m), may be the world's coldest spot (Figure 15.14). Here a record low of −127° F (−88.3° C) was observed. At the pole itself (Amundsen-Scott Station), July, August, and September of 1957 had averages about −76° F (−60° C) (Figure 15.14). Temperatures run roughly 40 F° (22 C°) higher, month for month, at Little America because it is located close to the Ross Sea and is at low altitude.

A remarkable feature of the high ice sheet interior is the intense chilling of air close to the snow surface. A strong temperature inversion develops in winter, so that the air near the surface may be 50 to 60 F° (28 to 33 C°) colder than air a few hundred feet higher. Downslope flow of this heavy, cold air layer causes blizzard winds to develop in favorable valley locations. Driving blizzard winds pack the sandlike snow into a hard mass, which then becomes etched into furrows (Figure 15.15).

The area of floating sea ice of the Arctic Ocean has a severe climate in which the mean temperature of every month is below freezing. However, the mean annual temperature is not nearly as low as on the high ice sheets. The major environmental difference between the Arctic Ocean and the ice sheets lies in the nature of the floating ice layer itself.

Sea ice and icebergs We can distinguish _sea ice_, formed by direct freezing of ocean water, from _icebergs_, which are bodies of land ice broken free from tide-level glaciers. Aside from differences in origin, a major difference between sea ice and floating masses of land ice is thickness. Sea ice is limited in thickness to about 15 ft (5 m).

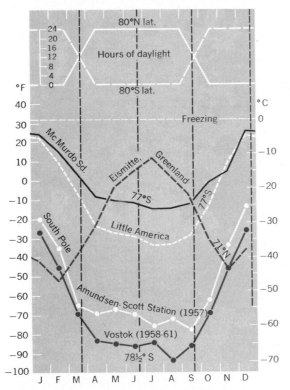

figure 15.14
Temperature graphs for five ice sheet stations.

figure 15.15
Wind eroded furrows on the Greenland Ice Sheet, 5 mi (8 km) from the ice margin, lat. 76° N. (L. H. Nobles.)

Pack ice is sea ice that completely covers the sea surface (Figure 15.16). Under the forces of wind and currents, pack ice breaks up into individual patches, called ice floes. The narrow strips of open water between such floes are known as leads. Where ice floes are forcibly brought together by winds, the ice margins buckle and turn upward into pressure ridges resembling walls or irregular hummocks. Travel on foot across the polar sea ice is extremely difficult because of such obstacles. The surface zone of sea ice is composed of fresh water.

Icebergs are formed by the breaking off, or calving, of blocks from a valley glacier or tongue of an ice sheet. An iceberg may be as thick as several hundred feet. Being only slightly less dense than seawater, the iceberg floats very low in the water, about five sixths of its bulk lying below water level (Figure

figure 15.16
The icebreaker Northwind forces a passage through arctic ice floes in mid-August. To the left is an open lead. Cutting across from left to right is a rugged zone of pressure ridges. (U.S. Coast Guard.)

figure 15.17
This great iceberg in the east Arctic Ocean dwarfs the icebreaker Eastwind. (U.S. Coast Guard.)

15.17). The ice is fresh, of course, since it is formed of compacted and recrystallized snow.

In the northern hemisphere, icebergs are derived largely from glacier tongues of the Greenland Ice Sheet. They drift slowly south with the Labrador and Greenland currents and may find their way into the North Atlantic in the vicinity of the Grand Banks of Newfoundland. Icebergs of the antarctic are

distinctly different. Whereas those of the North Atlantic are irregular in shape and present rather peaked outlines above water, the antarctic icebergs are commonly tabular in form, with flat tops and steep clifflike sides (Figure 15.18). This is because tabular bergs are parts of ice shelves, which are great, floating platelike extensions of the continental ice sheet. In dimensions, a large tabular berg of the antarctic may be tens of miles broad and over 2000 ft (600 m) thick, with an ice wall rising 300 ft (90 m) above sea level.

Highland climates

Because air temperature falls with increased altitude, we might infer that the climates encountered in climbing a high mountain range in the equatorial zone would simply duplicate the sucession of climates found on a sea-level traverse to the poles (Figure 15.19). There are two good reasons why this concept is not valid. First, air density decreases upward, so that there is progressively less air to absorb, radiate, and conduct heat. In other words, the balances of radiation and heat change greatly as we approach the alpine summits. Second, going upward does not bring a change in the annual cycle of solar radiation. On a summit at 20,000 ft in the Andes range on the equator, the sun's path in the sky is not far from the

figure 15.18
A tabular iceberg near the Bay of Whales, Little America, Antarctica. (U.S. Coast Guard.)

figure 15.19
Zone of perpetual snow within a few degrees of latitude of the equator. White Range, near Ancash, Peru. (Aero Service Division, Western Geophysical Co. of America.)

zenith at any time of year. This path is just the same for a place near sea level in the adjacent rainforest of the Amazon basin. In contrast, a station on the arctic sea ice, near the pole, would experience 6 months of day and 6 months of night.

At a high altitude the thinner atmosphere has relatively less carbon dioxide, water vapor, and dust. This air absorbs and reflects less solar energy, permitting a higher intensity of insolation at the ground.

Increasing intensity of insolation at higher altitude has a profound influence on daily extremes of air temperature. Surfaces exposed to sunlight heat rapidly and intensely; shaded surfaces are quickly and severely cooled. This

same effect results in rapid air heating during the day and rapid cooling at night at high-mountain locations.

This principle shows up in the set of daily-temperature graphs of Figure 15.20. The data cover two weeks in July at several stations on the west side of South America at lat. 15° S. The stations range in altitude from sea level to a maximum of over 14,000 ft (4.3 km). Not only does the average temperature of the month decrease with higher altitude, but the daily temperature range becomes much larger at higher altitude.

Look next at the annual cycle of monthly mean air temperatures for two stations in the tropical zone (Figure 15.21). New Delhi, the capital

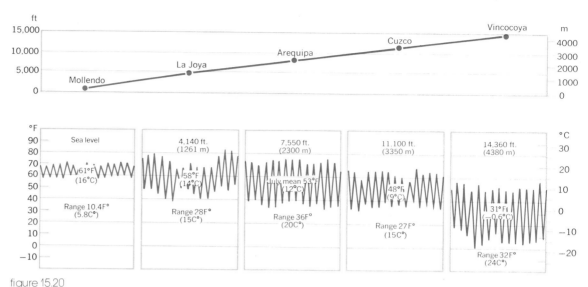

figure 15.20
As altitude increases, the monthly mean temperature declines, and the daily range becomes greater. (Data of Mark Jefferson.)

city of India, lies in the Ganges lowland; Simla, a mountain refuge from the hot weather, is located about 6500 ft (2000 m) higher in altitude, in the foothills of the Himalayas. When hot-season averages are over 90° F (18° C), Simla is enjoying a pleasant 65°F (18°C) which is a full 30 F° (17 C°) cooler. Notice, however, that the two temperature cycles are quite similar in phase, with the minimum month being January for both.

The general effect of increased altitude is first to bring an increase in precipitation, at least for the first few thousand feet of altitude increase. This change is due to the production of orographic rainfall, generated by the forced ascent of air masses (see Chapter 6).

The altitude effect shows nicely in the monthly rainfalls of New Delhi and Simla (Figure 15.21). New Delhi shows the typical rainfall pattern of the tropical wet-dry climate, with monsoon rains peaking in July and August. Simla has the same pattern, but the amounts are larger in every month, and the monsoon peak is very strong. Simla's annual total is well

figure 15.21
Climographs for New Delhi and Simla, both in northern India. Simla is a welcome refuge from the intense heat of the Gangetic Plain in May and June.

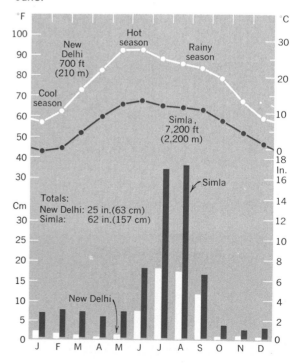

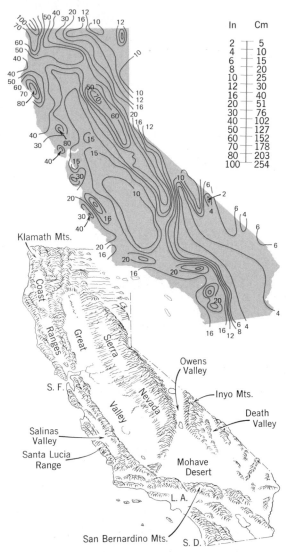

figure 15.22
The effect of mountain topography on rainfall is shown well by the state of California. Isohyets in inches.

figure 15.23
This tree limb is completely encased in ferns, growing as epiphytes. Montane forest in the Kenya uplands, Africa. (American Museum of Natural History.)

over twice that of New Delhi. Generally, mountains are humid climate zones. This is particularly striking in an arid or semiarid region, where the mountains form islands of humid climate surrounded by desert or steppe.

The influence of mountains on precipitation is strikingly illustrated by the state of California

(Figure 15.22). Mountain ranges alternate with valleys in an arrangement generally paralleling the Pacific coastline. The barriers lie across the flow of the prevailing westerlies, which bring moist maritime polar (mP) air masses inland in the winter season. Notice that in the Great Valley and Death Valley precipitation is less than 10 in. (25 cm) but, on the western slopes of the Sierra Nevada and Klamath Mountains, it is over 70 in. (175 cm).

From the standpoint of river flow and floods, mountain climates are of greatest importance in midlatitudes. The higher ranges serve as snow storage areas, keeping back the precipitation until early or midsummer, releasing it slowly through melting. In this way a continuous river flow is maintained. As melting proceeds to successively higher levels, the meltwater is supplied to the drainage basin. Among the snow-fed rivers of the western United States are the Columbia, Snake, Missouri, Platte, Arkansas, and Colorado.

Vegetation and life zones When ascending a high mountain range you will pass through distinct vegetation zones. The succession of zones is in some ways like that encountered in traveling from low latitudes to the arctic zone, but with important differences, particularly in the equatorial highlands.

The succession of vegetation forms you would encounter in ascending the Ruwenzori Range of central Africa, or the central mountain range of New Guinea, would go about as follows. Equatorial rainforest extends high up the mountain slopes but, between 4000 and 6000 ft (1200 and 1800 m) gradually changes to

montane forest Tree crowns are lower and less densely packed than in the equatorial rainforest. Tree ferns and bamboos are conspicuous, while epiphytes (air plants) become very abundant. Ferns are an important group of epiphytes (Figure 15.23). With increasing altitude, the trees become smaller, and their limbs and trunks are covered by mosses. In this zone of persistent mists and high relative humidities we find the mossy forest. Near its upper limit the forest trees

figure 15.24

Altitude zoning of climates in the equatorial Andes of southern Peru, lat. 10° to 15° S. (After Isaiah Bowman, The Andes of Southern Peru.)

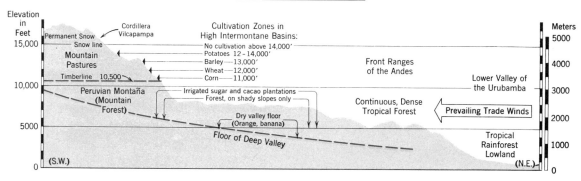

figure 15.25

Altitude zoning of mountain and plateau climates in the arid southwestern United States. Grand Canyon-San Francisco Mountain district of northern Arizona. (Data of G. A. Pearson, C. H. Merriam, and A. N. Strahler.)

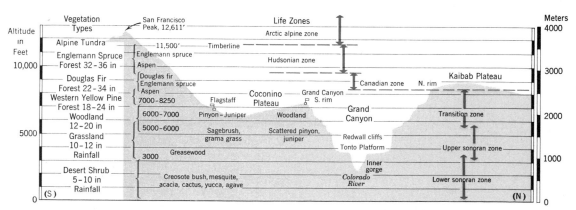

become dwarfed, giving elfin forest. Above the forest limit, which is about 12,000 ft (3600 m), we encounter a treeless vegetation that may be alpine tundra or grassland. Still higher, we reach the snowline, with its perpetual snow banks, and perhaps also living glaciers. The snowline in the equatorial zone lies in the range of 15,000 to 16,000 ft (4600 to 5000 m).

Figure 15.24 shows the altitude zoning of vegetation and agriculture in the equatorial Andes of Peru, a few degrees south of the equator. The high intermontane basins above timberline support subsistence farming of corn, wheat, barley, and potatoes.

In midlatitudes, where steppe or desert exists at low altitudes, zonation is particularly striking. Figure 15.25 shows the vegetation zones of the Colorado Plateau region of northern Arizona. Zone names, altitudes, dominant forest trees, and annual precipitation are given in the figure. Ecologists have set up a series of *life zones*, whose names suggest the similarities of these zones with latitude zones encountered in poleward travel on a meridian. Desert vegetation of the hot, dry Inner Gorge of the Grand Canyon gives way to grassland and woodland, then to a forest of western yellow pine (Figure 15.26). Still higher is the Hudsonian zone, 9500 to 11,500 ft (2900 to 3500 m). It bears a needleleaf forest quite similar to subarctic needleleaf forest. As the upper limit of forest, or timberline, is approached, the coniferous trees take on a stunted appearance and decrease in height to low shrub-like forms.

A vegetation zone of alpine tundra lies above tree line and resembles in many ways the arctic tundra (Figure 15.27). The snowline is encountered at about 12,000 to 13,000 ft (3700 to 4000 m) in these midlatitudes, which is, of course, much lower than at the equator.

figure 15.26

A forest of western yellow pine of the Kaibab Plateau of northern Arizona. The altitude here is about 7500 ft (2250 m). These pine forests have been a major timber resource in many highlands of the arid southwestern United States. (U.S. Forest Service.)

Poleward, the snowline decreases in altitude, eventually reaching sea level in the vicinity of the arctic circle.

As you might expect, soils also change their character with increasing altitude, responding to changes in climate and vegetation. In midlatitudes of the western United States, soil profiles change with life zones, as shown in Figure 15.28. From desert soils in low basins, the profile changes in succession to brown soils, through chestnut, chernozem, and prairie soils, up to an altitude of about 8000 ft (2400 m). Here, under the needleleaf forest, podzols are like those of the subarctic climate. Still higher you would find a tundra soil like that of the arctic tundra.

figure 15.27
Alpine tundra in the center of this view gives way at the left to needleleaf forest at a lower altitude. The bushlike plants are dwarf willow, barely able to grow under the impact of blizzard winds and driving snow. Arapahoe National Forest, Colorado. (U.S. Forest Service.)

figure 15.28
This schematic diagram shows the gradation of soils from a dry steppe-climate basin (left) to a cool, humid climate (right) as one ascends the west slopes of the Bighorn Mountains, Wyoming. (After J. Thorp.)

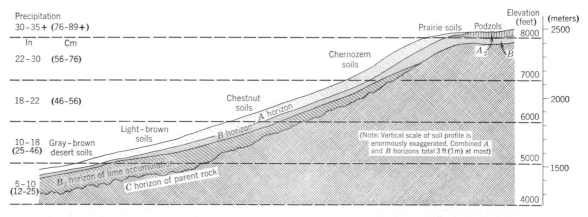

Global environments in review

In three chapters we have surveyed the principal environmental regions and their food resources. Each type of climate, together with its characteristic soils and natural vegetation, comprises a unique natural environmental region. Perhaps the impression most deeply implanted by this survey is the enormous range in environmental quality, judged in terms of plant resources and the potential for production of plant food for human consumption. It may be a bit disconcerting to consider that vast areas of the continents are almost total deserts, either too dry or too cold to support human populations. In the desert wastes of the Sahara or the snow-covered plateau of Antarctica, the average American family might stay alive for at most a few days without a supply of water or heat, or both, brought in from more favorable environments.

Another distinct impression our survey might leave is that the human family has already done a remarkably thorough job of exploiting the favorable environments of the globe. Through centuries of trial and error, inhabitants of the Old World discovered which soils are productive and which are not. They discovered which soil-water budgets will allow them to grow crops and which will not. These painfully won adaptations reflect the best that Man could do with only the power of his own body and that of domesticated animals. These distributions reflect the limits of naturally available fertilizers — mostly the body wastes of Man himself and his animals. From painful experience (if mass starvation be included) the Old World farmer learned that diversification of crops on his pitifully small farm plots was a wise choice, since crop failure from drought, plant disease, and insect pests would not likely take everything he had planted in any one season.

When the New World was opened up to European Man's occupation, the earliest patterns of land use closely followed the practices of the Old World. We have learned that the same climate, soils, and water budgets are approximately duplicated in widely separated parts of the world. Settlers to the New World found a humid continental climate like that of western Europe; they found tall-grass prairies like those of Hungary; they found steppes with black soils, like those of the Ukraine. In California they found an environment like that of the Mediterranean lands. Crop farming, dairying, cattle grazing, or grape and citrus culture fitted the analogous environments rather well, all things considered.

For each environmental region we have made an assessment of the prospects for greatly increased food production, using massive inputs of irrigation water, fertilizers, fuels, and pesticides in concert with the newest in genetic strains of staple crops and the best in management techniques.

With this chapter we conclude the study of three of the four key topics of traditional physical geography: climate, soil, and vegetation. A fourth remains; it is embodied in the word landform. In its broadest sense, in the context of physical geography, landform encompasses the solid platform or substrate on which soils and vegetation rest and from which the organic realm derives its inorganic nutrients. Landforms, the plural form of the word, refer specifically to surface configurations of the lithosphere, that is, the geometrical configurations of the upper surface of the bedrock and its regolith. Landforms come in all scales and in a great array of shapes. Landforms interact with climate, soils, and vegetation in many ways and on many scales. The remaining chapters of this book deal with landforms and with the lithosphere from which they are shaped.

Earth Materials

Chapter 16

WITH THIS CHAPTER we turn to the lithosphere, or solid mineral realm of our planet. Which materials and structures of the lithosphere are most important to Man and to the processes of the life layer? The inorganic mineral matter of the lithosphere is a vitally important nutrient reservoir for life processes, as we have seen in earlier chapters on soils and materials cycling in the biosphere. Rocks exposed at the earth's surface are the primary source of these nutrient materials.

The lithosphere is also a platform for life on the lands. The platform consists basically of the continents. Continents and ocean basins are global features of first order of magnitude. What materials make up the continents? Why are they shaped and distributed in such an odd way over the globe? These are important questions in the background of physical geography. The surfaces of the continents are shaped into a remarkable variety of surface configurations, called landforms. Landforms strongly influence the distribution of ecosystems; landforms exert strong controls over Man's occupation of the lands.

Our plan in this concluding block of eight chapters is first to examine the lithosphere on a large scale to determine its composition and structure and to interpret the major features of the continents. This section deals first with *geology*, the science of the solid earth. Then we investigate landforms in terms of their origin and environmental qualities. In following this plan, we will start with sources of energy and matter arising from deep within the planet, but we will end up with surface features shaped by solar energy through activity of the atmosphere and hydrosphere. In this way the final concepts will relate to the familiar theme of interaction within the life layer, and we will have traveled the full circle of physical geography.

Composition of the earth's crust

From the standpoint of the environment of Man, the really significant zone of the solid earth is the thin outermost layer — the earth's crust. This mineral skin, averaging about 10 mi (17 km) thickness for the globe as a whole, contains the continents and ocean basins and is the source of soil and other sediment vital to life, of salts of the sea, of gases of the atmosphere, and of all free water of the oceans, atmosphere, and lands.

Figure 16.1 displays in order the eight most abundant elements of the earth's crust in terms of percentage by weight. Oxygen, the predominant element, accounts for about half the total weight. It occurs in combination with silicon, the second most abundant element.

Aluminum and iron are third and fourth on the list. These metals are of primary importance in Man's industrial civilization, and it is most fortunate that they are comparatively abundant elements. Four metallic elements follow that we have already identified as nutrient bases in soils: calcium, sodium, potassium, and magnesium. All four are on the same order of abundance (2 to 4 percent). Their importance in soil fertility has already been stressed in Chapter 10.

If we were to extend the list, the ninth-place element would be titanium, followed in order by hydrogen, phosphorus, barium, and strontium. Phosphorus is one of the essential elements in plant growth.

Rocks and minerals

The elements of the earth's crust are organized into compounds that we recognize as minerals. *Rock* is broadly defined as any aggregate of minerals in the solid state. Rock comes in a very wide range of compositions, physical characteristics, and ages. A given rock is

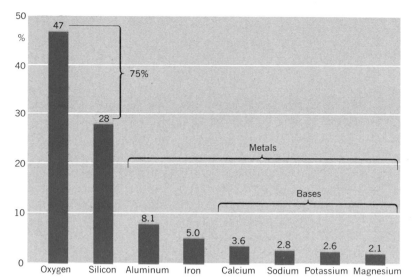

figure 16.1
The average composition of the earth's crust is given here in terms of the percentage by weight of the eight most abundant elements.

usually composed of two or more minerals, and usually many minerals are present; however, a few rock varieties consist almost entirely of one mineral. Most rock of the earth's crust is extremely old in terms of human standards, the times of formation ranging back many millions of years. But rock is also being formed at this very hour as a volcano emits lava that solidifies on contact with the atmosphere.

A mineral is perhaps easier to define than a rock. A _mineral_ is a naturally occurring, inorganic substance, usually possessing a definite chemical composition and characteristic atomic structure. A vast number of minerals exist, together with a great number of their combinations into rocks. We must generalize and simplify this discussion so as to refer to rocks and minerals in a way meaningful in terms of their environmental properties and value as natural resources.

Rocks of the earth's crust fall into three major classes. (1) _Igneous rocks_ are solidified from mineral matter in a high-temperature molten state. (2) _Sedimentary rocks_ are layered accumulations of mineral particles derived in various ways from preexisting rocks. (3)

Metamorphic rocks are igneous or sedimentary rocks that have been physically and chemically changed by application of heat and pressure during mountain-making activities. Although we have listed these classes in a conventional sequence, it will become obvious in later pages that no one class has first place in terms of origin. Instead, they form a continuous circuit through which the crustal minerals have been recycled during many millions of years of geologic time. We begin with igneous rock for simplicity, not because this is the original class of rock.

The silicate minerals

The great bulk of igneous rock consists of _silicate minerals_, which are all compounds containing silicon atoms in combination with oxygen atoms in a close linkage. In the crystal structure of silicate minerals one atom of silicon is linked with four atoms of oxygen as the unit building block of the compound. Most of the silicate minerals also have one, two, or more of the metallic elements listed in Figure 16.1 (i.e., aluminum, iron, calcium, sodium, potassium, and magnesium). We are simplifying the list of silicate minerals to seven; these are listed in Figure 16.2

Among the most common minerals of all the rock classes is *quartz*; its composition is silicon dioxide. Two silicate-aluminum compounds called *feldspars* follow. One type, *potash feldspar*, contains potassium as the dominant metal besides aluminum. A second type, *plagioclase feldspar*, is rich in sodium, calcium, or both.

Quartz and the feldspars are light in color (white, pink or grayish) and low in density compared with the other silicate minerals. For this reason they form a silicate mineral group described as *felsic* ("fel" for feldspar; "si" for silicate).

The next three silicate minerals are actually mineral groups, with a number of mineral varieties in each. They are the *mica group* (example: biotite), *amphibole group* (example: hornblende), and *pyroxene group* (example: augite). All three are compounds of silicon, aluminum, magnesium, iron, and potassium or calcium. The seventh mineral, *olivine*, is a

figure 16.2

Silicate minerals and igneous rocks. Only the most important silicate mineral groups are listed, along with four common igneous rock types.

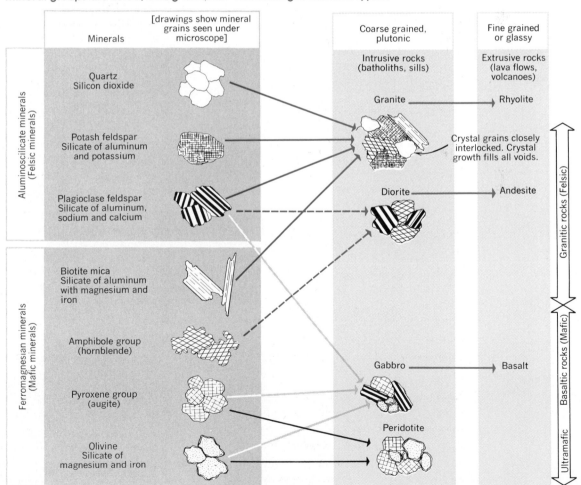

silicate of only magnesium and iron; it lacks aluminum. Altogether, these minerals are described as *mafic* ("ma" for magnesium; "f" from the chemical symbol for iron, Fe). The mafic minerals are dark in color (usually black) and are denser than the felsic minerals.

The igneous rocks

The silicate minerals occur in a high-temperature molten state as *magma.* From pockets at least several miles below the earth's surface, magma makes its way upward through older solid rock and eventually solidifies as igneous rock. No single igneous rock is made up of all seven silicate minerals that we have listed but, instead, of three or four of those minerals as the major components.

Figure 16.2 lists four common igneous rocks and the major silicate minerals of which they are composed. These four serve only as examples; there are many other important mineral combinations. The first example is *granite*. The bulk of granite consists of quartz (27 percent), potash feldspar (40 percent), and plagioclase feldspar (15 percent). The remainder is mostly biotite and amphibole. Because most of the volume of granite is of felsic minerals, we classify granite as a *felsic igneous rock*. As Figure 16.3 shows, granite is a mixture of white, grayish or pinkish, and black grains, but the overall appearance is a light gray or pink color. The mineral grains are very tightly interlocked, and the rock is exceedingly strong.

Diorite, the second igneous rock on the list, lacks quartz. It consists largely of plagioclase feldspar (60 percent) and secondary amounts of amphibole and pyroxene. Diorite is a light-colored felsic rock, only slightly denser than granite.

The third igneous rock is *gabbro*, in which the major mineral is pyroxene (60 percent). A substantial amount of plagioclase feldspar (20

figure 16.3

Seen close up, this granite, a coarse-grained, plutonic intrusive rock, proves to be made of tightly interlocking crystals of a few kinds of minerals. (A. N. Strahler.)

to 40 percent) is present and, in addition, there may be some olivine (0 to 20 percent). Gabbro is classed as a *mafic igneous rock*; it is dark in color and denser than the felsic rocks. The fourth igneous rock, *peridotite*, has a predominance of olivine (60 percent), and the rest is mostly pyroxene (40 percent). Peridotite is classed as an *ultramafic igneous rock*, denser even than the mafic types.

Our purpose in describing four common kinds of igneous rocks is to show that there is a range

figure 16.4
Igneous rock bodies.

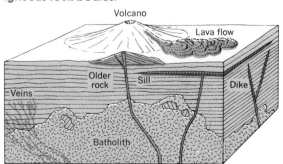

from felsic, through mafic, to ultramafic types, increasing in density from one end of the scale to the other. This arrangement is important because it is duplicated on a grand scale in the principal rock layers comprising the earth. We shall mention this arrangement again in describing the earth's interior structure.

Intrusive and extrusive igneous rocks

Magma that solidifies below the earth's surface and is surrounded by older, preexisting rock is called *intrusive igneous rock*. Where magma reaches the surface, it emerges as *lava*, which solidifies to form *extrusive igneous rock*.

Although both an intrusive rock and extrusive rock can come from a single type of magma, their outward appearance is quite different. Intrusive igneous rocks cool very slowly and, as a result, develop large mineral crystals. These coarsely grained intrusive rocks are classified as *plutonic rocks*. A good example is the granite pictured in Figure 16.3. In a plutonic rock the individual mineral crystals can easily be distinguished with the unaided eye.

Intrusive igneous rock typically accumulates in enormous subterranean bodies, called *batholiths*. Figure 16.4 shows the relationship of a batholith to the overlying rock. As it made its way upward, the magma made room for its bulk by dissolving and incorporating the older rock above it. Batholiths are several miles in depth and may extend beneath an area of several thousand square miles.

Figure 16.4 shows two other common forms of intrusive rock bodies. One is a *sill*, a platelike layer formed when magma forced its way between two horizontal rock layers, lifting the overlying rock to make room. A second is the *dike*, a near-vertical wall-like body formed by the spreading apart of a vertical rock fracture. Figure 16.5 shows a dike of mafic rock cutting

figure 16.5
A basaltic dike cutting granite, Cohasset, Massachusetts. (John A. Shimer.)

through a granite batholith. The dike rock is fine textured because of rapid cooling. Dikes are commonly the conduits by means of which magma reaches the surface. Magma entering small fractures in the overlying rock solidifies in a branching network of thin veins.

Extrusive igneous rocks, emerging as lava, cool very quickly in contact with the air or water at the earth's surface (Figure 16.6). Quick cooling gives lava a very fine-grained texture and sometimes even a glassy texture. Figure 16.7 shows two kinds of extrusive texture. One is a frothy, bubble-filled lava, called scoria (or sometimes pumice). The other is a natural volcanic glass. Most lava solidifies simply as a dense uniform rock of dull surface appearance in which the mineral grains are too small to be distinguished with the unaided eye.

Figure 16.2 names three extrusive rocks; each one is of the same mineral composition as a plutonic rock. *Rhyolite* is the name for lava of the same composition as granite; *andesite* is a lava of the composition of diorite. *Basalt* is lava of the composition of gabbro. Rhyolite and

figure 16.6
A freshly solidified basalt lava flow with ropy surface texture. Craters of the Moon National Monument, Idaho. (George A. Grant, U.S. Department of Interior.)

andesite are pale grayish or pink in color; basalt is black.

Lava flows, along with particles of solidified lava blown explosively from narrow vents, build large surface structures, called volcanoes. These are described in Chapter 17 in connection with process of mountain building.

figure 16.7
A frothy, gaseous lava solidifies into a light, porous scoria (left). Rapidly cooled lava may form a dark volcanic glass (right). (A. N. Strahler.)

Igneous processes and ore deposits

Igneous rocks are the primary source of many metals indispensable to our industrial society. Over spans of many millions of years igneous processes have concentrated certain rare metals into compact rock bodies, called ores. An _ore_ is a natural mineral concentration that can be profitably extracted and refined.

To obtain an appreciation of the scarcity of certain of the essential metals, study the figures in Table 16.1. The table gives estimated abundance, as percentage by weight, of each metal in average rock of the earth's crust. Notice that aluminum and iron are comparatively abundant elements. In contrast, many of the industrial metals that we regard as quite commonplace — zinc, copper, and lead, for example — are present in very small percentages in terms of the average rock composition. In the case of mercury, for which there is no adequate substitute at present, the abundance is on the order of about one part in 10 million.

table 16.1
Average abundance of metals in crustal rock

Element	Percent by Weight
Aluminum	8.1
Iron	5.0
Magnesium	2.1
Titanium	0.44
Manganese	0.10
Vanadium	0.014
Chromium	0.010
Nickel	0.0075
Zinc	0.0070
Copper	0.0055
Cobalt	0.0025
Lead	0.0013
Tin	0.00020
Uranium	0.00018
Molybdenum	0.00015
Tungsten	0.00015
Antimony	0.00002
Mercury	0.000008
Silver	0.000007
Platinum	0.000001
Gold	0.0000004

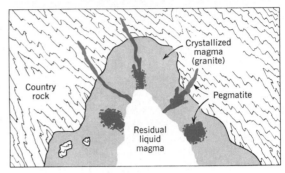

figure 16.8
This schematic cross section shows the relationship of pegmatite bodies to an intrusive igneous body and its surrounding country rock. (From A. N. Strahler, 1972, Planet Earth; Harper & Row, New York.)

contain enormous crystals of the mineral spodumene, a silicate of lithium (Figure 16.9). Single crystals weigh over 30 tons apiece. This and other pegmatite localities are a principal source of the light metal, lithium, which has many important uses in industry. Other vital

We will only briefly touch on the subject of how ores of metals accumulate as a result of igneous processes. Two examples of ores of this origin will illustrate the concept.

In the final stages of crystallization of a granite magma, a watery magma, rich in silicon, remains. Under high pressure, this solution leaves the main body of solidfying magma and penetrates the surrounding rock mass in small chambers and narrow passageways to crystallize in the form of *pegmatite*. Pegmatite bodies consist mostly of large mineral crystals. Pegmatite bodies take the form of irregular masses within the parent plutonic rock and occur as dikes and veins (Figure 16.8).

While the bulk of all pegmatites consists of quartz, feldspar, micas, and other common minerals of felsic rocks, unusual concentrations of rarer minerals occur. For example, pegmatites of the Black Hills of South Dakota

figure 16.9
Large spodumene crystals in the Etta pegmatite, Pennington County, South Dakota. The hammer rests on a single large spodumene crystal. (U.S. Geological Survey.)

industrial metals obtained from pegmatites include tantalum, columbium, and beryllium; the last is an important component in high-strength alloys of copper, cobalt, nickel, and aluminum.

Another type of ore deposit is produced by the effects of high-temperature solutions known as _hydrothermal solutions_. These solutions leave a magma during the final stages of its crystallization. Minerals carried by these solutions are deposited in fractures to produce mineral veins. Where veins occur in exceptional thicknesses, they constitute a lode.

Hydrothermal solutions also produce another important type of ore accumulation, the disseminated deposit, in which the ore is distributed throughout a very large mass of rock. Certain great copper deposits are of this type. One of the most celebrated of these is at Bingham Canyon, Utah (Figure 16.10).

Hydrothermal solutions rise toward the surface, making vein deposits in a shallow zone and even emerging as hot springs. Many valuable ores of gold and silver are deposits of the

shallow type. Particularly interesting is the occurrence of mercury ore in the form of the mineral cinnabar as a shallow hydrothermal deposit. Most renowned are the deposits of the Almaden district in Spain, where mercury has been mined for centuries and has provided most of the world's supply of that metal.

Although mineral resources are not a central topic in physical geography, they are a subject of major interest to geographers as one of several basic categories of natural resources distributed over the globe. The important point we make here is that metallic ores represent extremely rare concentrations of elements, and no more will be forthcoming once those now in the crust are used up.

Chemical alteration of igneous rocks

As a transition to the second class of rocks, the sedimentary rocks, it is important that we investigate the ways in which the silicate minerals of igneous rocks are changed into a

figure 16.10
This enormous open-pit copper mine supplies copper-bearing minerals disseminated through a large body of igneous rock. Bingham Canyon, Utah. (Kennecott Copper Corporation.)

new set of minerals upon exposure at the earth's surface.

The surface environment of the lands is poorly suited to the preservation of plutonic rocks formed under conditions of high pressure and high temperature. Most silicate minerals do not last long, geologically speaking, in the low temperatures and pressures of atmospheric exposure, particularly since free oxygen, carbon dioxide, and water are abundantly available. Chemical change in a response to the changed environment is called *mineral alteration*.

Rock surfaces are also acted on by physical forces of disintegration, breaking up igneous rock into small fragments and separating the component minerals, grain from grain. Fragmentation is essential for the chemical reactions of mineral alteration, since it results in a great increase in mineral surface area exposed to chemically active solutions. The processes of physical disintegration of rocks are discussed in Chapter 18. At this point, we are concerned with the chemical nature of mineral alteration as a process leading to production of sedimentary rock.

The presence of dissolved oxygen in water in contact with mineral surfaces leads to oxidation beneath the soil surface. *Oxidation* is the combination of oxygen with the metallic elements, such as calcium, magnesium, and iron, abundant in the silicate minerals. At the same time, carbon dioxide in solution forms carbonic acid, a weak acid in soil water. *Carbonic acid action* is capable of dissolving certain minerals, especially calcium carbonate. In addition, where decaying vegetation is present, soil water contains complex organic acids, capable of reacting with mineral compounds. Certain common minerals, such as rock salt (sodium chloride), dissolve directly in water, but simple solution is not particularly effective for the silicate minerals.

Water itself combines with silicate mineral compounds in a reaction known as *hydrolysis*. This process is not merely a soaking or wetting of the mineral, but a true chemical change producing a different compound and a different mineral. The reaction is not readily reversible under atmospheric conditions, so that the products of hydrolysis are stable and long lasting, as are the products of oxidation. In other words, these changes represent a permanent adjustment of mineral matter to a new surface environment of low pressures and temperatures.

Certain of the alteration products of silicate minerals are clay minerals. A *clay mineral* is one that has plastic properties when moist, because it consists of thin flakes of colloidal size, lubricated by layers of water molecules. We explained the ion-holding property of these clay colloids in Chapter 10.

Potash feldspar undergoes hydrolysis to become *kaolinite*, a white mineral with a greasy feel. Kaolinite becomes plastic when moistened. It is an important ceramic mineral used to make chinaware, porcelain, and tile.

Bauxite is an important alteration product of feldspars, occurring typically in warm climates of tropical and equatorial zones where rainfall is abundant year-round or in a rainy season. The dominant constituent of bauxite is sesquioxide of aluminum in combination with water. As we explained in Chapter 10, this is an unusually stable compound. Unlike kaolinite, which is a true clay, bauxite forms massive rocklike layers of laterite below the soil surface.

Another important clay mineral is *illite*, formed as an alteration product of feldspar and mica. Illite is a silicate of aluminum and potassium with water and is an abundant mineral in sedimentary rocks. It occurs as minute, thin flakes of colloidal dimensions and is carried long distances in streams (see Figure 10.5). *Montmorillonite* is another common clay mineral (more correctly a group of minerals) derived from alteration of feldspar, mafic minerals, or volcanic ash.

A most important alteration product of the mafic minerals is *limonite,* consisting of sesquioxide of iron with water. This stable form of iron oxide is found widely distributed in rocks and soils. It is closely associated with bauxite in laterites of low latitudes. Limonite supplies the typical reddish to chocolate-brown colors of soils and rocks. Some shallow accumulations of limonite were formerly mined as a source of iron.

Sediments and sedimentary rocks

The second great class of rocks are the sedimentary rocks. Their substance is derived both from preexisting rock of any origin and from newly formed organic matter. Igneous rock is the most important source of the inorganic mineral matter that goes to make up sedimentary rock. This explains why the sedimentary rocks are best studied after the igneous rocks.

In the process of mineral alteration, solid rock is softened and fragmented, yielding particles of many sizes. When transported in a fluid medium — air, water, or ice — these particles are known collectively as *sediment*. Used in its broadest sense, sediment includes both inorganic and organic matter. Dissolved mineral matter in the form of ions in solution is also included.

Streams carry sediment to lower levels and to locations where accumulation is possible. Wind and glacial ice also transport sediment, but not necessarily to lower elevations or to places suitable for accumulation. Usually the most favorable sites of sediment accumulation are in shallow seas bordering the continent, but they may also be inland seas and large lakes. Thick accumulations of sediment may become deeply buried under newer sediments. Over long spans of time the sediments undergo physical or chemical changes; they become compacted and hardened, forming sedimentary rock.

There are three major classes of sediment: (1) *clastic sediment* consists of mineral particles

figure 16.11
Planes of stratification show clearly in these clays and silts exposed in the South Dakota Badlands. (Douglas Johnson.)

derived by breakage from a parent rock source. Examples are the materials in a sand bar on a river bed or on a beach. (2) *Chemically precipitated sediment* consists of inorganic mineral compounds precipitated from a salt water solution or as hard parts of organisms. One example is a layer of rock salt; another is the white lime rock of a coral reef. (3) *Organic sediment* consists of the tissues of plants and animals, accumulated and preserved after the death of the organism. An example is a layer of peat in a bog. From the wide range of examples, it is obvious that the sedimentary rocks include a wide range of varieties in terms of both physical and chemical properties. We can only touch on a few of the important kinds of sedimentary rocks.

Despite their great variety, the sedimentary rocks share some physical features in common. The sediment accumulates in nearly horizontal layers, called *strata* (or simply "beds") (Figure 16.11). Strata are separated by stratification planes or "bedding planes," which allow one layer to be easily removed from the next. Strata of widely different compositions can occur one above the next, so that the eroded strata of a great accumulation of sedimentary rocks show a diversity of bands and ledges. A fine example is seen in the upper walls of the Grand Canyon, in Arizona (see Figure 21.8).

The clastic sedimentary rocks

Clastic sediments are derived from any one of the rock groups — igneous, sedimentary, and metamorphic — which gives a very wide range of parent minerals. One sediment source is from the silicate minerals and the alteration products of those minerals. Because quartz is hard and is immune to alteration, it is usually the most important single component of the clastic sediments (Figure 16.12). Second in abundance are fragments of unaltered fine-grained parent rocks. Feldspar and mica are

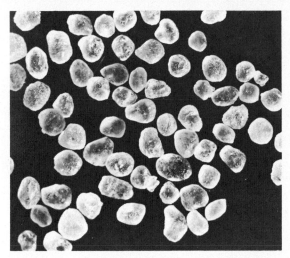

figure 16.12
Rounded quartz grains from an ancient sandstone. The grains average about 1 mm (0.04 in.) in diameter. (Andrew McIntyre, Columbia University.)

also commonly present. Clay minerals, particularly kaolinite, illite, and montmorillonite, are major constituents of the very fine clastic sediments.

The natural range of particle sizes in clastic sediment determines the ease and distance of travel of particles in transport by water currents. Obviously, the finer the particles, the more easily they are held in suspension in the fluid; the coarser particles tend to settle to the bottom of the fluid layer. In this way a separation of grades, called *sorting*, occurs. Sorting determines the texture of the sediment deposit and of the sedimentary rock derived from that sediment. Colloidal clays do not settle out unless they are made to clot together into larger groups. This clotting process, called *flocculation*, usually occurs when clays carried by rivers mix with the salt water of the ocean.

Compaction and cementation of layers of sediment leads to formation of sedimentary rocks. The following are important varieties. *Conglomerate* consists of pebbles or cobbles, usually of a very durable rock, set in a matrix

figure 16.13
Conglomerate is a mixture of pebbles and sand cemented into a hard rock. (A. N. Strahler.)

of sand (Figure 16.13). Conglomerate layers represent lithified beaches or stream-bed deposits. *Sandstone* is formed of the sand grades of sediment, cemented into a solid rock by silica (silicon dioxide) or calcium carbonate (Figure 16.14). The sand grains are commonly of quartz, such as those shown in Figure 16.12. *Siltstone* and *shale* are lithified layers of silt and clay particles, respectively (Figure 16.11).

figure 16.14
This cliff of sandstone shows jointing into large blocks. Colfax County, New Mexico. (U.S. Geological Survey.)

Shale is the most abundant of the sedimentary rocks. It is formed largely of the clay minerals, kaolinite, illite, and montmorillonite. The compaction of the original clay involves a considerable loss of volume as water is driven out.

Chemically precipitated sedimentary rocks

Under favorable conditions mineral compounds are deposited from the salt solutions of seawater and of salty inland lakes in desert climates. One of the most common sedimentary rocks formed in this way is *limestone*, composed largely of the mineral *calcite*. Calcite is calcium carbonate ($CaCO_3$) and is the same mineral we have encountered in the grassland soils, where it occurs as grains and nodules in the *B* horizon and in heavy layers as lime-crust. Limestone strata formed the sea floor — marine limestones — have accumulated in great thicknesses in many ancient seaways in past geologic eras (Figure 16.15). A closely related rock is *dolomite*, composed of calcium-magnesium carbonate. Limestone and dolomite are grouped together as the *carbonate rocks*. They are dense rocks, sometimes white, sometimes pale gray, and occasionally black.

Many varieties of limestones are formed of the mineral hard parts of organisms, either of plants or of animals. For example, chalk, a pure white rock, is formed of countless skeletons of marine algae. Reef limestone is formed from the hard parts of corals. Invertebrate shells, crushed and sorted by wave action and heaped into beaches, make up another variety of limestone, called "coquina" (Figure 16.16).

Seawater also yields sedimentary layers of silica (silicon dixoide) in hard, noncrystalline form called *chert*. Chert is a variety of sedimentary rock, but also commonly occurs

figure 16.15
These limestone strata in Oklahoma were steeply tilted long after being laid down. (Photo by Lofman. Courtesy of Exxon Corporation.)

combined with limestone (cherty limestone). Another chemical precipitate is *phosphate rock*, a marine sediment rich in the mineral phosphorus. Its principal use is in phosphate fertilizers, essential to modern agriculture.

Where evaporation is sustained and intense, shallow water bodies acquire a very high level of salinity. Shallow bays and estuaries in coastal deserts are one such saline body; another type is the playa lake of inland desert basins. Sedimentary minerals and rocks deposited from such concentrated solutions are called *evaporites*. Ordinary rock salt, the mineral halite (sodium chloride), has accumulated in this way into important sedimentary rock units.

Hydrocarbon compounds in sedimentary rocks

Hydrocarbon compounds (compounds of carbon, hydrogen, and oxygen) form a most important type of organic sediment. These

substances occur both as solids (peat and coal) and as liquids and gases (petroleum and natural gas), but only coal qualifies physically as a rock.

Peat, a soft, fibrous substance of brown to black color, accumulates in a bog environment where the continual presence of water inhibits decay and oxidation of plant remains. One form of peat is of freshwater origin and represents the filling of shallow lakes. Thousands of such peat bogs are found in North America and Europe; they occur in depressions remaining after recession of the great ice sheets of the Pleistocene Epoch (Chapter 22). This peat has been used for centuries as a low-grade fuel. Peat of a different sort is formed in the saltwater environment of tidal marshes (Chapter 23).

At various times and places in the geologic past, conditions were favorable for the large-scale accumulation of plant remains, accompanied by subsidence of the area and

figure 16.16
This specimen of coquina shows the mixture of shell fragments of which the rock is formed. (A. N. Strahler.)

10 cm

2 in.

figure 16.17

A coal seam 8 ft (2.4 m) thick exposed in a river bank, Dawson County, Montana. (U.S. Geological Survey.)

burial of the compacted organic matter under thick layers of inorganic sediments. *Coal* is the end result of this process. Coal seams are interbedded with shale, sandstone, and limestone strata (Figure 16.17).

Petroleum, or *crude oil*, as the liquid form is often called, includes many hydrocarbon compounds. *Natural gas*, found in close association with accumulations of liquid petroleum, is a mixture of gases. The principal gas is methane (marsh gas). Petroleum and natural gas are not classed as minerals, but they originated as organic compounds in sediments and are classed as mineral fuels.

It is generally agreed that petroleum and natural gas are of organic origin, but the nature of the process is hypothetical. A favored explanation for petroleum is that the oil originated within microscopic floating marine plants such as the diatoms. As each diatom died, it released a minute droplet of oil, which became enclosed in muddy bottom sediment. Eventually, the mud became a shale formation in which the oil was disseminated. Today we find oil shales holding petroleum in a dispersed state, and these give support to the organic

hypothesis. However, petroleum occurs in heavy concentrations in porous rock such as sandstone. This fact requires that the oil in some manner was forced to migrate from its source region to a porous reservoir rock.

The simplest arrangement of strata favorable to trapping petroleum and natural gas is an uparching of the type shown in Figure 16.18. Shale forms an impervious cap rock. A porous sandstone beneath the cap rock serves as a reservoir. Natural gas occupies the highest position, with the oil below it. As oil and gas are withdrawn, the ground water rises to fill the vacated pores.

From the standpoint of Man's energy resources, the outstanding concept relating to the hydrocarbon compounds within the earth — *fossil fuels*, they are collectively called — is that they have required hundreds of millions of years to accumulate, whereas they are being consumed at a prodigious rate by our industrial society. These fuels are nonrenewable resources. Once they are gone there will be no more, since the quantity produced in a thousand years by geologic processes is scarcely measurable in comparison to the quantity sorted through geologic time.

figure 16.18

Idealized cross section of an oil pool on a dome structure in sedimentary strata. Well A will draw gas; well B will draw oil; and well C will draw water. The cap rock is shale; the reservoir rock is sandstone. (From A. N. Strahler, 1972, Planet Earth, Harper & Row, New York.)

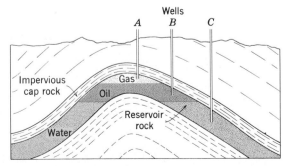

Metamorphic rocks

Any of the igneous or sedimentary rocks may be altered by the tremendous pressures and high temperatures that accompany mountain-building movements of the earth's crust. The resulting metamorphic rock is greatly changed in appearance and composition. Metamorphic rocks are harder and more compact than their parent types, except when the latter are igneous rocks. Moreover, the kneading action and heating that metamorphic rocks have undergone produce new structures and even new minerals. Each sedimentary and igneous rock has an equivalent metamorphic rock.

Shale, after being squeezed and sheared under mountain-making forces, is altered into *slate*. This gray or brick-red rock splits neatly into thin plates so familiar as roofing shingles and as flagstones of patios and walks.

With continued application of pressure and internal shearing, slate changes into *schist*, the most advanced grade of metamorphic rock. Schist has a structure called foliation, consisting of thin but rough and irregularly curved planes of parting in the rock. Schist is set apart from slate by the coarse texture of the mineral grains, the abundance of mica, and the presence of scattered large crystals of new minerals such as garnet. These crystals have grown during the process of internal shearing of the rock.

The metamorphic equivalent of conglomerate, sandstone, and siltstone is *quartzite*, formed by addition of silica to fill completely the interstices between grains. This process is carried out by the slow movement of underground waters carrying the silica into the sandstone, where it is deposited. Pressure and kneading of the rock is not essential in producing a quartzite.

Limestone, after undergoing metamorphism, becomes *marble*, a rock of sugary texture when freshly broken. During the process of internal shearing, the calcite mineral of the limestone has reformed into larger, more uniform crystals than before. Bedding planes are obscured, and masses of mineral impurities are drawn out into swirling streaks and bands.

Finally, the important metamorphic rock, *gneiss*, may be formed either from intrusive igneous rocks or from clastic sedimentary rocks that have been in close contact with intrusive magmas. A single description will not fit all gneisses, because they vary considerably in appearance, mineral composition, and structure. One conspicuous variety of gneiss is strongly banded into light and dark layers or lenses (Figure 16.19), which may be contorted into wavy folds. These bands, which have differing mineral compositions, are interpreted as the relics of sedimentary strata, such as shale and sandstone, to which new mineral matter has been added from nearby intrusive rocks.

figure 16.19
Outcrop of banded gneiss of Precambrian age, east coast of Hudson Bay, south of Povungnituk, Quebec. (Photograph G.S.C. No. 125221 by F. C. Taylor, Geological Survey of Canada, Ottawa.)

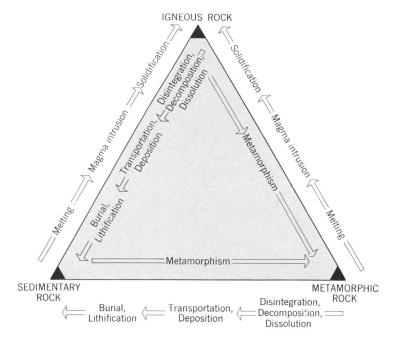

figure 16.20
Three major rock classes as corners of a triangle. (From A. N. Strahler, 1971, The Earth Sciences, 2nd ed., Harper & Row, New York.)

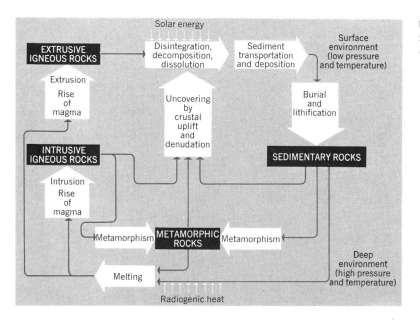

figure 16.21
Schematic diagram of the cycle of rock transformations.

The cycle of rock transformation

We can now bring together the formative processes of rocks into a single unified concept of recycling of matter through geologic time. A triangular diagram (Figure 16.20) shows the three major rock classes with the directions of transformation that are possible. From any one of the three corners change is possible in the direction of the other corners. Labels along the three sides explain the changes that occur in the direction of change.

Since the triangular diagram does not place the changes of rock in an environmental framework, we have set up a different sort of schematic diagram, shown in Figure 16.21. Here we distinguish between a surface environment of low pressures and temperatures and a deep environment of high pressures and temperatures. The deep environment is the realm of the igneous and metamorphic rocks.

The surface environment is one of rock alteration and sediment deposition.

Seen in its complete form, the total circuit of rock changes in response to environmental stress constitutes the *cycle of rock transformation*. This diagram emphasizes that mineral matter is continually recycled through the three major rock classes. Igneous rocks are by no means the "original" rocks of the earth's crust. Actually, there is no known record of the rocks that first formed the earth's crust; they were consumed and recycled long ago.

In the next chapter we shall expand our scale of vision to examine the earth's crust more broadly. How do the three great classes of rocks fit into the global patterns of continents and ocean basins and of mountain chains and low plains? The geographer is interested in these distributional patterns. Armed with a basic knowledge of minerals and rocks, we can effectively turn to these larger configurations of earth materials.

The Earth's Crust

Chapter 17

ENVIRONMENTAL REGIONS of the globe depend for their distribution on configurations of the earth's crust dictated by geologic processes. These processes are powered by energy sources deep within the earth. The internal earth forces have shaped the continents and ocean basins without conforming in the least to the orderly latitude zones of climate. We can think of latitude zones of temperature, winds, and precipitation as concentric color bands painted on a circular dinner plate as the potter's wheel spins. Now take that dinner plate and drop it on the floor; pick up the pieces and put them back into place. The fracture patterns cut across the circular color zones in a discordant and random pattern. This same unique combination of disorder on order characterizes the earth's surface environment. We find volcanoes erupting today in the cold desert of Antarctica as well as near the equator in Africa. Volcanoes are quite insensitive to the climate in which they rise.

An alpine mountain range has been pushed up in the cold subarctic zone of Alaska, where it trends east-west, but another lies astride the equator in South America and runs north-south. Evidently the geologic processes that create high mountains are quite insensitive to latitude and climate. But the reverse is not true, because climates respond to mountain ranges through the orographic effect. In this way, the chance configurations of the earth's relief features bring diversity to the global climate.

In this chapter we shall survey the major geologic features of our planet, starting with its deep interior as a layered structure. We then examine the outermost layer, or crust, comparing crustal forms of the continents with those of the ocean basins. Within only the past decade geologists have provided a unified theory to explain the differences between continents and ocean basins and to interpret the major forms of crustal unrest in a meaningful concept. Fortunately, we are reviewing geologic processes just at the moment in history when a new revolution in geology has accomplished a scientific upheaval. The revolutionary findings can now furnish us with a complete scenario of earth history on a grand scale of both time and spatial dimensions.

The earth's interior

Figure 17.1 is a cutaway diagram of the earth to show its major parts. The earth is an almost spherical body approximately 4000 mi (6400 km) in radius. The center is occupied by the _core_, a spherical zone about 2200 mi (3500 km) in radius. Because of the sudden change in behavior of earthquake waves upon reaching this zone, it has been concluded that the outer core has the properties of a liquid, in abrupt contrast to a solid mass that surrounds it. However, the innermost part of the core is probably in the solid state.

figure 17.1
Concentric zones make up the earth's interior.

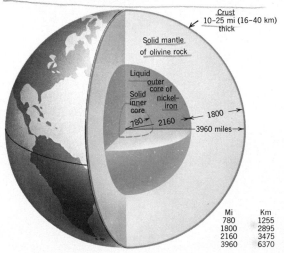

Crust
10–25 mi (16–40 km) thick

Solid mantle of olivine rock

Liquid outer core of nickel-iron

Solid inner core

780   2160   1800

3960 miles

Mi	Km
780	1255
1800	2895
2160	3475
3960	6370

Outside of the core lies the _mantle_, a layer about 1800 mi (2895 km) thick, composed of mineral matter in a solid state. Judging from the behavior of earthquake waves, the mantle is probably composed largely of the mineral olivine and resembles the ultramafic igneous rock peridotite.

The crust

Outermost and thinnest of the earth zones is the _crust_, a layer about 5 to 25 mi (8 to 40 km) thick, formed largely of igneous rocks. The base of the crust, where it contacts the mantle, is sharply defined. This contact is established from the way in which earthquake waves change velocity abruptly at that level (Figure 17.2_A_. The surface of separation between the crust and mantle is called the _Moho_, a simplification of the name of the seismologist who discovered it.

figure 17.2

The earth's crust is much thicker under continents than beneath the ocean basins.

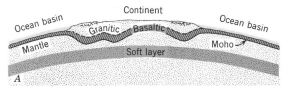

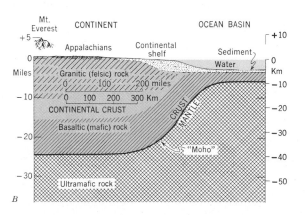

From a study of earthquake waves it is concluded that the crust consists of two layers: (1) a lower, continuous layer of basaltic (mafic) rock; (2) an upper layer of granitic (felsic) rock, which constitutes the bulk of continents. There is no sharply defined surface of separation between the granitic and basaltic layers. The granitic layer is absent beneath the ocean basins. Figure 17.2_B_ shows schematically a small part of the crust near the margin of a continent.

The crust of the continents is much thicker than the crust under the ocean basins. You can visualize the continents as vast icebergs floating in the sea with only a small part visible above the water and a great bulk deeply submerged. The rock of the earth's mantle has yielded by very slow flowage, permitting the less dense continental layer of the crust to come to rest like a floating iceberg.

Distribution of continents and ocean basins

The first-order relief features of the earth are the continents and oceans basins. Using a globe, we can compute that about 29 percent of the globe is land, 71 percent oceans. However, if the seas were to drain away, it would become obvious that broad areas lying close to the continental shores are actually covered by shallow water, less than 600 ft (180 m) deep. From these relatively shallow continental shelves the ocean floor drops rapidly to depths of thousands of feet. In a way, then, the ocean basins are brimful of water. The oceans have even spread over the margins of ground that would otherwise be assigned to the continents. If the ocean level were to drop by 600 ft (180 m) the surface area of continents would increase to 35 percent; the ocean basins would decrease to 65 percent. We can use these figures as representative of the true relative proportions.

Scale of the earth's relief features

Before turning to a description of the major subdivisions of the continents and ocean basins, we need to grasp the true scale of the earth's relief features in comparison with the earth as a sphere. Most relief globes and pictorial relief maps are greatly exaggerated in vertical scale.

For a true-scale profile around the earth we might draw a chalk-line circle 21 ft (6.4 m) in diameter, representing the earth's circumference on a scale of 1:2,000,000. A chalk line 3/8 in. (0.15 cm) wide would include within its limits not only the highest point on the earth, Mt. Everest (29,000 ft; 8840 m), but also the deepest known ocean trenches, somewhat deeper than 35,000 ft (10,700 m).

Figure 17.3 shows profiles correctly curved and scaled to fit a globe whose diameter is 21 ft (6.4 m). The surface profile is drawn to natural scale, without vertical exaggeration.

Although the most imposing landscape features of Asia and North America are shown, they are only trivial irregularities on the great global circle.

The geologic time scale

To place crustal rocks and structures in their positions in time, we need to refer to some major units in the scale of geologic time. Table 17.1 is a greatly abbreviated list of the major time divisions and their ages. All time older than 600 million years is Precambrian time. Three eras of time follow: Paleozoic, Mesozoic, and Cenozoic. These eras saw the evolution of life forms in the oceans and on the lands. The last era, the Cenozoic, is divided into five epochs, of which the youngest is the Pleistocene Epoch, known popularly as the "Ice Age." Most landforms of the continental surfaces were shaped by processes of erosion and deposition during late Cenozoic time. Man, as the genus *Homo*, evolved during the

figure 17.3
These profiles show the earth's great relief features in true scale. Sea-level curvature is fitted to a globe 21 ft (6.4 m) in diameter.

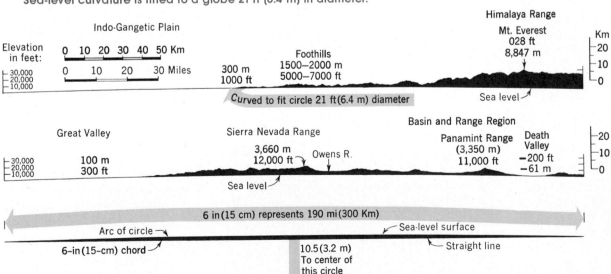

table 17.1
Table of geologic time

(Era)	(Epoch)	(Years before Present)
	Recent	0
		── 10,000
Cenozoic Era	Pleistocene (Glaciation of North America and Europe)	
		── 2 million
	Pliocene (Cascades and Sierras formed)	
	Miocene	
	Oligocene	
	Eocene	
		── 65 million
Mesozoic Era	(Rocky Mountains formed)	
		── 225 million
Paleozoic Era	(Appalachian Mountains formed)	
		── 600 million
Precambrian Time	(Canadian Shield rocks, 1 to 3 billions years)	

Oldest dated rocks 3.5 billion
Earth formed 4.6 billion
Age of universe 17-18 billion

late Pliocene Epoch and throughout the Pleistocene Epoch. As you can see, Man's time on earth has been an insignificant moment in the vast duration of geologic history.

Mountain-making belts

Broadly viewed, the continents consist of two basic subdivisions: (1) active belts of mountain making; (2) inactive regions of old rocks. At this point, we introduce the term *tectonic activity*, which means the bending and breaking of crustal rocks under internal earth forces. Along with *volcanism* (the formation of volcanic rocks by extrusion of magma), tectonic activity is responsible for the growth of mountain ranges and their internal structure.

Active mountain-making belts are narrow zones; most lie along continental margins. These belts are sometimes referred to as *alpine chains*, because they are characterized by high, rugged mountains such as the Alps of central Europe. These mountain belts were formed in the Cenozoic era, and activity has continued in many parts to the present day. The mountain belts are characterized by broadly curved instead of straight patterns on the map. Each curved section of an alpine chain is referred to as a *mountain arc*; the arcs are linked in sequence to form the two principal mountain belts. One is the *circum-Pacific belt*; it rings the Pacific Ocean basin (Figure 17.4). In North and South America, this belt is largely on the continents

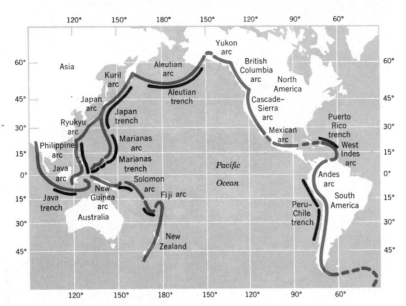

figure 17.4
The circum-Pacific ring of mountain arcs and island arcs. Deep trenches lie offshore in many parts of the ring.

and includes the Andes and Cordilleran ranges. In the western part of the Pacific basin the mountain arcs mostly lie well offshore from the continents and take the form of *island arcs*, running through the Aleutians, Kuriles, Japan, the Philippines, and many lesser islands. Between the large islands these arcs are represented by volcanoes rising above the sea as small, isolated islands.

The second chain of major mountain arcs forms the *Eurasian-Melanesian belt*, starting in the west at the Atlas Mountains of North Africa and running through the Near East and Iran to join the Himalayas. The belt then continues through Southeast Asia into Indonesia, where it joins the circum-Pacific belt in a T-junction. Figure 17.5 shows the location of this second belt. We shall return to the location of these active belts of mountain making later, explaining them in terms of global crustal activity. At this point we turn our attention to the different types of mountain landforms produced by volcanism and tectonic activity.

Mountain landforms of volcanic origin

We have identified volcanism as one of the forms of mountain building. The extrusion of magma builds landforms, and these collectively can accumulate both as volcanoes and as thick lava flows to make imposing mountain ranges. Most of these volcanic chains are within the circum-Pacific belt. The Cascade Mountains of northern California, Oregon, and Washington represent one such chain. The Aleutian Range of Alaska is another. Important segments of the Andes Mountains in South America and the island of Java in Indonesia consist of volcanoes.

Volcanoes are conical or dome-shaped structures built by the emission of lava and its contained gases from a restricted vent in the earth's surface (Figure 17.6). The magma rises in a narrow, pipelike conduit from a magma reservoir far below. Reaching the surface, igneous material may pour out in tonguelike lava flows or may be ejected under pressure of confined gases as solid fragments. Ejected solid

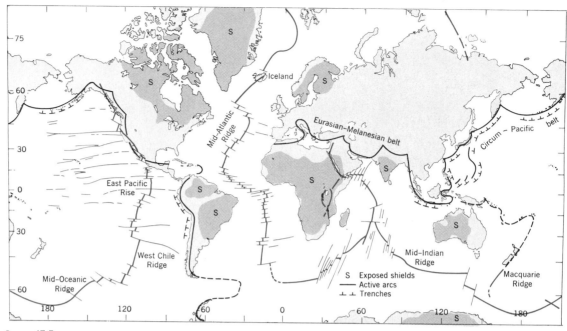

figure 17.5

A generalized world map of major crustal features. Major mountain belts and shield areas are shown on the continents. The Mid-Oceanic Ridge runs through the ocean basins.

fragments ranging in size from gravel and sand down to fine silt size are collectively called *tephra*. Form and dimensions of a volcano are quite varied, depending on the type of lava and the presence or absence of tephra. The nature of volcanic eruption, whether explosive or quiet, depends on the type of magma.

The important point is that the felsic lavas (rhyolite and andesite) have a high degree of viscosity (property of tackiness, resisting flowage) and hold large amounts of gas under pressure. As a result, these lavas produce explosive eruptions. In contrast, mafic lava (basalt) is highly fluid (low viscosity) and holds little gas, with the result that the eruptions are quiet and the lava can travel long distances to spread out in thin layers.

Tall, steep-sided volcanic cones are produced by felsic lavas. These cones usually steepen toward the summit, where a depression, the crater, is located. Tephra in these volcanic

eruptions takes the form of fine particles, described as volcanic ash, which falls on the area surrounding the crater and contributes to the structure of the cone (Figure 17.6). The interlayering of ash layers and lava streams

figure 17.6

An idealized cross section through a composite volcano and the magma chamber beneath. (From A. N. Strahler, 1971, The Earth Sciences, 2nd ed., Harper & Row, New York.)

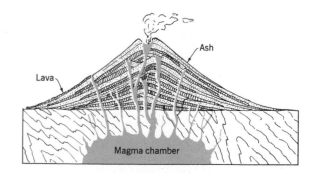

figure 17.7
Mount Shishaldin, an active composite volcano on Unimak Island in the Aleutian Islands, rises to an elevation just over 9300 ft (2800 m). A plume of condensed steam marks the summit crater. (U.S. Navy Department, from The National Archives)

produces a _composite volcano_. The world's lofty conical volcanoes, well known for their scenic beauty, are of the composite type. Examples are Mt. Hood in the Cascade Range, Fujiyama in Japan, Mt. Mayon in the Philippines, and Mt. Shishaldin in the Aleutian Islands (Figure 17.7). Tephra also includes volcanic bombs, solidified masses of lava ranging up to the size of large boulders, that fall close to the crater and roll down the steep slopes (Figure 17.8). Very fine volcanic dust rises high into the troposphere and stratosphere, where it remains suspended for years.

Another important form of emission from the explosive types of volcanoes is a cloud of incandescent gases and fine ash. This intensely hot cloud travels rapidly down the flank of the volcanic cone, searing everything in its path. On the island of Martinique, in 1902, a glowing cloud issued without warning from Mt. Pelée, sweeping down on St. Pierre, destroying the city and killing all but one of its 30,000 inhabitants.

Many of the world's active composite volcanoes lie in the circum-Pacific ring extending from the Andes in South America, through the Cascades and the Aleutians, into Japan; then south into the East Indies and New Zealand (Figure 17.4). There is also an important Mediterranean group, which includes active volcanoes of Italy and Sicily. Otherwise, Europe has no active volcanoes.

One of the most catastrophic of natural phenomena is a volcanic explosion so violent that it destroys the entire central portion of the volcano. There remains only a great central depression named a _caldera_. Although some of the upper part of the volcano is blown outward in fragments, most of it subsides into the ground beneath the volcano. Vast quantities of ash and dust are emitted and fill the atmosphere for many hundreds of square miles.

Krakatoa, a volcanic island in Indonesia, exploded in 1883, leaving a huge caldera. It is estimated that 18 cu mi (75 cu km) of rock disappeared during the explosion. Great seismic sea waves generated by the explosion killed many thousands of persons living on low coastal areas of Sumatra and Java.

A. Seen from a distance, Sakurajima shows a great cauliflower cloud of volcanic gases and condensed steam.

C. Reaching the sea, the hot lava makes clouds of steam.

B. A blocky lava flow is advancing over a ground surface littered with volcanic bombs and ash.

D. Volcanic ash buried this village. Torrential rains have carved rills into the soft ash.

figure 17.8
Sakurajima, a Japanese volcano, erupted violently in 1914. These pictures show various scenes from the eruption. (T. Nakasa.)

Flood basalts and shield volcanoes

Magma reaching the surface through rock fractures, called fissures, spreads out in thin tongues of lava. In some regions basalt lava has welled up in enormous volumes, accumulating layer upon layer to make a total thickness of thousands of feet and covering several thousand square miles. These accumulations are called *flood basalts*. An example is the Columbia Plateau of Washington and Oregon (Figure 17.9).

The quietly erupting, highly fluid basaltic lavas give rise to a second major group of volcanoes, known as *shield volcanoes*. The best examples are from the Hawaiian Islands, which consist entirely of lava domes (Figure 17.10).

Shield volcanoes are characterized by gently rising, smooth slopes that flatten near the top, producing a broad-topped volcano. The Hawaiian domes range to elevations up to 13,000 ft (4000 m) above sea level but, including the basal portion lying below sea level, they are more than twice that high. In

figure 17.9
Flood basalts of the Columbia Plateau region. The basalt layers have been eroded to produce steep cliffs, rimming broad, flat-topped mesas. Dry Falls, Grand Coulee, central Washington. (John S. Shelton.)

figure 17.10
This air view of Mauna Loa, Hawaii, shows a chain of pit craters leading up to the great central depression at the summit. (U.S. Army Air Force.)

width they range from 10 to 50 mi (16 to 80 km) at sea level and up to 100 mi (160 km) wide at the submerged base.

Shield volcanoes are built by repeated outpourings of lava. Explosive behavior and emission of tephra are not as important as for composite cones built of felsic magmas. The lava in the Hawaiian lava domes is of a dark basaltic type. It is highly fluid and travels far down the low slopes, which do not usually exceed angles of 4 or 5°.

Instead of the explosion crater, lava domes have a wide, steep-sided central depression that may be 2 mi (3.2 km) or more wide and several hundred feet deep (Figure 17.10). These large depressions are a type of caldera produced by subsidence accompanying the removal of molten lava from beneath. Molten basalt is actually seen in the floors of deep pit craters on the floor of the caldera or elsewhere over the surface of the lava dome (Figure 17.10). Most lava flows issue from fissures on the sides of the volcano.

At points where frothy basalt magma is ejected under high pressure from a narrow vent, tephra accumulate in a conical hill, called a *cinder cone*. Groups of cinder cones are commonly built in areas where basaltic lava flows are emerging from fissures. The cones rarely grow to heights over a few hundred feet.

Environmental impact of volcanic activity

The eruptions of volcanoes and lava flows are environmental hazards of the severest sort, often taking a heavy toll of plant and animal life and the works of Man. What natural phenomenon can compare with the Mt. Pelée disaster in which thousands of lives were snuffed out in seconds? Perhaps only an earthquake or storm surge of a tropical cyclone is equally disastrous — you can take your choice. Wholesale loss of life and destruction

of towns and cities are frequent in the history of peoples who live near active volcanoes. Loss occurs principally from sweeping clouds of incandescent gases that descend the volcano slopes like great avalanches; from lava flows whose relentless advance engulfs whole cities; from the descent of showers of ash, cinders, and bombs; from violent earthquakes associated with volcanic activity; and from mudflows of volcanic ash saturated by heavy rain. For habitations along low-lying coasts there is the additional peril of great seismic sea waves, generated elsewhere by explosive destruction of volcanoes.

Tectonic activity

Tectonic activity takes two basic forms in the active mountain-making belts. Figure 17.11 illustrates the principles. Starting with a crustal block (*A*) of a given width, *crustal compression* may occur, causing *folding*. Folding produces a series of wavelike undulations in sedimentary strata, called *folds*. Elsewhere, the crust may be

figure 17.11
Folding occurs under compression; rifting occurs where the crust is pulled apart.

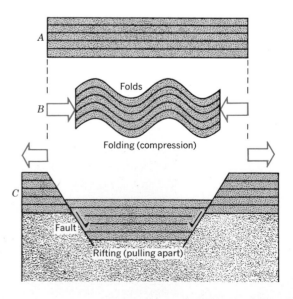

pulled apart; this activity is called *crustal spreading*. Being brittle, the crustal rock near the surface fractures into blocks, which slide past one another in the process of *faulting*. The break itself is called a *fault*: A *fault block*, a single crustal mass bounded by two faults, settles down to form a depression.

Faulting accompanying crustal spreading can be called *rifting*. Rifting is an important activity in certain active belts on the continents, although by no means as widespread globally as on the ocean floor. Rifting within the continental crust is a major activity in two widely separated world regions. One is the basin-and-range region of the western United States; another is the Red Sea and the East African Rift Zone (Figure 17.5).

Where crustal compression is extremely severe, rock folds are tightly compressed and pile up on one another, creating *alpine structure* (Figure 17.12). As tight folds are formed, the rock breaks into slices, the upper mass riding over the mass beneath. This form of breaking is called *overthrust faulting*. Rocks involved in alpine structure are thick sequences of sedimentary strata, usually including shales and limestones.

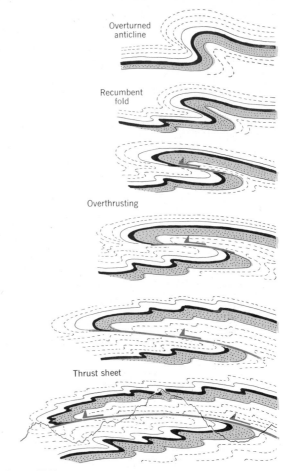

Overturned anticline

Recumbent fold

Overthrusting

Thrust sheet

figure 17.12

In alpine structure the tightly folded strata are broken by overthrust faults; one slice is heaped on the next. (From A. N. Strahler, 1971, The Earth Sciences, 2nd ed., Harper & Row, New York.)

Landforms made by faulting

Faulting is always accompanied by a slippage or displacement along the plane of breakage. Faults are often of great horizontal extent, so that the surface trace of a fault, or *fault line*, may run along the ground for many miles. Faulting occurs in sudden slippage movements that generate earthquakes. A particular fault movement may result in a slippage of as little as an inch or as much as 25 ft (8 m). Successive movements may occur many years apart, but can accumulate total displacements of hundreds or thousands of feet.

One common type of fault is the *normal fault*. The plane of slippage, or *fault plane*, is steep or nearly vertical. One side is raised or upthrown relative to the other, which is downthrown. A normal fault results in a steep, straight, clifflike feature called a *fault scarp* (Figure 17.13). Fault scarps range in height from a few feet to a few thousand feet (Figure 17.14). Their length is measurable in miles; often they attain lengths of 200 mi (320 km).

Normal faults rarely are isolated features. More often they occur in multiple arrangements, commonly as a parallel series of faults. This

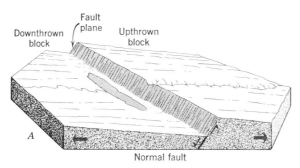

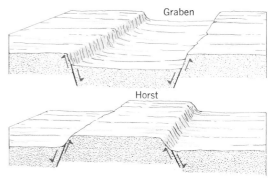

figure 17.15
Graben and horst.

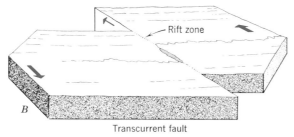

figure 17.13
Normal and transcurrent faults.

gives rise to a grain or pattern of rock structure and topography. A narrow block dropped down between two normal faults is a *graben* (Figure 17.15). A narrow block elevated

figure 17.14
This fresh fault scarp in alluvial materials formed during the Hebgen Lake earthquake of August 17, 1959, in Gallatin County, Montana. Displacement was about 19 ft (6 m) at the maximum point. Vehicle stands on the upthrown side of the fault. (U.S. Geological Survey.)

between two normal faults is a *horst*. Grabens make conspicuous topographic trenches, with straight, parallel walls. Horsts make blocklike plateaus or mountains, often with a flat top, but steep, straight sides.

In rifted zones, regions where normal faulting is on a grand scale, mountain masses called *block mountains* are produced. These faulted mountain blocks can be classed as tilted and lifted (Figure 17.16). A tilted block has one steep face, the fault scarp, and one gently sloping side. A lifted block, which is a type of horst, is bounded by steep slopes on both sides.

In a *transcurrent fault* the movement is predominantly in a horizontal direction (Figure 17.13B). No scarp, or a very low one at most, results. Instead, only a thin fault line is traceable across the surface. Streams sometimes turn and follow the fault line for a short distance. Sometimes a narrow trench, or rift, marks the fault.

figure 17.16
Fault block mountains may be of tilted type (left) or lifted type (right). (After W. M. Davis.)

Earthquake —
an environmental hazard

Everyone has read many news accounts about disastrous earthquakes and has seen pictures of their destructive effects. Californians know about severe earthquakes from firsthand experience, but many other areas in North America have experienced earthquakes, and a few of these have been severe. An *earthquake* is a motion of the ground surface, ranging from a faint tremor to a wild motion capable of shaking buildings apart and causing gaping fissures to open up in the ground.

The earthquake is a form of energy of wave motion transmitted through the surface layer of the earth in widening circles from a point of sudden energy release — the focus. Like ripples produced when a pebble is thrown into a quiet pond, these *seismic waves* travel outward in all directions, gradually losing energy.

As we already noted, earthquakes are produced by sudden movements along faults;

commonly these are normal faults or transcurrent faults. Through the San Francisco Bay area passes the famed San Andreas fault. The devastating earthquake of 1906 resulted from slippage along this fault, which is of the transcurrent type. The fault is 600 mi (965 km) long and extends into southern California, passing about 40 mi (60 km) inland of the Los Angeles metropolitan area. In places the fault is expressed as a rift valley (Figure 17.17). Associated with the San Andreas fault are several important transcurrent branch faults, all capable of generating severe earthquakes.

We shall not go into the details of mechanics of faults and how they produce earthquakes. It must be enough to say that rock on both sides of the fault is slowly bent over many years as tectonic forces are applied in the movement of large crustal masses. Energy accumulates in the bent rock, just as it does in a bent crossbow. When a critical point is reached, the strain is relieved by slippage on the fault and a large quantity of energy is instantaneously released

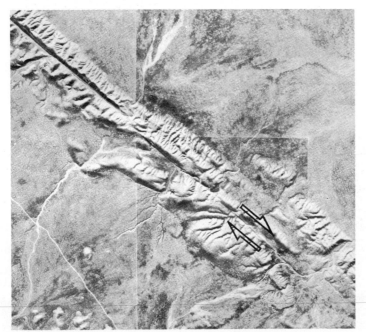

figure 17.17
In this vertical air view, the nearly straight trace of the San Andreas fault contrasts sharply with the sinuous lines of stream channels. San Bernardino County, California. (Litton Industries, Western Geophysical Company of America.)

figure 17.18
This road in the Santa Cruz Mountains of California was offset by fault movement during the San Francisco earthquake of 1906. Lateral displacement was about 4 ft (1.2 m). (Sketched from a photograph by E. P. Carey.)

in the form of seismic waves. Slow bending of the rock takes place over many decades. Its release then causes offsetting of features that formerly crossed the fault in straight lines, for example, a roadway or fence (Figure 17.18). Faults of this type can also show a slow, steady displacement known as fault creep, which

tends to reduce the accumulation of stored energy.

The Richter scale of earthquake magnitudes was devised in 1935 by the distinguished seismologist, Charles F. Richter, to indicate the quantity of energy released by a single earthquake. Scale numbers range from 0 to 9, but there is no upper limit except for nature's own limit of energy release. A value of 8.6 was the largest observed between 1900 and 1950.

One of the great earthquakes of recent times was the Good Friday Earthquake of March 27, 1964, with the surface point of origin located about 75 mi (120 km) from Anchorage, Alaska. Its magnitude was 8.4 to 8.6 on the Richter scale, approaching the maximum known. Of particular interest in connection with earthquakes as environmental hazards are the secondary effects. At Anchorage most of the damage was from secondary effects, since buildings of wooden frame construction often experience little damage when built on solid rock areas. Damage was largely from earth movements in weak clays underlying the city (Figure 17.19). These clays developed liquid

figure 17.19
Slumping and flowage of unconsolidated sediments, resulting in property destruction at Anchorage, Alaska, Good Friday earthquake of March 27, 1964. (U.S. Army Corps of Engineers.)

properties upon being shaken and allowed great segments of ground to subside and to pull apart in a succession of steps, tilting and rending houses. Other secondary effects were from the rise of water level and landward movement of large waves, destroying shipping and low-lying structures.

Another important secondary effect of a major earthquake is the *seismic sea wave*, or *tsunami* as it is known to the Japanese. A train of these waves is often generated in the ocean at a point near the earthquake source by a sudden movement of the sea floor. The waves travel over the ocean in ever-widening circles, but they are not perceptible at sea in deep water. However, when a wave arrives at a distant coastline, the effect is to cause a rise of water level. Wind-driven waves, superimposed on the heightened water level, allow the surf to attack places inland that are normally above the reach of waves. For example, the particularly destructive seismic sea wave of 1933 in the Pacific Ocean caused waves to attack ground as high as 30 ft (9 m) above normal tide level, causing widespread destruction and many deaths by drowning in low-lying coastal areas. It is thought that coastal flooding that occurred in Japan in 1703, with an estimated life loss of 100,000 persons, may have been caused by seismic sea waves.

Continental shields and mountain roots

As we noted earlier, mountain-making belts account for only a small portion of the continental crust. The remainder consists of inactive regions of older rock. Within these inactive regions we recognize two structural types of crust: (1) shields; (2) roots of older mountain belts. *Shields* are low-lying continental surfaces beneath which lie igneous and metamorphic rocks in a complex arrangement (Figure 17.20). The rocks are very old, mostly of Precambrian age, and have had a very involved geologic history. For the most part, the shields are regions of low hills and low plateaus, although there are some exceptions where large crustal blocks have been uplifted. Many thousands of feet of rock have been eroded from the shields during their exposure throughout a half-billion years.

Large areas of the continental shields are under a cover of younger sedimentary layers, ranging in age from Paleozoic through Cenozoic eras. These strata accumulated at times when the shields subsided and were inundated by shallow seas. Marine sediments were laid down on the ancient shield rocks in thickness ranging from hundreds to thousands of feet. These shield areas were then arched up and again became land surfaces.

figure 17.20
A schematic cross section through the upper continental crust showing rocks and structures of the active belts and inactive regions. In general, the geologic age becomes younger from right to left.

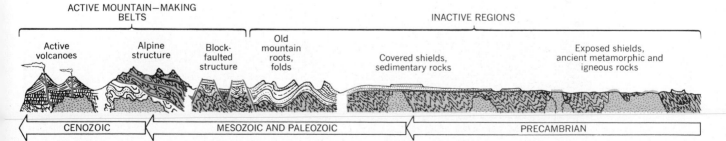

Erosion by running water has since removed large amounts of the sedimentary cover, but it remains intact over vast areas. We refer to such areas as *covered shields*, to distinguish them from the *exposed shields* in which the Precambrian rocks lie bare. Figure 17.20 shows the covered and exposed shields in relation to the older mountain belts and the various kinds of structures in the active tectonic belts. An example of an exposed shield is the Canadian Shield of North America. Exposed shields are also extensive in Scandinavia, South America, Africa, penisular India, and Australia (Figure 17.5).

Roots of older mountain belts run within the shields in many places. These *mountain roots* are mostly formed of Paleozoic and Mesozoic sedimentary rocks that have been intensely deformed and locally changed into meta- morphic rocks. The Appalachian Mountains of eastern North America are a good example. The Highlands of Scotland offer another example. Thousands of feet of overlying rocks have been removed from these old tectonic belts, so that only root structures remain. Roots appear as chains of long, narrow ridges, rarely rising over a few thousand feet above sea level. These landforms are described in Chapter 21.

The crust beneath the ocean basins

Crustal rock beneath the ocean floors consists almost entirely of basalt, covered over large areas by a comparatively thin accumulation of sediments. Age determinations of the basalt and its sediment cover show that the oceanic crust is very young, geologically speaking. Most of that crust was formed during the Cenozoic era and is less than 60 million years old. In a few places the age is late Mesozoic. When we consider that the bulk of the continental crust is of Precambrian age — mostly over a billion years old — the very young age of the oceanic crust is all the more remarkable. We will need to fit this fact into the general theory of global tectonics.

In overall plan the ocean basins are char- aracterized by a central ridge structure that divides the basin about in half. This feature is shown by schematic diagram in Figure 17.21. The *Mid-Oceanic Ridge* consists of submarine hills rising gradually to a mountainous central zone. Precisely in the center of the ridge, at its highest point, is an *axial rift*, which is a trenchlike feature. The form of this rift suggests that the crust is being pulled apart along the line of the rift axis. On either side of the Mid-Oceanic Ridge are broad deep plains and

figure 17.21
This schematic block diagram shows the ocean basins as symmetrical elements on a central axis. The model applies particularly well to the North and South Atlantic oceans.

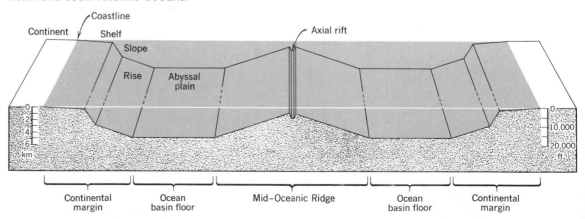

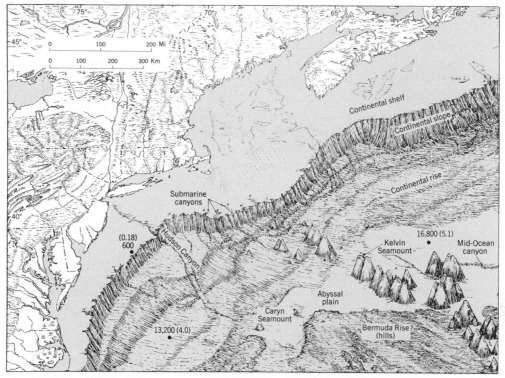

figure 17.22

Features of the continental margin and ocean-basin floor off the coast of the northeastern United States. Depth in feet; kilometers in parentheses. (Portion of Physiographic Diagram of the North Atlantic Ocean, 1968, revised, by B. C. Heezen and M. Tharp, Boulder, Colo., Geol. Soc. of Amer., reproduced by permission.)

hill belts belonging to the *ocean basin floor*. Their average depth is about 17,000 ft (5 km). The flat surfaces are called *abyssal plains*; they are extremely smooth because they have been built up of fine sediment.

Nearing the continental margins, the ocean floor begins to slope gradually upward, forming the *continental rise*. The floor then steepens greatly in the *continental slope* (Figure 17.22). At the top of this slope we arrive at the brink of the *continental shelf*, a gently sloping platform some 75 to 100 mi (120 to 160 km) wide along the eastern margin of North America. Water depth is about 600 ft (180 m) at the outer edge of the shelf.

The continental shelf is underlain by a great thickness of sedimentary strata derived from the continent. These strata form a large wedge-shaped deposit, thinning landward and feathering out over the continental shield. Figure 17.23 is a block diagram showing the wedge of continental shelf sediments. These sediments are brought from the land by rivers and spread over the shallow sea floor by currents. A great deal of attention is now being paid to the continental shelf wedge as a potential source of rich petroleum accumulations, reached only from offshore drilling platforms. Below the continental rise and its adjacent abyssal plain is another thick sediment deposit, formed of deep-sea sediments carried down the continental slope by swift muddy currents.

Sea-floor spreading

The Mid-Oceanic Ridge and its axial rift can be traced through the ocean basins for a total distance of about 40,000 mi (64,000 km). Figure 17.5 shows the extent of the ridge. From the South Atlantic, the ridge turns east and enters the Indian Ocean. Here, one branch penetrates Africa, while the other continues east between Australia and Antarctica, then swings across the South Pacific. Nearing South America, it turns north and penetrates North America at the head of the Gulf of California.

The axial rift of the Mid-Oceanic Ridge is the origin of numerous earthquakes, indicating continued tectonic activity. Now unmistakable evidence has been accumulated to show that the crust is spreading apart along the rift. The rate of separation is from 1 to 2 in. (2.5 to 5 cm) per year. As this *sea-floor spreading* occurs, basaltic lava rises from beneath the rift, solidifying and forming new oceanic crust. This continuing process explains why the ocean basin crust is so young in geologic age. The axial rift is broken across at right angles by numerous fractures, which are a type of fault. Segments of the rift are offset along these fractures, as shown on the map (Figure 17.5).

Oceanic trenches

The circum-Pacific belt is characterized by a chain of deep *oceanic trenches* lying offshore from the mountain arcs and running close to the island arcs. The positions of these trenches are shown in Figure 17.4. Floors of the narrow trenches reach depths of 24,000 ft (7.5 km) and even more (Figure 17.24). Many lines of evidence show that the oceanic crust is sharply downbent to form these trenches. Numerous earthquakes, many of them very intense, occur in the trenches and adjacent mountain arcs, attesting to continuing tectonic activity.

Lithospheric plates

Crustal spreading along the axial rift and downbending of the crust in the oceanic trenches involves more than just the earth's crust. Motion extends into the upper mantle, so that both the crust and upper mantle move as a single rigid layer, called the *lithosphere* (Figure 17.25). This use of the word lithosphere is somewhat more restricted than its meaning as the entire solid earth. Immediately below the lithosphere is a soft layer,

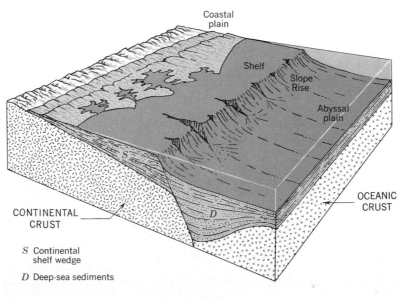

figure 17.23
This block diagram shows an inner wedge of sediments beneath the continental shelf and an outer wedge of deep-sea sediments beneath the continental rise and abyssal plain.

S Continental shelf wedge

D Deep-sea sediments

figure 17.24
The Peru-Chile Trench, off the west coast of South America. (Portion of <u>Physiographic Diagram of the South Atlantic Ocean</u>, 1961, by B. C. Heezen and M. Tharp, Boulder, Colo., Geol. Soc. of Amer., reproduced by permission.)

called the *asthenosphere*. In this layer the mantle rock is heated close to its melting point. The soft layer yields by flowage, allowing the lithosphere above to glide slowly over the globe, much as pack ice of the Arctic Ocean drifts slowly under the force of currents and winds.

The rigid lithosphere, which is approximately 40 mi (65 km) thick, is broken into a number of

figure 17.25
The lithosphere, a rigid upper plate, glides over the soft asthenosphere. While one edge of a lithospheric plate is being formed during seafloor spreading on an axial rift, the opposite edge is disappearing into the mantle by subduction.

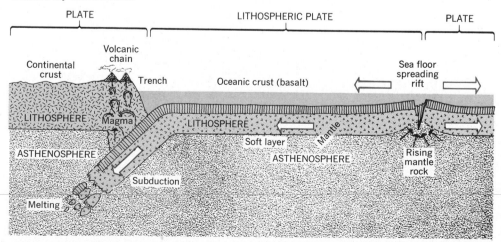

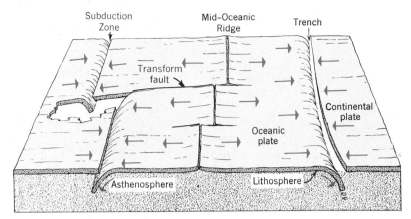

figure 17.26
This schematic diagram shows major features of plate tectonics. Earth curvature has been removed. (From A. N. Strahler, 1971, <u>The Earth Sciences</u>, 2nd ed., Harper & Row, New York.)

enormous segments, called *lithospheric plates* (Figure 17.25). Sea-floor spreading at the axial rift of the Mid-Oceanic Ridge represents a line along which two lithospheric plates are moving apart. In contrast, a trench represents a line along which two lithospheric plates are colliding. In this case, one plate is forced to plunge down into the mantle, a process called *subduction*. Figure 17.26 is a three-

dimensional representation of lithospheric plates and their relative motions.

Besides coming in contact along lines of sea-floor spreading and subduction, lithospheric plates can slide past one another along simple faults, called *transform faults* (Figure 17.26). Motion on a transform fault is horizontal and always in the same direction as the fault line. Consequently, on a transform fault, the plates do not separate, nor does one plate pass beneath the other. On the sea floor, transform faults appear as long, straight walls or narrow trenches. Many of the faults offsetting the Mid-Oceanic Ridge are transform faults (Figure 17.5).

figure 17.27
Two stages in the evolution of an tectonic belt. (<u>A</u>) A volcanic island arc contributes sediment to a geosyncline between the arc and the mainland. (<u>B</u>) The geosyncline is crumpled by intense impact of the moving lithospheric plate, giving rise to alpine structure, metamorphic rocks, and batholiths.

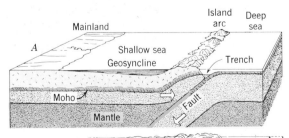

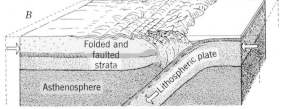

Mountain making and lithospheric plates

Mountain making can now be fitted into the concept of lithospheric plates undergoing subduction near the margin of a continent. Figure 17.27 shows two stages. In the upper block diagram the trench is bordered by a chain of volcanoes. Magma comes from great depth, where melting has affected the down-plunging plate. Between the island arc and the mainland lies a shallow sea. Sediments accumulating in this sea comprise a *geosyncline*. Thousands of feet of shales and limestones accumulate in the geosyncline, which gradually subsides under the load of the sediment.

In a second stage of tectonic activity, shown in the lower diagram, the down-plunging lithospheric plate has pushed strongly against the opposing plate, crumpling the strata of the geosyncline into tight folds. A zone of alpine structure has been formed. Farther inland, folding is less intense and a belt of open, wavelike folds is produced. At great depth beneath the alpine structure heat and pressure have converted the sediments into metamorphic rock. Large masses of magma, produced by melting of the lithospheric plate, have moved up as plutonic rock, forming batholiths deep beneath the alpine mountain chain.

Global plate tectonics

The culmination of the new revolution in geology has been the recognition of a global system of lithospheric plates. Their physical interaction goes under the general term *plate tectonics.* Figure 17.28 shows the major plates of the globe. The American plate includes the North American and South American continental crust and all of the oceanic crust as far east as the Mid-Atlantic Ridge. The Pacific plate is the only unit bearing only oceanic crust. It occupies all of the Pacific region west and north of the Mid-Oceanic Ridge. It undergoes subduction beneath the American plate along the Alaskan-British Columbia coastal zone. The Antarctic plate occupies the globe south of the Mid-Oceanic Ridge system. The African plate consists of the African continental crust and a zone of surrounding oceanic crust limited by the Mid-Oceanic Ridge. A single Eurasian plate, which consists largely of continental crust, is bounded on the east and south by subduction zones of the great alpine mountain chains and island arcs. The Australian plate consists of continental

crust of India and Australia, as well as oceanic crust of the Indian Ocean and a part of the southwestern Pacific. Other smaller plates are shown and named on the map.

While the driving force for plate motions is not known, a leading hypothesis attributes the motion to convection currents deep in the mantle (Figure 17.29). In some respects this model convection current system in the mantle resembles the atmospheric circulation at low latitudes within the Hadley cell (Chapter 5). Unequal heating of mantle rock leads to differences in rock density, and this, in turn, may set in motion very slow rising and sinking motions.

It is postulated that horizontal mantle motion between the zone of rising mantle rock (Mid-Oceanic Ridge) and sinking rock (beneath the subduction zones) exerts a dragging force on the lithospheric plate. Many variations of the convection hypothesis have been set forth and various other energy sources have been proposed as well. While the reality of plate motions seems well established, we are far from understanding the impelling forces.

Evolution of the continents

Through geologic time, as lithospheric plates have drifted apart, vast areas of ocean basins with a thin basaltic crust have come into existence. At the same time, tectonic processes acting in the subduction zones have gradually created thick continental crust with its felsic (granitic) upper zone. In this way the continents have evolved and increased in size over a 3- to 4-billion year period.

As early as the late nineteenth century, the proposal had been made that the continents of North and South America, Europe, Africa, Australia, and Antarctica, and the subcontinent of India (south of the Himalayas) along with Madagascar, had originated as a single supercontinent, named *Pangaea* (Figure

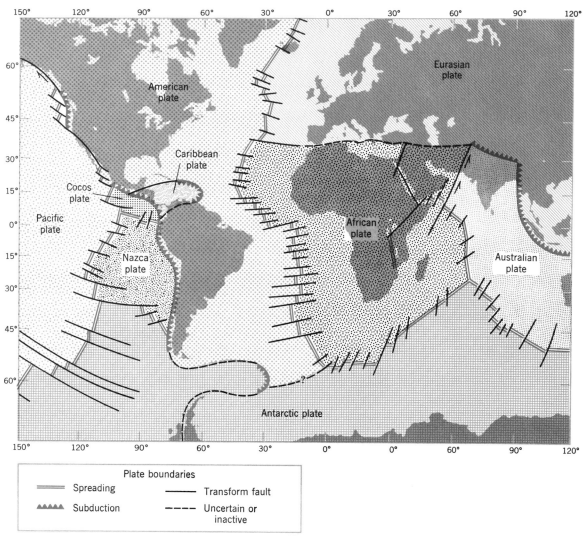

figure 17.28

World map of lithospheric plates.

17.30). Pangaea began to break apart about 200 million years ago, during the Mesozoic Era. The fragments slowly separated and underwent some horizontal rotation as well.

The Atlantic Ocean is envisioned as the area opened up by drifting apart of the Americas from Africa and Europe. This hypothesis of *continental drift* was generally rejected or skeptically regarded by most of the geological fraternity, particularly the Americans, until about 1960, when sound evidence of crustal spreading came to light. Now a majority of geologists have accepted continental drift, much as it was originally outlined, and have fitted it into the global theory of plate tectonics. Figure 17.30 shows postulated stages in the separation of the continents, according to a recent version.

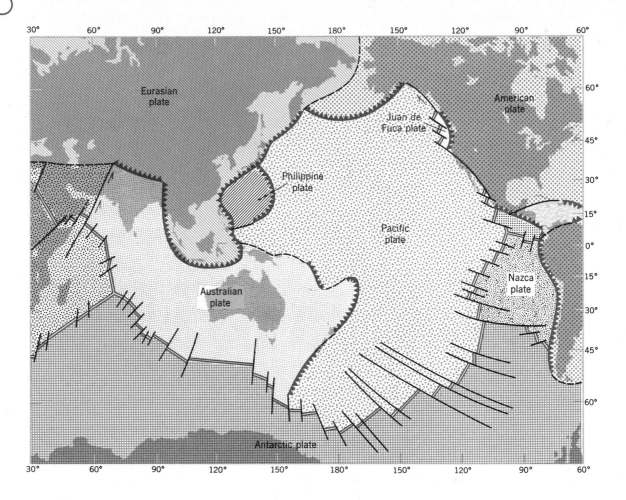

The rock cycle in a new light

Plate tectonics has placed a new understanding on the rock transformation cycle we described in Chapter 16. As Figure 17.31 shows, a moving lithospheric plate bearing oceanic crust can be visualized as a great conveyor belt. The oceanic sediment deposited on the oceanic crust is carried toward a subduction zone, where it is scraped off, along with the continental sediment. The sediment is transformed into igneous and metamorphic rock deep within the subduction zone and is added to the continental crust. The descending plate is melted and its substance returned to the mantle.

figure 17.29

Simplified model of a convection system in the mantle causing motion of the overlying lithospheric plates.

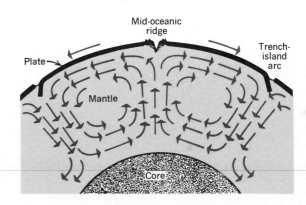

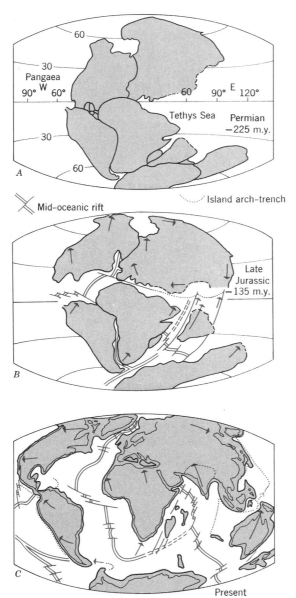

figure 17.30

The breakup of Pangaea is shown in three stages. Inferred motion of lithospheric plates is indicated by arrows. (Redrawn and simplified from maps by R. S. Dietz and J. C. Holden, 1970, Jour. Geophys. Research, Vol. 75, pp. 4943-4951, Figures 2 to 6.)

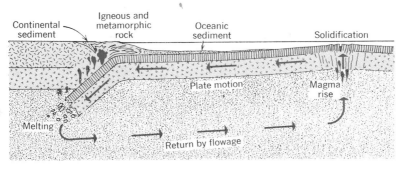

figure 17.31

The moving lithospheric plate works like a conveyor belt, bringing sediment to a subduction zone where it is transformed into new igneous and metamorphic rock.

The earth's crust in review

In two chapters we have made a survey of the composition and structure of the earth's crust. One underlying concept is that of a rock cycle, continuously in operation over some 3 or more billions of years of geologic time. The recycling of crustal mineral matter has taken place through the mechanism of plate tectonics. The continents have gradually grown in extent through accumulation of granitic rock produced in subduction zones, and in this way, the first-order relief features of the globe have come into existence.

Active tectonic and volcanic belts are dominant environmental features of the continents. Cutting across the belts of prevailing winds, high mountain chains induce orographic precipitation and, at the same, create rain-shadow deserts. Landforms produced by volcanic activity and by faulting are important in these new mountain chains. These activities also pose environmental hazards for those populations who live nearby. However, great parts of the continental surfaces — those inactive regions of the shields and their ancient mountain roots — consist of landforms shaped by an entirely different set of processes. It is to these landforms that we turn in the remaining chapters.

Weathering and Mass Wasting

Chapter 18

NOW THAT WE have completed a review of the earth's crust, its mineral composition, and its tectonic and volcanic landforms, we can focus on the shallow life layer itself. At this interface, the externally acting, solar-driven energy systems of the atmosphere and oceans mesh with the internally driven geologic system that has created and raised the continental masses. The geologic processes have caused varied rock types to become exposed to the surface environment. Our study of the interaction of these two great planetary systems began in Chapter 16 with a study of the chemical alteration of rock and the production of sediment. This is a process essential to the rock transformation cycle and to growth of the continental crust. We will now look at the role of rock alteration and related processes in shaping the surface of the lands.

Landforms are the distinctive configurations of the land surface: mountains, hills, valleys, plains, and the like. Landforms are environmentally significant because they influence the place-to-place variation in ecological factors such as water availability and exposure to radiant solar energy. Through varying height and degrees of inclination of the ground surface, landforms directly influence hydrologic and soil-forming processes; these, in turn, act as direct controls on ecosystems.

Geomorphology

Study of the life layer of the continents brings together two major branches of the earth sciences: geomorphology and hydrology. *Geomorphology* is the study of landforms, including their history and processes of origin. Geomorphology deals largely with the action of *fluid agents* which erode, transport, and deposit mineral and organic matter. The four fluid agents are running water in surface and underground flow systems, waves, acting with currents in oceans and lakes, glacial ice,

moving sluggishly in great masses, and wind, blowing over the ground.

Of the four agents, three are forms of water. Consequently, the science of hydrology is inseparably interwoven with geomorphology. One might be not far wrong in saying simply that the hydrologist is preoccupied with "where water goes," the geomorphologist with "what water does." Hydrology concerns itself with the hydrologic cycle in an attempt to calculate the water balance and to measure rates of flow of water in all parts of that cycle (Chapter 9). Geomorphology concerns itself with geologic work that the water in motion performs on the land.

Denudation is a useful term for the total action of all processes by which the exposed rocks of the continents are worn away and the resulting sediments are transported to the sea by the fluid agents. Denudation is an overall lowering of the land surface. Denudation tends toward reducing the continents to nearly featureless sea-level surfaces and, ultimately, through wave action, to submarine surfaces. If it had not been repeatedly counteracted by tectonic activity throughout geologic time, denudation would have eliminated all terrestrial environments.

The important point that emerges as we look back through geologic time is that terrestrial life environments have been in constant change, even as plants and animals have undergone their evolutionary development. The varied denudation processes have produced, maintained, and changed a wide variety of landforms, which have been the habitats for evolving life forms. In turn, the life forms have become adapted to those habitats and have diversified to a degree that matches the diversity of the landforms themselves.

Geomorphic and hydrologic systems operating in the life layer have long been subjected to quite radical modification by the works of Man. Agriculture has for centuries altered the surface properties of areas of subcontinental size. Agriculture has modified the action of running water and the water balance, to say nothing of radically changing the character of the soil. Urbanization is an even more radical alteration, seriously upsetting hydrologic processes. Engineering and mining activities, such as strip mining and the construction of highways, dams, and canals, not only upset hydrologic systems but can completely destroy or submerge entire assemblages of landforms. Of the four fluid agents of landform sculpture, only glaciers of ice have so far successfully resisted changes of activity imposed by Man.

Invariably, Man's attempts to control the action of running water, waves, and currents produce unpredicted and undesirable side effects, some of which are physical, others ecological. An important reason to study geomorphology is to predict the consequences of man-made changes and so to plan wisely for management of the environment.

Weathering and mass wasting

The processes by which rock is physically disintegrated and chemically decomposed during exposure to atmospheric influences are referred to collectively as *weathering processes*. In addition to the chemical and physical changes in mineral matter that result from weathering, there is a continued agitation in the soil and regolith because of changes in temperature and water content. These daily and seasonal rhythms of change continue endlessly.

The spontaneous downward movement of soil, regolith, and rock under the influence of gravity (but without the dynamic action of moving fluids) is included under the general term *mass wasting*. In Chapter 1 we emphasized the role of gravity as a pervasive environmental factor. All processes of the life layer take place in the earth's gravity field, and

all particles of matter tend to respond to gravity. Movement to lower levels takes place when the internal strength of a mass of soil or rock declines to a critical point below which the force of gravity cannot be resisted. This failure of strength under the ever-present force of gravity takes many forms and scales, and we shall see that Man is a major agent in causing several forms of mass wasting.

Bedrock, regolith and soil

When you examine a freshly cut cliff, such as that in a new highway cut or quarry wall, you will probably find several kinds of earth materials, shown in Figure 18.1. Solid, hard rock that is still in place and relatively unchanged is called *bedrock*. It grades upward into a zone where the rock has become decayed and has disintegrated into clay, silt, and sand particles. This zone is the *residual regolith*. On top is the soil layer.

One or both of these zones may be missing. In some places everything is stripped off down to the bedrock, which then appears at the surface as an *outcrop* (Figure 18.1). Sometimes, following cultivation or forest fires, the soil is partly eroded away. Severe erosion exposes the regolith. The thickness of soil and regolith is quite variable. Although the soil is rarely more than a few feet thick, the regolith of decayed and fragmented rock may extend down tens or even hundreds of feet. Formation of regolith is greatly aided by the presence of innumerable bedrock cracks, called *joints* (Figure 18.1). Water can move easily through joints to promote rock decay.

Another variety of unconsolidated mineral matter that may be found covering the bedrock is *transported regolith*. It consists of materials such as stream-laid gravels and sands, floodplain silts, clays of lake bottoms, beach and dune sands, or rubble left by a melting glacier. All types have in common a history of

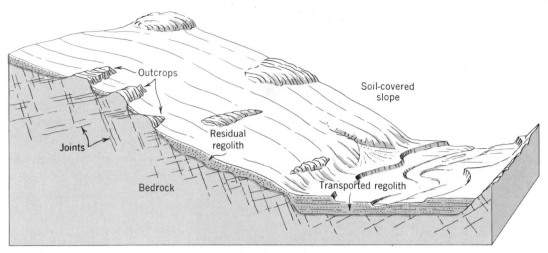

figure 18.1

Residual regolith on the hillside is derived from bedrock beneath. Transported regolith of the valley bottom has been brought from some distance away by a stream.

having been transported by streams, ice, waves, or wind. Residual regolith is formed in place by disintegration of bedrock below and is of local origin. In contrast, transported regolith consists of rocks and minerals from distant sources, and these materials may be quite unlike the underlying bedrock. Figure 18.1 shows stream valley deposits, called alluvium, that represent transported regolith. Once deposited, transported regolith may remain undisturbed for many thousands of years, allowing a soil layer to be formed in its uppermost layer.

Physical weathering

Physical weathering produces fine particles from massive rock by the action of forces strong enough to fracture the rock. One of the most important physical weathering processes in cold climates is *frost action*, the repeated growth and melting of ice crystals in the pore spaces of soil and in rock fractures. As water in joints freezes, it forms needlelike ice crystals extending across the openings. As these ice needles grow, they exert a powerful force against the confining walls and can easily pry apart the joint blocks.

Freezing water strongly affects soil and rock in all midlatitude and high-latitude regions having a cold winter season, but its effects are most striking in high mountains, above the timberline, and in arctic regions. Here the separation and shattering of joint blocks produces surfaces littered with angular blocks. Frost action on cliffs of bare rock in high mountains detaches rock fragments that fall to the cliff base. Where production of fragments is rapid, they accumulate to form *talus slopes*. Most cliffs are notched by narrow ravines that funnel the fragments into separate tracks, so as to produce conelike talus bodies arranged side by side along the cliff (Figure 18.2). Fresh talus slopes are unstable, so that the disturbance created by walking across the slope or dropping a large rock fragment from the cliff above will easily set off a sliding of the surface layer of particles.

Closely related to the growth of ice crystals is the weathering process of rock disintegration

figure 18.2
Talus cones at the base of a frost-shattered cliff. Moraine Lake in the Canadian Rockies. (Ray Atkeson.)

because there the ground water seeps outward to reach the rock surface (Figure 18.3). In the southwestern United States, many of the deep niches or cavelike recesses formed in this way were occupied by Indians. Their cliff dwellings gave protection from the elements and safety from armed attack (Figure 18.4).

Salt crystallization also acts adversely on masonry buildings and highways. Brick and concrete in contact with moist soil are highly susceptible to grain-by-grain disintegration from this cause. The salt crystals can be seen as a soft, white, fibrous layer on basement floors and walls. Man has added to these destructive effects by spreading deicing salts on streets and highways. Sodium chloride (rock salt), widely used for this purpose, is particularly destructive to concrete pavements and walks, curbstones, and other exposed masonry structures.

Most rock-forming minerals expand when heated and contract when cooled. Where rock surfaces are exposed daily to the intense heating of the sun alternating with nightly cooling, the resulting expansion and contraction exerts powerful disruptive forces on the rock. Although firsthand evidence is lacking, it seems likely that temperature

by growth of salt crystals. This process operates extensively in dry climates and is responsible for many of the niches, shallow caves, rock arches, and pits seen in sandstone formations. During long drought periods, ground water is drawn to the surface of the rock by capillary force. As evaporation of the water takes place in the porous outer zone of the sandstone, tiny crystals of salts are left behind. The growth force of these crystals is capable of producing grain-by-grain breakup of the sandstone, which crumbles into a sand and is swept away by wind and rain. Especially susceptible are zones of rock lying close to the base of a cliff,

figure 18.3
In dry climates there is a slow seepage of water from the cliff base. Here niches develop through rock weathering.

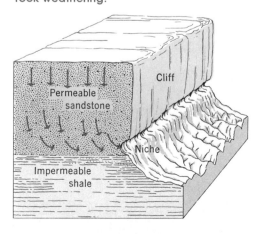

figure 18.4
White House Ruin occupies a deep niche in the sandstone wall of Canyon de Chelly, Arizona. (Ray Atkeson.)

changes cause breakup of rock already weakened by other agents of weathering.

The wedging of plant roots also deserves consideration as a mechanism causing joint blocks to be separated. You may have seen a tree whose lower trunk and roots are firmly wedged between two great joint blocks of massive rock (Figure 18.5). Whether the tree has actually been able to spread the blocks farther apart or has merely occupied the available space is open to question. However, it is certain that pressure exerted by growth of tiny rootlets in joint fractures causes the loosening of countless small rock scales and grains.

A widespread process of rock disruption related to physical weathering results from _unloading_, the relief of confining pressure of overlying rock. Unloading occurs as rock is brought nearer to the earth's surface through the erosional removal of overlying rock. Rock

formed at great depth beneath the earth's surface (particularly igneous and metamorphic rock) is in a slightly compressed state because of the confining pressure of overlying rock. On

figure 18.5
Jointing in sandstone resembles pavement blocks at Artists View, Catskill Mountains, New York. A tree has grown up between two joint blocks. (A. N. Strahler.)

figure 18.6
Sheeting of granite facilitates quarrying operations. (Light Quarry Division of Rock of Ages Corporation, Barre, Vermont.)

figure 18.7
North Dome and Basket Dome in Yosemite National Park, California, are exfoliation domes developed from huge masses of solid igneous rock. (Douglas Johnson.)

being brought to the surface, the rock expands slightly in volume. Expansion causes thick shells of rock to break free from the parent mass below. The new surfaces of fracture are a form of jointing called *sheeting structure*. The rock sheets show best in massive rocks such as granite and marble.

Sheeting structure is well developed in granite quarries, where it greatly facilitates the removal of rock (Figure 18.6). Individual sheets sometimes break free and arch upward with explosive violence. A similar phenomenon in deep mines and tunnels is the explosive breaking away of ceilings. Known as "popping rock," this spontaneous rock disintegration is a hazard to miners.

Where sheeting structure has formed over the top of a single large body of massive rock, an *exfoliation dome* is produced (Figure 18.7). Domes are among the largest of the landforms caused primarily by weathering. In Yosemite Valley, California, where domes are spectacularly displayed, the individual rock shells may be as thick as 50 ft (15 m).

Forms produced by chemical weathering

We investigated chemical weathering processes in Chapter 16 under the heading of chemical alteration. Recall that the dominant processes of chemical change affecting silicate minerals are oxidation, carbonic acid action, and hydrolysis. Feldspars and the mafic minerals are very susceptible to chemical

figure 18.8
A felsic dike (sharp-cornered blocks in center) in deeply altered mafic rock (rounded masses and rough gray rock surface). Sangre de Cristo Mountains, New Mexico. (A. N. Strahler.)

figure 18.9
Spheroidal weathering, shown here, has produced many thin concentric shells in a basaltic igneous rock. Lucchetti Dam, Puerto Rico. (U.S. Geological Survey.)

decay (Figure 18.8). On the other hand, quartz is a highly stable mineral, almost immune to decay.

As chemical alteration penetrates the bedrock, joint blocks are attacked from all sides. Decay progresses inward to make concentric shells of soft rock (Figure 18.9). This form of change is called *spheroidal weathering* and produces onionlike bodies from which thin layers can be peeled away from a spherical core.

Decomposition by hydrolysis and oxidation changes strong rock into very weak regolith. This change allows erosion to operate with great effectiveness, wherever the regolith is exposed. Weakness of the regolith also makes it susceptible to natural forms of mass wasting.

In warm, humid climates of the equatorial, tropical, and subtropical zones hydrolysis and oxidation often result in the decay of igneous and metamorphic rocks to depths as much as 300 ft (90 m). Geologists who first studied this deep rock decay in the Southern Appalachians named the rotted layer *saprolite* (literally "rotten rock"). To the civil engineer, deeply weathered rock is of major importance in constructing highways, dams, or other heavy structures. Although the saprolite is soft and can be removed by power shovels with little blasting, there is serious danger of failure of foundations under heavy loads. This regolith also has undesirable plastic properties because of a high content of clay minerals.

The hydrolysis of granite is accompanied by *granular disintegration*, the grain-by-grain breakup of the rock. This process creates many interesting boulder and pinnacle forms by rounding of angular joint blocks (Figure 18.10). These forms are particularly conspicuous in arid regions. There is ample moisture in most deserts for hydrolysis to act, given sufficient time. The products of grain-by-grain breakup form a coarse desert gravel, which consists largely of grains of quartz and partially decomposed feldspars.

figure 18.10
Egg-shaped granite boulders are produced from joint blocks by granular disintegration in a semiarid climate near Prescott, Arizona. (A. N. Strahler.)

Atmospheric carbon dioxide is dissolved in all surface waters of the lands, including rainwater, soil water, and stream water. The solution of carbon dioxide in water produces a weak acid, called carbonic acid. Carbonate sedimentary rocks (limestone, marble) are particularly susceptible to the acid action. Mineral calcium carbonate is dissolved, yielding calcium ions and bicarbonate ions, both of which are carried away in solution in water of streams.

Carbonic acid reaction with limestone produces many interesting surface forms, mostly of small dimensions. Outcrops of limestone typically show cupping, rilling, grooving, and fluting in intricate designs (Figure 18.11). In a few places the scale of

deep grooves and high wall-like rock fins reaches proportions that prevent passage of people and animals.

Mass wasting

Everywhere on the earth's surface, gravity pulls continually downward on all materials. Bedrock is usually so strong and well supported that it remains fixed in place but, where a mountain slope becomes too steep, bedrock masses break free, falling or sliding to new positions of rest. In cases where huge masses of bedrock are involved, the result can be catastrophic in loss to life and property in towns and villages in the path of the slide. Such slides are a major form of environmental hazard in mountainous regions. Soil and regolith, being poorly held together, are much more susceptible to gravity movements than bedrock. There is abundant evidence that on most slopes at least a small amount of downhill movement is going on constantly. Much of this motion is imperceptible, but sometimes the regolith slides or flows rapidly.

figure 18.11
Solution rills in limestone, west of Las Vegas, Nevada. Scale is indicated by a pocket knife in center. (John S. Shelton.)

Taken altogether, the various kinds of downhill movements occurring under the pull of gravity, collectively called mass wasting, constitute an important process in denudation of the continental surfaces. Man has added to the natural forms of mass wasting by moving enormous volumes of rock and soil on construction sites of dams, canals, highways, and buildings.

Soil creep

On almost any steep, soil-covered slope, you can find evidence of extremely slow downhill movement of soil and regolith, a process called *soil creep*. Figure 18.12 shows some of the evidence that the process is going on. Joint blocks of distinctive rock types are found moved far downslope from the outcrop. In some layered rocks such as shales or slates, edges of the strata seem to bend in the downhill direction. This is not true plastic bending, but is the result of slight movement on many small joint cracks (Figure 18.13). Fence posts and telephone poles lean down-slope and even shift measurably out of line. Retaining walls of road cuts lean and break outward under pressure of soil creep from above.

figure 18.13
Slow creep has caused this downhill bending of steeply dipping sandstone layers. (Ward's Natural Science Establishment, Inc., Rochester, New York.)

What causes soil creep? Heating and cooling of the soil, growth of frost needles, alternate drying and wetting of the soil, trampling and burrowing by animals, and shaking by earthquakes all produce some disturbance of the soil and mantle. Because gravity exerts a downhill pull on every such rearrangement, the particles are urged slowly downslope.

Earthflow

In humid regions having steep hillsides, masses of water-saturated soil, regolith, or weak shale may slide downslope during a period of a few hours. This movement is an *earthflow*. Figure 18.14 is a sketch of an earthflow, showing how the material slumps away from the top, leaving a steplike terrace. The plastic mass then flows outward to form a bulging toe. Shallow earthflows, affecting only the soil and regolith, are common on sod-covered and forested slopes that have been saturated by heavy rains.

Earthflows are a common cause of blockage of highways and railroad lines. Because the rate of flowage is slow, these flows are not often a

figure 18.12
The slow, downhill creep of soil and regolith shows up in many ways on a hillside. (After C. F. S. Sharpe.)

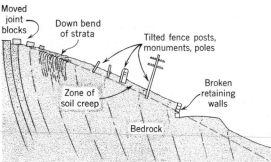

figure 18.14
Earthflows in a mountainous region. (After W. M. Davis.)

figure 18.15
Solifluction lobes cover this Alaskan mountain slope in the tundra climate. (U.S. Geological Survey.)

threat to life. However, property damage to buildings, pavements, and utility lines is often costly where construction has taken place on unstable soil slopes.

A special variety of earthflow characteristic of the tundra climate is *solifluction* (from Latin words meaning "soil" and "flow"). In early summer, when thawing has penetrated the upper foot or so, the soil is fully saturated. Soil water cannot escape downward because of the underlying impermeable frozen mass (permafrost). Flowing almost imperceptibly, this saturated soil forms terraces and lobes that give the mountain slope a stepped appearance (Figure 18.15).

Mudflows and debris floods

One of the most spectacular forms of mass wasting and one that is potentially a serious environmental hazard is the *mudflow*. This is mud stream of fluid consistency that pours down canyons in mountainous regions (Figure 18.16). In deserts, where vegetation does not protect the mountain soils, local convectional storms produce rain much faster than it can be absorbed by the soil. As the water runs down the slopes, it forms a thin mud, which flows down to the canyon floors. Following stream courses, the mud continues to flow until it becomes so thick it must stop. Great boulders are carried along, buoyed up in the mud. Roads, bridges, and houses in the canyon floor are engulfed and destroyed. Where the mudflow

figure 18.16
Thin, streamlike mudflows issue occasionally from canyon mouths in arid regions. The mud spreads out on the piedmont slopes in long, narrow tongues.

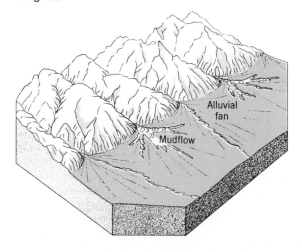

Alluvial fan

Mudflow

(left) **Frost-shattered blocks of quartzite on the summit of the Snowy Range, Wyoming, elevation 12,000 ft (3700 m). The expanse of broken rock is a felsenmeer (rock-sea). (Arthur N. Strahler)**

(left) **Frost heaving above timberline. Needle ice, attached to the underside of a rock fragment, lifted the mass as the crystals lengthened. Below is the cavity in which the ice formed. (Mark A. Melton)**

(below) **Mafic igneous rock undergoing decay in place to produce a thick regolith. White dikes of felsic rock are little affected. Sangre de Cristo Mts., New Mexico. (Arthur N. Strahler)**

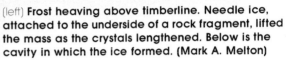

(above) **Granular disintegration of a boulder of coarse granite is forming hollows beneath a hardened surface layer covered by desert varnish. Sacaton Mountains, Arizona. (Mark A. Melton)**

Weathering Processes

(above) **Spheroidal weathering in mafic igneous rock, Spanish Peaks, Colorado. (Orlo E. Childs)**

(right) **Joint blocks of pink granite, rounded by granular disintegration in a semiarid climate. Granite Dells, near Prescott, Arizona. (Arthur N. Strahler)**

(above) **Cavernous weathering in cliffs of consolidated volcanic ash (tuff), Frijoles Canyon, New Mexico.** (Arthur N. Strahler)

Weathering Forms

(below) **A double rock arch formed by weathering of a massive sandstone layer (Entrada formation), Arches National Monument, Utah.** (Arthur N. Strahler)

(above) **Montezuma Castle in the Verde Valley, Arizona, is an Indian cliff dwelling occupying niches weathered from limy strata of an ancient lake deposit.** (Donald L. Babenroth)

(above) **Half Dome, Yosemite Valley, California, a large exfoliation dome with thick granite shells (sheeting structure).** (Orlo E. Childs)

(left) **The White House Ruin, a former Indian habitation, occupies a great niche in sandstone in the lower wall of Canyon de Chelly, Arizona.** (Mark A. Melton) (right) **Limestone showing cavernous weathering by solution. These boulders were removed by power shovel from an enclosing regolith of reddish clay, near Roanoke, Virginia. (Arthur N. Strahler)**

(below) **Cavities and pits produced by carbonic acid action on limestone, Everglades National Park, Florida. (Arthur N. Strahler)**

(left) **This large sinkhole in limestone, its floor choked with soil, has been used as a small cornfield, near Hilton, Virginia. (Arthur N. Strahler)**

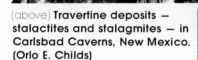

Limestone Solution and Caverns

(above) **Travertine deposits — stalactites and stalagmites — in Carlsbad Caverns, New Mexico. (Orlo E. Childs)**

(below) **Travertine deposits, Howe's Cavern, New York. Stalactites and drip curtains hang from joint fractures in the ceiling rock. (Orlo E. Childs)**

(left) **A shallow sinkhole on the Kaibab Plateau, Arizona. Clay impedes the drainage of water, allowing a pond to form. (Arthur N. Strahler)**

(left) **Talus cones of angular quartzite boulders built against the steep headwall of a glacial cirque. Snowy Range, Wyoming. (Arthur N. Strahler)**

(above) **A small earthflow made up of saturated glacial till, Vermont. The flow occurred during heavy summer rains. (Orlo E. Childs)**

(left) **Upper part of a coastal slide set off by an earthquake, San Pedro, California. Large blocks slumped down to different levels, displacing the roadway. (Ned L. Reglein)**

Mass Wasting

(one above) **This mudflow issued from a steep mountain canyon and buried a highway near Farmington, Utah. The boulders were carried many miles, suspended in the soft mud. (Orlo E. Childs)**

(above) **This great landslide occurred in the Gros Ventre River canyon in 1925, blocking the river and creating a large lake. The mass descended about 2000 ft (600 m). (Orlo E. Childs)**

(right) **A large, prehistoric landslide underlies the hummocky ground in the middle distance. Pavilion, British Columbia. (Mark A. Melton)**

emerges from the canyon and spreads across a piedmont plain, their may be severe property damage and even loss of life.

Mudflows also occur on the slopes of erupting volcanoes. Freshly fallen volcanic ash and dust is turned into mud by heavy rains and flows down the slopes of the volcano. Herculaneum, a city at the base of Mt. Vesuvius, was destroyed by a mudflow during the eruption of 79 A.D. At the same time, the neighboring city of Pompeii was buried under volcanic ash.

Mudflows show varying degrees of consistency, from a mixture about like the concrete that emerges from a mixing truck to thinner consistencies that are little different than in turbid stream floods. The watery type of mudflow is called a *debris flood* in the western United States, and particularly in southern California, where it occurs commonly and with disastrous effects.

Landslide and debris avalanche

Landslide is the rapid sliding of large masses of bedrock. Wherever mountain slopes are steep there is a possibility of large, disastrous landslides. In Switzerland, Norway, or the Canadian Rockies, for example, villages built on the floors of steep-sided valleys have been destroyed and their inhabitants killed by the sliding of millions of cubic yards of rock, set loose without any warning.

A great landslide disaster occurred in 1959 in Montana when a severe earthquake (Hebgen Lake earthquake) caused an entire mountainside to slide into the Madison River gorge, killing 27 persons (Figure 18.17). The Madison Slide, as it is known, formed a debris dam over 200 ft (60 m) high and produced a new lake. Volume of this slide was about 35 million cu yd (27 million cu m). The lake has

figure 18.17
Seen from the air, the Madison Slide forms a great dam of rubble across the Madison River Canyon. The rock mass which slid away from the mountain summit at the left was 2000 ft (600 m) long and 1000 ft (300 m) high. Momentum carried the tumbled rock debris over 400 ft (120 m) up the opposite side of the canyon. (U.S. Geological Survey.)

now been made permanent by construction of a protected spillway. Severe earthquakes in mountainous regions are a major immediate cause of landslides and earthflows.

Aside from occasional great catastrophes, landslides have rather limited environmental influence because of their sporadic occurrence in thinly populated mountainous regions. However, small slides can repeatedly block or break an important mountain highway or railway line.

In high alpine mountain chains, glacial erosion has produced extremely steep valley gradients. Large quantities of glacial rock rubble (moraine) and relict glaciers are perched precariously at high positions. Here the *alpine debris avalanche* occurs, a sudden rolling of a mixture of rock waste and glacial ice. The tongue of mixed debris and ice travels downvalley at very high speed. A recent disaster of this type occurred in 1970 in the high Andes of Peru. A severe earthquake, which caused widespread death and destruction, set off the fall of a large snow cornice from a high peak. After a free fall of 3000 ft (900 m), the snow mass was partially melted by impact and incorporated a great quantity of loose rock to become a debris avalanche. Traveling downvalley at speeds reaching 300 mi (480 km) per hour, the avalanche wiped out the town of Yungay and several smaller villages. The death toll, which included earthquake casualties, was estimated in the thousands.

Man-induced mass wasting

Man's activities induce mass wasting in forms ranging from mudflow and earthflow to landslide. These activities include (1) piling up of waste soil and rock into unstable accumulations that fall spontaneously, and (2) removal of support by undermining natural masses of soil, regolith, and bedrock.

At Aberfan, Wales, a major disaster occurred when a hill 600 ft (180 m) high, built of rock waste (culm) from a nearby coal mine, spontaneously began to move and quickly developed into a mudflow of thick consistency. The waste pile had been constructed on a steep hillslope and on a spring line, as well, making a potentially unstable configuration. The debris tongue overwhelmed part of the town below, destroying a school and taking over 150 lives (Figure 18.18). Phenomena of this type are often called ''mudslides'' in the news media.

In Los Angeles County, California, real estate development has been carried out on very steep hillsides and mountainsides by the process of bulldozing roads and homesites out of the deep regolith. The excavated regolith is

figure 18.18

This debris flow at Aberfan, Wales, buried a school and killed many children. (From A. N. Strahler, 1972, Planet Earth, Harper & Row, New York.)

pushed out into adjacent embankments where its instability poses a threat to slopes and stream channels below. When saturated by heavy winter rains, these embankments can give way, producing earthflows, mudflows, and debris floods that travel far down the canyon floors and spread out on the piedmont surfaces, burying streets and yards in bouldery mud. Many debris floods of this area are also produced by heavy rains falling on undisturbed mountain slopes denuded by fire of vegetative cover in the preceding dry summer. Some of these fires are set by Man, whether inadvertently or deliberately. Disturbance of slopes by construction practices is simply an added source of debris and serves to enhance what is already an important environmental hazard.

Man as an agent of land scarification

Man now possesses enormous machine power and explosives, capable of moving great masses of regolith and bedrock from one place to another for two basic purposes: first, to extract mineral resources; second, to reorganize terrain into suitable configurations for highway grades, airfields, building foundations, dams, canals, and various other large structures. Both activities involve removal of earth materials, which destroys entirely the preexisting ecosystems and habitats of plants and animals. The same activities include building up of new land on adjacent surfaces, using those same earth materials. This process destroys ecosystems and habitats by burial. What distinguishes man-made forms of mass wasting from the natural forms is that Man can use machinery to raise earth materials against the force of gravity. Also, Man's use of explosives in blasting can bring to bear in a rock mass disruptive forces that exceed by many orders of magnitude the forces of physical weathering.

Scarification is a general term for excavations and other land disturbances produced for purposes of extracting mineral resources; it includes accumulation of waste matter as spoil or tailings. Among the forms of scarification are open-pit mines, strip mines, quarries for structural materials, borrow pits along highway grades, sand and gravel pits, clay pits, phosphate pits, scars from hydraulic mining, and stream gravel deposits reworked by dredging. Open-pit mining of low-grade copper ores is illustrated in Figure 16.10.

Scarification is on the increase. Demands for coal to meet energy requirements are on the rise. There are also increased demands for industrial minerals used in manufacturing and construction. At the same time, as the richer and more readily available mineral deposits are consumed, industry turns to poorer grades of ores and to less easily accessible coal deposits. As a result, the rate of scarification is further increased.

Strip mining and mass wasting

Where coal seams lie close to the surface or actually outcrop along hillsides, the *strip mining* method is used. Here, earthmoving equipment removes the covering strata (overburden) to bare the coal, which is lifted out by power shovels. There are two kinds of strip mining, each adapted to the given relationship between ground surface and coal seam.

Area strip mining is used in regions of flat land surface under which the coal seam lies horizontally (Figure 18.19A). After the first trench is made and the coal removed, a parallel trench is made, the overburden of which is piled as a spoil ridge into the first trench. In this way the entire seam is gradually uncovered, and a series of parallel spoil ridges remains.

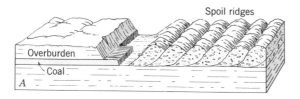

figure 18.19

(A) Area strip mining. (B) Contour strip mining.

figure 18.20

Contour strip mining in Wise County, Virginia.
(Kenneth Murray, Kingsport, Tennessee, from
Nancy Palmer Photo Agency, Inc.)

The contour strip mining method is used where
a coal seam outcrops along a steep hillside
(Figure 18.19B). The coal is uncovered as far
back into the hillside as possible, and the
overburden is dumped on the downhill side.
There results a bench bounded on one side by
a freshly cut rock wall and on the other by a
ridge of loose spoil with a steep outer slope
leading down into the valley bottom. The
benches form sinuous patterns following the
plan of the outcrop (Figure 18.20). Strip mining
is carried to depths as great as 100 ft (30 m)
below the surface.

The spoil bank produced by contour strip
mining is unstable and is a constant threat to
the lower slope and valley bottom below.
When saturated by heavy rains and melting
snows, the spoil generates earthflows and
mudflows; these descend on houses, roads,
and forest. The spoil also supplies sediment
that clogs stream channels far down the
valleys.

Weathering and
mass wasting in review

In this chapter we have compared natural
processes of wasting of the continental surfaces
with man-induced changes of a similar nature.
While the processes of weathering and
soil creep are for the most part slow-acting
and produce effects that are visible only
when accumulated over centuries, the
natural mass-wasting processes also include
catastrophic events. Indeed, some large
landslides dwarf anything that Man has
accomplished in equal time by earth-moving
with machines and explosives. But Man works
constantly with enormous quantities of energy,
and the cumulative environmental damage and
destruction continues to mount, almost
unchecked. Only now are the conflicts of
interest beginning to emerge and to be
squarely faced.

Ground Water and Surface Water

Chapter 19

IN THIS CHAPTER we return to the hydrologic cycle. Chapter 9 dealt with a phase of the hydrologic cycle in which soil water is recharged by precipitation and returned directly to the atmosphere by evapotranspiration. Recall that many soil-water budgets show a substantial water surplus, to be disposed as runoff. Our investigation now deals with that water surplus and the paths it follows as subsurface water. There are two basic paths of escape for surplus water.

First, surplus water may percolate through the soil, traveling downward under the force of gravity to become a part of the underlying ground-water body. Following subterranean flow paths, this water emerges to become surface water, or it may emerge directly in the shore zone of the ocean.

Second, surplus water may flow over the ground surface as runoff to lower levels. As it travels, the dispersed flow becomes collected into streams, which eventually conduct the runoff to the ocean. In this chapter, we trace both the subsurface and surface pathways of flow of surplus water. In so doing, we will complete the hydrologic cycle.

Surplus water, as runoff, is a vital part of the environment of terrestrial life forms and of Man in particular. Surface water in the form of streams, rivers, ponds, and lakes constitutes one of the distinctive environments of plants and animals.

Our heavily industrialized society requires enormous supplies of fresh water for its sustained operation. Urban dwellers consume water in their homes at rates of 50 to 100 gallons per person per day. Huge quantities of water are used for cooling purposes in air-conditioning units and power plants.

In view of projections based on existing rates of increase in water demands, we shall be hard put in the future to develop the needed supplies of pure fresh water. Water pollution also tends to increase as populations grow and urbanization advances over broader areas. A disconcerting concept is that the available resource of pure fresh water is shrinking while demands are rising. Knowledge of hydrologic processes enables us to evaluate the total water resource, to plan for its management, and to protect it from pollution.

Ground water

Ground water is that part of the subsurface water that fully saturates the pore spaces of the bedrock and regolith. The ground water occupies the _saturated zone_ (Figure 19.1). Above it is the _unsaturated zone_ in which water does not fully saturate the pores. Water is held in the unsaturated zone by capillary force in tiny films adhering to the mineral surfaces.

Ground water is extracted from wells dug or drilled to reach the ground water zone. In the ordinary shallow well, water rises to the same height as the _water table_, or upper boundary of the saturated zone (Figure 19.1)

Where wells are numerous in an area, the position of the water table can be mapped in detail by plotting the water heights and noting the trend of change in altitude from one well to

figure 19.1
Zones of subsurface water.

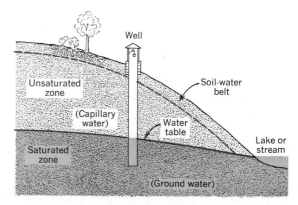

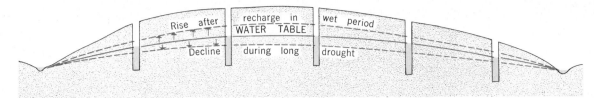

figure 19.2
Configuration of the water-table surface conforms with the land surface above it.

the other. The water table is highest under the highest areas of land surface: hilltops and divides. The water table declines in altitude toward the valleys where it appears at the surface close to streams, lakes, or marshes (Figure 19.2). The reason for this water table configuration is that water percolating down through the unsaturated zone tends to raise the water table, whereas seepage into streams tends to draw off ground water and to lower its level.

Because ground water moves extremely slowly, a difference in water table level, or *hydraulic head*, is built up and maintained between high and low points on the water table. In periods of abnormally high precipitation, this head is increased by a rise in the water table under divide areas; in periods of water deficit, occasioned by drought, the water table falls (Figure 19.2).

The subsurface phase of the hydrologic cycle is completed when the ground water emerges in places where the water table intersects the ground surface. Such places are the channels of streams and the floors of marshes and lakes. By slow seepage and spring flow the water emerges fast enough to balance the rate at which water enters the ground water table by percolation.

Figure 19.3 shows paths of flow of ground water. Flow takes paths curved concavely upward. Water entering the hillside midway between divide and stream flows rather directly. Close to the divide point on the water

table, however, the flow lines go almost straight down to great depths from which they recurve upward to points under the streams. Progress along these deep paths is incredibly slow; that near the surface is much faster. The most rapid flow is close to the place of discharge in the stream, where the arrows are shown to converge.

Ponds, lakes, and bogs

The various events of geologic history, such as faulting, or erosion and deposition by wind and by ice sheets, have created many natural depressions with floors below the altitude of the water table. Water stands in these depressions in the form of ponds and lakes. In humid climates water level coincides closely with the water table in the surrounding area.

figure 19.3
Paths of ground water movement under divides and valleys. (After M. K. Hubbert.)

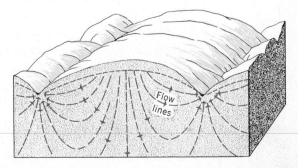

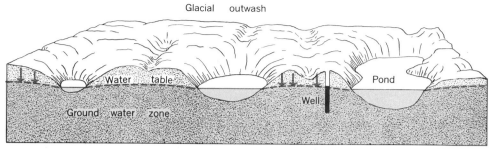

figure 19.4

Freshwater ponds in sandy glacial deposits on Cape Cod, Massachusetts. (From A Geologist's View of Cape Cod, Copyright © 1966 by Arthur N. Strahler. Reproduced by permission of Doubleday & Co.)

Seepage of ground water, as well as direct runoff of precipitation, maintains these free water surfaces permanently throughout the year. Examples of such freshwater ponds are found widely distributed in North America and Europe where plains of glacial sand and gravel contain natural pits and hollows left by the melting of stagnant ice masses (see Chapter 22). Figure 19.4 is a block diagram showing small freshwater ponds on Cape Cod. The surface elevation of these ponds coincides closely with the level of the surrounding water table.

Many former freshwater water-table ponds have become partially or entirely filled by the organic matter from growth and decay of water-loving plants. The result is a bog with a surface close to the water table.

Aquifers, aquicludes, and artesian wells

Where rock layers lie nearly horizontal, or in gently inclined positions, the ground water relations may be quite different from the simple pattern illustrated in Figure 19.3. Suppose, for example, that the region is one of sedimentary strata with beds of sandstone alternating with beds of shale (Figure 19.5). Sandstone is usually porous, providing a large ground water storage reservoir through which water may move easily. A water-transmitting rock body is termed an *aquifer*. By contrast, a shale bed is dense and lacks open pores; it impedes the flow of ground water and is called an *aquiclude*. In the particular case shown in Figure 19.5, a thin bed of shale has effectively blocked the downward percolation of water to the main water table below, creating a perched water table. Where the perched water table meets the valley side a line of springs emerges. Most natural springs are mere trickles of water, unseen and unnoticed under a cover of dense vegetation. A few springs discharge enormous volumes of water where an unusually good aquifer is exposed (Figure 19.6).

figure 19.5

A perched water table.

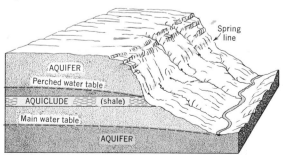

figure 19.6
Thousand Springs, Idaho, emerges from the north side of the Snake River Canyon, opposite the mouth of the Salmon River. The spring issues from a porous basalt layer. (U.S. Geological Survey.)

Where strata are inclined a favorable situation may exist for development of an *artesian well*, one in which the water flows upward toward the surface through its own pressure. In Figure 19.7 we see a highly diagrammatic representation of such conditions; the vertical exaggeration is very great merely to show the principle. The eroded edge of a sandstone aquifer is exposed to intake of water at a high

figure 19.7
Ground water in a sandstone aquifer is held under pressure beneath a capping aquiclude of shale. Water rises above the ground surface from an artesian well.

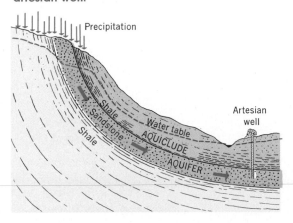

position. Water entering here passes deep underground to a position below the valley floor. Here the water is under a strong pressure, or hydraulic head, from the weight of the overlying water. This pressure is sufficient to force water up to the surface in a well drilled down through the shale aquiclude into the aquifer. Flow as an artesian spring may occur naturally if there are faults in the strata that permit water to leak upward through the shale layer.

Use of ground water as a resource

Man's withdrawal of ground water has begun to make serious impacts on the environment in many places. The drilling of vast numbers of wells, from which water is forced out in great volumes by powerful pumps, has profoundly altered nature's balance of ground water recharge and discharge. Increased urban populations and industrial developments require larger water supplies, needs that cannot always be met from construction of new surface water reservoirs.

In agricultural lands of the semiarid and desert climates, heavy dependence is placed on irrigation water from pumped wells, especially since many of the major river systems have already been fully developed for irrigation from surface supplies. Wells can be drilled within the limits of a given agricultural or industrial property and can provide immediate supplies of water without any need to construct expensive canals or aqueducts.

Formerly the small well needed to supply domestic and livestock needs of a home or farmstead was actually dug by hand as a large cylindrical hole, lined with masonry where required. By contrast, the modern well put down to supply irrigation and industrial water is drilled by powerful machinery that may bore a hole 16 in. (40 cm) or more in diameter to depths of 1000 ft (300 m) or more. Drilled wells are sealed off by metal casings that exclude impure near-surface water and prevent clogging of the tube by caving of the walls. Near the lower end of the hole, where it enters the aquifer, the casing is perforated to admit the water. The yields of single wells range from as low as a few gallons per day in a domestic well to many millions of gallons per day for large, deep industrial or irrigation wells.

As water is pumped from a well, the level of water in the well drops. At the same time, the surrounding water table is lowered in the shape of a conical surface, termed the _cone of depression_. The difference in height between the cone base and the original water table is the _drawdown_ (Figure 19.8). By producing a steeper gradient of the water table, the flow of ground water toward the well is also increased, so that the well will yield more water. This holds only for a limited amount of drawdown, beyond which the yield fails to increase. The cone of depression may extend as far out as 10 mi (16 km) or more from a well where very heavy pumping is continued. Where many wells are in operation, their intersecting cones produce a general lowering of the water table.

Depletion often greatly exceeds the rate at which the ground water of the area is recharged by percolation from rain or from the beds of streams. In an arid region, much of the ground water for irrigation is from wells driven into thick sands and gravels. Recharge of these deposits depends on the seasonal flows of water from streams heading high in adjacent mountain ranges. The extraction of ground water by pumping can greatly exceed the recharge by stream flow. Cones of depression deepen and widen; deeper wells and more powerful pumps are then required. Overdrafts of water accumulate, and the result is exhaustion of a natural resource not renewable except by long lapses of time.

In humid areas where a large annual water surplus exists, natural recharge is by general percolation over the entire ground area surrounding the well. Here the prospects of achieving a balance of recharge and withdrawal are highly favorable through the control of pumping. An important recycling measure is the return of waste waters or stream waters to the ground water table by means of recharge wells in which water flows down rather than up.

figure 19.8
Drawdown and cone of depression in a pumped well.

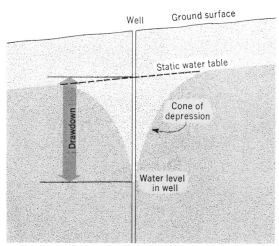

Pollution of ground water

Disposal of solid wastes poses a major environmental problem in the United States because our advanced industrial economy produces an endless source of garbage and trash. Traditionally, these waste products have been trucked to the town dump, and there burned in continually smoldering fires that emit foul smoke and gases. The partially consumed residual waste is then buried under earth.

In recent years, a major effort has been made to improve solid-waste disposal methods. One method is high-temperature incineration.

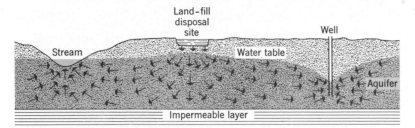

figure 19.9
Leachate from a waste disposal site moves toward a supply well (right) and a stream (left). (From A. N. Strahler, 1972, Planet Earth, Harper & Row, New York.)

Another is the _sanitary land-fill_ method in which waste is not allowed to burn. Instead, the waste is continually covered by protective overburden, usually sand or clay available on the land-fill site. The waste is thus buried in the unsaturated zone. Here it is subject to reaction with percolating rainwater infiltrating the ground surface. This water picks up a wide variety of ions from the waste body and carries these down as _leachate_ to the water table. Once in the water table, the leachate follows the flow paths of the ground water.

As shown in Figure 19.9, a mounding of the water table develops beneath the disposal site. Loose soil of the disposal area facilitates infiltration of precipitation, while lack of vegetation reduces the evapotranspiration. Consequently, the recharge here is greater than elsewhere, and the mound is maintained. After leachate has moved vertically down by gravity percolation to the water table mound, it moves radially outward from the mound to surrounding lower points on the water table. As shown in Figure 19.9, a supply well with its cone of depression draws ground water from the surrounding area. Linkage between outward flow from the waste disposal site and inward flow to the well can bring leachate into the well, polluting the ground water supply.

Limestone caverns

Most of you are familiar with the names of famous caverns, such as Mammoth Cave or Carlsbad Caverns. Millions of Americans have visited these famous tourist attractions. _Caverns_

figure 19.10
Cavern development in the ground water zone, followed by travertine deposition in the unsaturated zone. (From A. N. Strahler, 1971, The Earth Sciences, 2nd ed. Harper & Row, New York.)

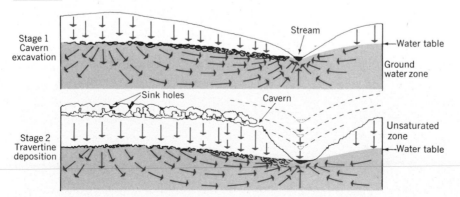

are interconnected subterranean cavities in the bedrock, formed by the action of circulating ground water on limestone. Figure 19.10 suggests how caverns may develop. In the upper diagram the action of carbonic acid is shown to be particularly concentrated in the saturated zone just below the water table. Products of solution are carried along in the ground water flow paths to emerge in streams. In a later stage, shown in the lower diagram, the stream has deepened its valley and the water table has been correspondingly lowered to a new position. The cavern system previously excavated is now in the unsaturated zone (Figure 19.11). Evaporation of percolating water on exposed rock surfaces in the caverns now begins the deposition of carbonate matter, known as *travertine*. Encrustations of travertine take many beautiful forms — stalactites, stalagmites, columns, drip curtains, and terraces (Figures 19.12 and 19.13).

The environmental importance of caves is felt in several ways. Throughout Man's early development, caves were an important habitation. Now we find the skeletal remains of these people, together with their implements and cave drawings, preserved through the centuries in caves in many parts of the world. Today, caverns are being used as storage facilities, living quarters, and factories.

Caverns have provided some valuable deposits of guano, the excrement of birds or bats, which is rich in nitrates. Guano deposits have been used in the manufacture of fertilizers and explosives. Bat guano was taken from Mammoth Cave for making gunpowder during the war of 1812.

Karst landscapes

Where limestone solution is very active, we find a landscape with many unique landforms. This is especially true along the Dalmatian coastal area of Yugoslavia, where the land-

figure 19.11
The gaping entrance to Jenolan Caves, Australia, is at the base of a high limestone cliff. A subterranean river (foreground) makes its exit from the cavern system at this point. (New South Wales Government Printer.)

scape is called *karst*. The term may be applied to the topography of any limestone area where sinkholes are numerous and small surface streams are nonexistent. A *sinkhole* is a surface depression in limestone of a cavernous region. Some sinkholes are filled with soil washed from nearby hillsides. Others are steep-sided, deep holes.

Development of a karst landscape is shown in Figure 19.14. In an early stage, funnellike sinkholes are numerous. Later, the caverns collapse, leaving open, flat-floored valleys. Some important regions of karst or karstlike topography are the Mammoth Cave region of Kentucky, the Yucatan Peninsula, and parts of Cuba and Puerto Rico.

Forms of overland flow

We have now traced the subsurface movements of surplus water beneath the lands.
We turn next to trace the surface flow paths of
surplus water.

Runoff that flows down the slopes of the land
in broadly distributed sheets is referred to as
overland flow (Figure 19.15). We distinguish
overland flow from *stream flow*, in which the
water occupies a narrow channel confined by
lateral banks. Overland flow can take several
forms. It may be a continuous thin film, called
sheet flow, where the soil or rock surface is
smooth. Flow may take the form of a series of
tiny rivulets connecting one water-filled hollow
with another, where the ground is rough or
pitted. On a grass-covered slope, overland flow
is subdivided into countless tiny threads of
water, passing around the stems. Even in a

figure 19.13
**A travertine terrace in the Jenolan Caves. (New
South Wales Government Printer.)**

figure 19.12
**A drip curtain of travertine, Jenolan Caves. (New
South Wales Government Printer.)**

figure 19.14
**Features of a karst landscape. (A) Rainfall enters
the cavern system through sinkholes in the
limestone. (B) Extensive collapse of caverns
reveals surface streams flowing on shale beds
beneath the limestone. The flat-floored valleys
can be cultivated. (Drawn by E. Raisz.)**

A

B

figure 19.15
Overland flow on this sloping corn field is being generated during a summer thunderstorm. The flow is being received by a narrow stream channel in the foreground and conducted rapidly away toward the left. (Soil Conservation Service.)

heavy and prolonged rain, you might not notice overland flow in progress on a sloping lawn. On heavily forested slopes, overland flow may pass entirely concealed beneath a thick mat of decaying leaves.

Stream flow

Overland flow eventually contributes to a stream, which is a much deeper, more concentrated form of runoff. We define a *stream* as a long, narrow body of flowing water occupying a trenchlike depression, or channel, and moving to lower levels under the force of gravity.

The *channel* of a stream is a narrow trough, shaped by the forces of flowing water to be most effective in moving the quantities of water and sediment supplied to the stream (Figure 19.16). Channels may be so narrow that a person can jump across them, or as wide as 1 mi (1.5 km) for great rivers such as the Mississippi.

The size of a stream channel can be stated in terms of the area of cross section, A, which is the area in square feet between the stream surface and the bed, measured in a vertical slice across the stream (Figure 19.16). The rate of fall in altitude of the stream surface in the downstream direction is the *gradient*. As a stream flows under the influence of gravity, the water encounters resistance — a form of friction — with the channel walls. As a result, water close to the bed and banks moves slowly; that in the deepest and most centrally located zone flows fastest. Figure 19.16 indicates by arrows the speed of flow at various points in the stream. The single line of maximum velocity is located in midstream, where the channel is straight and symmetrical, but about one third of the distance down from surface to bed.

Our statement about velocity needs to be qualified. Actually, in all but the most sluggish streams, the water is affected by *turbulence*, a system of innumerable eddies that are continually forming and dissolving. A particular molecule of water, if we could keep track of it, would describe a highly irregular, corkscrew path as it is swept downstream. Motions include upward, downward, and sideward directions. Turbulence in streams is

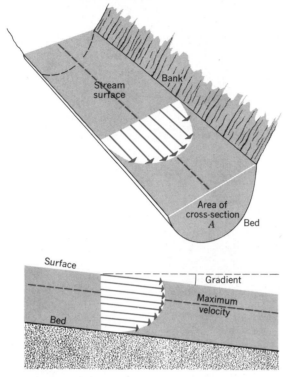

figure 19.16
Stream flow within a channel is most rapid near the center and just below the water surface.

extremely important because of the upward elements of flow that lift and support fine particles of sediment. The murky, turbid appearance of streams in flood is ample evidence of turbulence, without which sediment would remain near the bed. Only if we measure the water velocity at a certain fixed point for a long period of time, say several minutes, will the average motion at that point be downstream and in a line parallel with the surface and bed. Average values are shown by the arrows in Figure 19.16.

Because the velocity at a given point in a stream differs greatly according to whether it is being measured close to the banks and bed or out in the middle line, a single value, the _mean velocity_, is needed. Mean velocity is computed for the entire cross section to express the activity of the stream as a whole.

Stream discharge

A most important measure of stream flow is _discharge_, Q, defined as the volume of water passing through a given cross section of the stream in a given unit of time. Commonly, discharge is stated in cubic feet per second (abbreviated to cfs). In metric units discharge is stated in cubic meters per second (cms). Discharge may be obtained by taking the mean velocity, V, and multiplying it by cross-sectional area, A. This relationship is stated by the important equation, $Q = AV$.

We realize that water will flow faster in a channel of steep gradient than one of gentle gradient, since the component of gravity acting parallel with the bed is stronger for the steeper gradient. As shown in Figure 19.17, velocity increases quickly where a stream passes from a wide pool of low gradient to a steep stretch of rapids. As velocity V increases, cross-sectional area A must decrease; otherwise their product, AV, would not be held constant. In the pool, where velocity is low, cross-sectional area is correspondingly increased.

Drainage systems

Seeking to escape to progressively lower levels and eventually to the sea, runoff becomes organized into a _drainage system_. The system

figure 19.17
Schematic diagram of the relationships among cross-sectional area, mean velocity, and gradient. (From A. N. Strahler, 1971, The Earth Sciences, 2nd ed., Harper & Row, New York.)

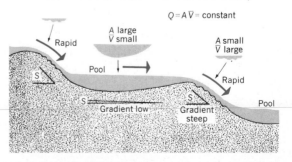

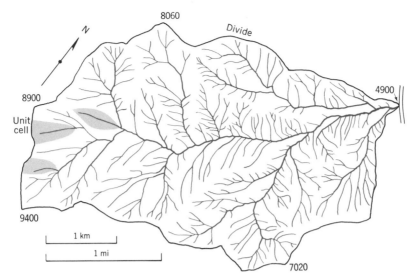

figure 19.18
Channel network of a small drainage basin. (Data of U.S. Geological Survey and Mark A. Melton.)

consists of a branched network of stream channels, as well as the sloping ground surfaces that contribute overland flow to those channels. The entire system is bounded by a drainage divide, outlining a more-or-less pear-shaped _drainage basin_. The drainage system is adjusted to dispose as efficiently as possible of the runoff and its contained load of mineral particles.

A typical stream network contributing to a single outlet is shown in Figure 19.18. Note that each fingertip tributary receives runoff from a small area of land surface surrounding the channel. This area may be regarded as the unit cell of the drainage system. The entire surface within the outer divide of the drainage basin constitutes the watershed for overland flow. A drainage system is a converging mechanism funneling and integrating the weaker and more diffuse forms of runoff into progressively deeper and more intense paths of activity.

Stream gauging

An important activity of the U.S. Geological Survey is the measurement, or _gauging_, of stream flow in the United States. In

figure 19.19
Idealized diagram of a stream gauging installation. The operator rides in the cable car.

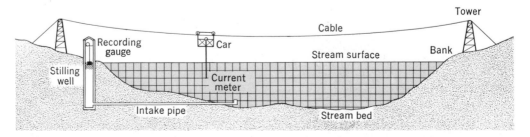

cooperation with states and municipalities this organization maintains over 6000 gauging stations on principal streams and their tributaries. Information on daily discharge and flood discharges is essential for planning the distribution and development of surface waters as well as for design of flood-protection structures and for the prediction of floods as they progress down a river system.

A stream gauging station requires a device for measuring the height of the water surface, or *stage* of the stream. Simplest to install is a staff gauge, which is simply a graduated stick permanently attached to a post or bridge pier. This must be read directly by an observer whenever the stage is to be recorded. More useful is an automatic-recording gauge, which is mounted in a stilling tower built beside the river bank (Figure 19.19). The tower is simply a hollow masonry shaft filled by water that enters through a pipe at the base. By means of a float connected by a wire to a recording mechanism above, a continuous ink-line record of the stream stage is made on a graph paper attached to a slowly rotating drum.

To measure stream discharge, we must determine both the area of cross section of the stream and the mean velocity. This requires that a *current meter* (Figure 19.20) be lowered into the stream at closely spaced intervals so that the velocity can be read at a large number of points evenly distributed in a grid pattern through the stream's cross section (Figure 19.19). A bridge often serves as a convenient means of crossing over the stream; otherwise, a cable car or small boat is used. The current meter has a set of revolving cups whose rate of turning is proportional to current velocity. As the velocities are being measured from point to point, a profile of the river bed is also made by sounding the depth. From these readings a profile is drawn and the cross-sectional area is measured from the profile. Mean velocity is computed by summing all individual velocity

figure 19.20

In gauging large rivers, the current meter is lowered on a cable by a power winch. Earphones, connected by wires to the meter, receive a series of clicks whose frequency indicates water velocity. (U.S. Geological Survey.)

readings and dividing by the number of readings. Discharge can then be computed using the formula $Q = AV$.

Stream flow and precipitation

It seems obvious that the discharge of a stream will increase in response to a period of heavy rainfall or snowmelt. The response is delayed, of course, but the length of delay depends on a number of factors. The most important factor is the size of the drainage basin feeding the

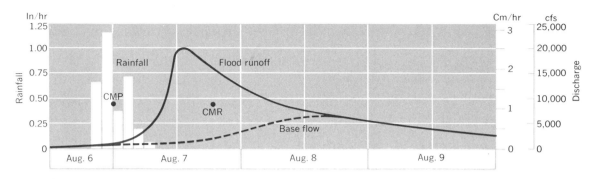

figure 19.21

Four days of flow of Sugar Creek, Ohio. (Data from Hoyt and Langbein, Floods.)

stream above the place where the gauge is located. The relationship between stream discharge and precipitation is best studied by means of a simple graph, called a *hydrograph*.

Figure 19.21 is a hydrograph for a drainage basin about 300 sq mi (800 sq km) in area, located in Ohio within the humid continental climate. The graph gives data for a two-day summer storm. Rainfall is shown by a bar graph giving the number of inches of precipitation in each two-hour period. Also plotted on the graph (smooth line), is the discharge of Sugar Creek, the trunk stream of the drainage basin. The average total rainfall over the watershed of Sugar Creek was about 6 in. (15 cm); of this amount about half passed down the stream within three days' time. Some rainfall was held in the soil as soil water, some evaporated, and some infiltrated to the water table to be held in long-term storage in the ground water body.

Studying the rainfall and runoff graphs in Figure 19.21, we see that prior to the onset of the storm, Sugar Creek was carrying a small discharge. This flow, being supplied by the seepage of ground water into the channel, is termed *base flow*. After the heavy rainfall began, several hours elapsed before the stream gauge at the basin mouth began to show a rise in discharge. This interval, called the *lag time*,

indicates that the branching system of channels was acting as a temporary reservoir. The channels were at first receiving inflow more rapidly than it could be passed down the channel system to the stream gauge.

Lag time is measured as the difference between center of mass of precipitation (CMP) and center of mass of runoff (CMR), as labeled in Figure 19.21. The peak flow of Sugar Creek was reached almost 24 hours after the rain began; the lag time was about 18 hours. Note also that the rate of decline in discharge was much slower than the rate of rise.

In general, the larger a watershed, the longer is the lag time between peak rainfall and peak discharge, the more gradual is the rate of decline of discharge after the peak has passed. Notice that the flow of Sugar Creek showed a slow but distinct rise in the amount of discharge contributed by base flow.

Base flow and overland flow

In regions of humid climates, where the water table is high and normally intersects the important stream channels, the hydrographs of larger streams will show clearly the effects of two sources of water: (1) base flow and (2) overland flow. Figure 19.22 is a hydrograph of

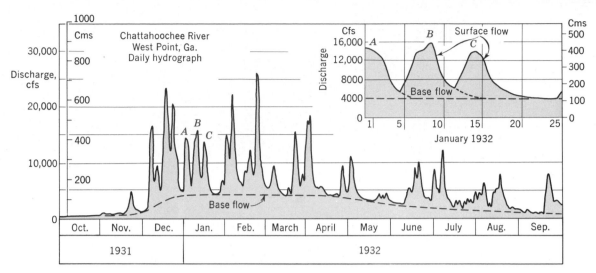

figure 19.22
Flow peaks of the Chattahoochee River. (Data from E. E. Foster, _Rainfall and Runoff_.)

the Chattahoochee River, Georgia, a large river draining a watershed of some 3350 sq mi (8700 sq km), much of it in the humid southern Appalachian Mountains. The sharp, abrupt fluctuations in discharge are produced by overland flow following rain periods of one to three days' duration. These are each similar to the hydrograph of Figure 19.21, except that they are here shown much compressed by the time scale.

After each rain period the discharge falls off rapidly but, if another storm occurs within a few days, the discharge rises to another peak. The enlarged inset graph shows details for the month of January. Where a long period intervenes between storms, the discharge falls to a low value, the base flow, where it levels off.

Throughout the year the base flow, which represents ground water inflow into the stream, undergoes a marked annual cycle. During the period of recharge (winter and early spring), water table levels are raised and the rate of inflow into streams is increased. For the Chattahoochee River, the rate of base flow

during January, February, March, and April holds uniform at about 4000 cfs (110 cms). The base flow begins to decline in spring, as heavy evapotranspiration losses reduce soil water and therefore cut off the recharge of ground water. The decline continues through the summer, reaching a low of about 1000 cfs (30 cms) by the end of October.

Floods

Everyone has seen enough news photos of river floods to have a good idea of the appearance of flood waters and the havoc wrought by their erosive power and by the silt and clay that they leave behind. Even so, it is not easy to define the term _flood_. Perhaps it is enough to say that a condition of flood exists when the discharge of a river cannot be accommodated within the margins of its normal channel, so that the water spreads over the adjoining ground on which crops or forests are able to flourish.

Most larger streams of humid climates have a _floodplain_, a belt of low flat ground bordering the channel on one or both sides inundated by

figure 19.23
The city of Hartford was partly inundated by the Connecticut River flood of March 1936. The river channel is to the left, its banks marked by a line of trees. (Official Photograph, 8th Photo Section, A. C., U.S. Army.)

stream waters about once a year. This flood usually occurs in the season when abundant supplies of surface water combine with effects of a high water table to supply more runoff than can stay within the channel. Such an annual inundation is considered a flood, even though its occurrence is expected and does not prevent the cultivation of crops after the flood has subsided. The seasonal inundation does not interfere with the growth of dense forests, which are widely distributed over low, marshy floodplains in all humid regions of the world. Still higher discharges of water, the rare and disastrous floods that may occur as infrequently as three to five decades, inundate

ground lying well above the floodplain (Figure 19.23).

For practical purposes, the National Weather Service, which provides a flood-warning service, designates a particular stage of gauge height at a given place as the _flood stage_, implying that the critical level has been reached above which overbank flooding may be expected to set in.

Flood prediction

The National Weather Service operates a River and Flood Forecasting Service through 85 offices located at strategic points along major river systems of the United States. Each office issues river and flood forecasts to the communities within the associated district, which is laid out to cover one or more large watersheds. Flood warnings are publicized by every possible means. Close cooperation is maintained with various agencies to plan evacuation of threatened areas and the removal or protection of vulnerable property.

Graphs of flood stages tell the likelihood of occurrence of given stages of high water for each month of the year. Figure 19.24 shows expectancy graphs for two rivers. The meaning of the strange-looking bar symbols is explained in the key. The Mississippi River at Vicksburg illustrates a great river responding largely to spring floods so as to yield a simple annual cycle. All floods have occurred in the first 6 months of the year; none in the second 6 months. The Sacramento River shows nicely the effect of the Mediterranean climate, with its winter wet season and long severe summer drought. Winter floods are caused by torrential rainstorms and snowmelt in the mountain watersheds of the Sierra Nevada and southern Cascades. By midsummer river flow has shrunken to a very low stage.

Hydrologic effects of urbanization

Hydrology of a watershed is altered in two ways by urbanization. First, an increasing percentage of the surface is rendered impervious to infiltration by construction of roofs, driveways, walks, pavements, and parking lots. It has been estimated that in residential areas, for a lot size of 15,000 sq ft (1400 sq m), the impervious area amounts to about 25 percent, whereas for a lot size of

figure 19.24
The highest water stage that occurred in each month is given in terms of percentages on these graphs. (After National Weather Service.)

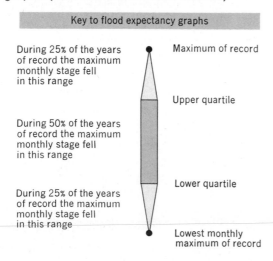

Key to flood expectancy graphs

During 25% of the years of record the maximum monthly stage fell in this range — Maximum of record

Upper quartile

During 50% of the years of record the maximum monthly stage fell in this range

Lower quartile

During 25% of the years of record the maximum monthly stage fell in this range — Lowest monthly maximum of record

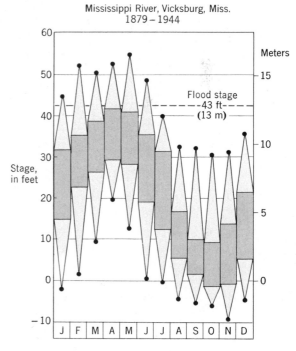

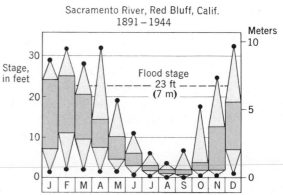

6000 sq ft, the impervious area amounts to about 80 percent.

An increase in proportion of impervious surface reduces infiltration and increases overland flow generally from the urbanized area. An important result is to increase the frequency and height of flood peaks during heavy storms. There is also a reduction of recharge to the ground water body beneath, and this reduction, in turn, decreases the base flow contribution to channels in the area. Thus the full range of stream discharges, from low stages in dry periods to flood stages, is made greater by urbanization.

A second change caused by urbanization is the introduction of storm sewers that allow storm runoff from paved areas to be taken directly to stream channels for discharge. Runoff travel time to channels is shortened at the same time that the proportion of runoff is increased by expansion in impervious surfaces. The two changes together conspire to reduce the lag time, as shown by the schematic hydrographs in Figure 19.25.

Many rapidly expanding suburban communities are now finding that certain low-lying residential areas, formerly free of flooding, are being subjected to inundation by over-bank flooding of a nearby stream. The need for careful terrain study and land-use planning is obvious in such cases to protect the unwary homebuyer from locating in a flood-prone neighborhood.

Water pollution

In this chapter we have examined the flow of water itself, largely from the hydrologist's point of view. Streams, lakes, bogs, and marshes are specialized habitats of plants and animals; their ecosystems are particularly sensitive to changes induced by Man in the water balance and in water chemistry. Not only does our

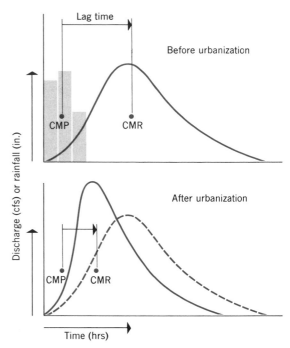

figure 19.25

These schematic hydrographs show the effect of urbanization on lag time and peak discharge. Urbanization increases the peak flood flow and diminishes the lag time. Points CMP and CMR are centers of mass of rainfall and runoff, respectively, as in Figure 19.21. (After L. B. Leopold, U.S. Geological Survey.)

industrial society make radical physical changes in water flow by construction of engineering works (dams, irrigation systems, canals, dredged channels), but we also pollute and contaminate our surface waters with a large variety of wastes. Some of these wastes are in the form of ions in solution. The subject of water quality is appropriate to discuss in this chapter. Introduction of mineral sediment into surface waters as a result of land disturbances is a suitable topic for the next chapter.

Chemical pollution by direct disposal into streams and lakes of wastes generated in industrial plants is a phenomenon well known to the general public and can be seen firsthand in almost any industrial community in the

United States. Direct outfall of sewage, whether raw (untreated) or partially treated, is another form of direct pollution of streams and lakes.

In urban and suburban areas pollutant matter entering streams and lakes includes deicing salt, lawn conditioners (lime and fertilizers), and sewage effluent. In agricultural regions, important sources of pollutants are fertilizers and the body wastes of livestock. Major sources of water pollution are associated with mining and processing of mineral deposits. In addition to chemical pollution, there is thermal pollution from discharge of heated water from nuclear generating plants. The possibility of contamination by radioactive substances released from nuclear processing plants also exists.

Among the common chemical pollutants of both surface water and ground water are sulfate, nitrate, phosphate, chloride, sodium, and calcium ions. Sulfate ions enter runoff both by fallout from polluted urban air and as sewage effluent. Important sources of nitrate ions are fertilizers and sewage effluent. Excessive concentrations of nitrate in water supplies are highly toxic and, at the same time, their removal is difficult and expensive. Phosphate ions are contributed in part by fertilizers and by detergents in sewage effluent. Phosphate and nitrate are plant nutrients and can lead to excessive growth of algae in streams and lakes. This process is called *eutrophication*. Chloride and sodium ions are contributed both by fallout from polluted air and by deicing salts used on highways. Instances are recorded in which water supply wells close to highways have become polluted from deicing salts.

A particular form of chemical pollution of surface water goes under the name of *acid mine drainage*. It is an important form of environmental degradation in parts of Appalachia where abandoned coal mines and strip mine workings are concentrated (Chapter 18). Ground water emerges from abandoned mines and as soil water percolating through strip mine spoil banks. This water is charged with sulfuric acid and various salts of metals, particularly of iron. The sulfuric acid is formed by reaction of water with iron sulfides, particularly the mineral pyrite (Fe_2S), which is a common constituent of coal seams. Acid of this origin in stream waters can have adverse effects on animal life. In sufficient concentrations it is lethal to certain species of fish and has at times caused massive fish kills. The acid waters also cause corrosion to boats and piers. Government sources have estimated that nearly 6000 mi (10,000 km) of streams in this country, together with almost 40 sq mi (100 sq km) of reservoirs and lakes are seriously affected by acid mine drainage. One particularly undesirable by-product of acid mine drainage is precipitation of iron to form slimy red and yellow deposits in stream channels.

Sulfuric acid may also be produced from drainage of mines from which sulfide ores are being extracted, and from tailings produced in plants where the ores are processed. Such chemical pollution also includes salts of various toxic metals, among them zinc, lead, arsenic, copper, and aluminum. Yet another source of chemical pollution of streams is by radium derived from the tailings of uranium ore processing plants.

Toxic metals, among them mercury, along with pesticides and a host of other industrial chemicals are introduced into streams and lakes in quantities that are locally damaging or lethal to plant and animal communities. In addition, sewage introduces live bacteria and viruses that are classed as biological pollutants; these pose a threat to health of Man and animals.

Thermal pollution is a term applied generally to the discharge of heat into the environment from combustion of fuels and from nuclear energy conversion into electric power. We have described thermal pollution of the atmosphere and its effects in Chapter 4. Thermal pollution of water is different in its environmental effects because it takes the form of heavy discharges of heated water locally into streams, estuaries, and lakes. The thermal environmental impact may thus be quite drastic in a small area.

It is appropriate here to point out that the forms of pollution we have listed here often do severe damage to the esthetic qualities of landscapes of streams and lakes.

Fresh water as a natural resource

Fresh water is a basic natural resource essential to Man's varied and intense agricultural and industrial activities. Runoff held in reservoirs behind dams provides water supplies for great urban centers, such as New York City and Los Angeles; diverted from large rivers, it provides irrigation water for highly productive lowlands in arid lands, such as the Imperial Valley of California and the Nile Valley of Egypt. To these uses of runoff are added hydroelectric power, where the gradient of a river is steep, or routes of inland navigation, where the gradient is gentle.

Unlike ground water, which represents a large water storage body, fresh surface water in the liquid state is stored only in small quantities. (An exception is the Great Lakes system.) Referring back to Figure 9.3, note that the quantity of available ground water is about 30 times as large as that stored in freshwater lakes, while the water held in streams is only about 1/100 that in lakes. Because of small natural storage capacities, surface water can be drawn only at a rate comparable with its annual renewal through precipitation. Dams are built

to develop useful storage capacity for surplus runoff that would otherwise escape to the sea but, once the reservoir has been filled, water use must be scaled to match the natural supply rate averaged over the year. Development of surface water supplies brings on many environmental changes, both physical and biological, and these must be taken into account in planning for future water developments.

Studies of the total United States water resource have led to the conclusion that for the nation as a whole there is an adequate freshwater supply for at least several decades into the future. For individual regions of the country, however, we can anticipate severe shortages that can be met only by transfer of water from regions of surplus. The enormous concentration of persons and industry into a small area as, for example, Los Angeles County and the New York City region, require development of surface water facilities in distant uplands, along with elaborate water transfer systems using expensive and vulnerable aqueducts. It is only by continual and costly expansion of these facilities that shortages can be avoided. Compounding distribution problems are those of water pollution. If the quality of our fresh water supplies continues to be degraded, it will be necessary to resort more and more to expensive water treatment procedures necessary to keep the water usable.

In recent years, much emphasis has been placed on alternatives to massive water transfers to meet rising demands. The basic alternative is a reduction in water use, particularly reductions in wasteful use in urban systems and in the huge demands for irrigation in arid lands. It is obvious that problems of water resource management involve a broad spectrum of economic, political, and cultural factors.

Landforms Made by Running Water

Chapter 20

IN THIS CHAPTER we will deal with the activity of overland flow and stream flow as agents of landform sculpture. But, before we embark on our study of the work of the fluid agents, it is important to look at landforms in general in broad perspective. The configuration of continental surfaces reflects the balance of power, so to speak, between internal earth forces, acting through volcanic and tectonic processes, and external forces, acting through the agents of denudation. Seen in this perspective, landforms in general fall into two basic categories.

Landforms produced directly by volcanic and tectonic activity are _initial landforms_ (Figure 20.1). We have investigated these landforms in Chapter 17. They include volcanoes and lava flows, fault blocks and rift valleys, and mountains of alpine structure elevated in active zones of collision between lithospheric plates. The energy for lifting molten rock and rigid crustal masses to produce the initial landforms has an internal heat source, most probably natural radioactivity in rock of the crust and mantle.

figure 20.1
Initial and sequential landforms.

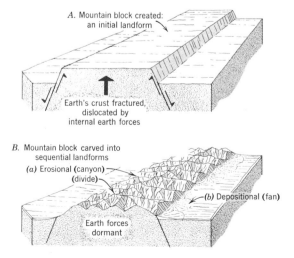

A. Mountain block created: an initial landform

Earth's crust fractured, dislocated by internal earth forces

B. Mountain block carved into sequential landforms
(a) Erosional (canyon)
(divide)
(b) Depositional (fan)
Earth forces dormant

Landforms shaped by processes and agents of denudation belong to the class of _sequential landforms,_ meaning that they follow in sequence after the initial landforms are created and the landmass has been raised to its elevated position. As shown in Figure 20.1, a single uplifted crustal block raised by faulting (itself an initial landform) is set on by agents of denudation and carved up into a large number of sequential landforms.

Any landscape is really nothing more than the existing stage in a great contest. As lithospheric plates collide or rift apart, the internal earth forces spasmodically elevate parts of the crust to create initial landforms. The external agents persistently keep wearing these masses down and carving them into vast numbers of smaller sequential landforms.

All-stages of this struggle can be seen in various parts of the world. High alpine mountains and volcanic chains exist where the internal earth forces have recently dominated. Rolling plains of the continental shields reflect the ultimate victory of agents of denudation. All intermediate stages can be found. Because the internal earth forces act repeatedly, new landmasses keep coming into existence as old ones are subdued.

Fluvial processes and landforms

Landforms shaped by running water are conveniently described as _fluvial landforms_ to distinguish them from landforms made by the other fluid agents — glacial ice, wind, and waves. Fluvial landforms are shaped by the _fluvial processes_ of overland flow and stream flow. Weathering and the slower forms of mass wasting, such as soil creep, operate hand in hand with overland flow, and cannot be separated from the fluvial processes.

Fluvial landforms and fluvial processes dominate the continental land surfaces the world over. Throughout geologic history, glacial ice has been present only in comparatively small global areas located in the polar zones and in high mountains. Landforms made by wind action occupy only trivially small parts of the continental surfaces, while landforms made by waves and currents are restricted to a very narrow contact zone between oceans and continents. In terms of area the fluvial landforms are dominant in the environment of terrestrial life and are the major source areas of Man's food resources through the practice of agriculture. Almost all lands in crop cultivation and almost all grazing lands have been shaped by fluvial processes.

Fluvial processes perform the geological activities of erosion, transportation, and deposition. Consequently, there are two major groups of fluvial landforms: erosional landforms and depositional landforms (Figure 20.2). Valleys are formed where rock is eroded away by fluvial agents. Between the valleys are ridges, hills, or mountain summits representing unconsumed parts of the crustal block. All such sequential landforms shaped by progressive removal of the bedrock mass are *erosional landforms*.

Rock and soil fragments that are removed from the parent mass are transported by the fluid agent and deposited elsewhere to make an entirely different set of surface features, the *depositional landforms*. Figure 20.2 illustrates

the two groups of landforms. The ravine, canyon, peak, spur, and col are erosional landforms; the fan, built of rock fragments below the mouth of the ravine, is a depositional landform. The floodplain, built of material transported by a stream, is also a depositional landform.

Normal and accelerated slope erosion

Fluvial action starts on the uplands of drainage basins. Overland flow, by exerting a dragging force over the soil surface, picks up particles of mineral matter ranging in size from fine colloidal clay to coarse sand or gravel, depending on the speed of the flow and the degree to which the particles are bound by plant rootlets or held down by a mat of leaves. Added to this solid matter is dissolved mineral matter in the form of ions produced by acid reactions or direct solution. This slow removal of soil is part of the natural geological process of landmass denudation; it is both inevitable and universal. Under stable natural conditions, the erosion rate in a humid climate is slow enough that a soil with distinct horizons is formed and maintained, enabling plant communities to maintain themselves in a stable equilibrium. Soil scientists refer to this state of activity as the *geologic norm*.

By contrast, the rate of soil erosion may be enormously speeded up by Man's activities or by rare natural events to result in a state of *accelerated erosion*, removing the soil much faster than it can be formed. This condition comes about most commonly from a forced change in the plant cover and in the physical state of the ground surface and uppermost soil horizons. Destruction of vegetation by clearing of land for cultivation or by forest fires sets the stage for a series of drastic changes. Interception of rain by foliage is ended; protection afforded by a ground cover of fallen

figure 20.2
Erosional and depositional landforms.

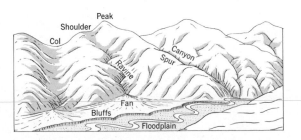

figure 20.3

A large raindrop (above) lands on a wet soil surface, producing a miniature crater (below). Grains of clay and silt are thrown into the air and the soil surface is disturbed. (Official U.S. Navy photograph.)

leaves and stems is removed. Consequently, the rain falls directly on the mineral soil.

Direct force of falling drops (Figure 20.3) causes a geyserlike splashing in which soil particles are lifted and then dropped into new positions, a process termed *splash erosion*. It is estimated that a torrential rainstorm has the ability to disturb as much as 100 tons of soil per acre (225 metric tons per hectare). On a sloping ground surface, splash erosion shifts the soil slowly downhill. A more important effect is to cause the soil surface to become much less able to infiltrate water because the

natural soil openings become sealed by particles shifted by raindrop splash. Reduced infiltration permits a much greater proportion of overland flow to occur from a given amount of rain. Increased overland flow intensifies the rate of soil removal.

Another effect of destruction of vegetation is to reduce greatly the resistance of the ground surface to the force of erosion under overland flow. On a slope covered by grass sod, even a deep layer of overland flow causes little soil erosion because the energy of the moving water is dissipated in friction with the grass stems, which are tough and elastic. On a heavily forested slope, countless check dams made by leaves, twigs, roots, and fallen tree trunks take up the force of overland flow. Without such vegetative cover the eroding force is applied directly to the bare soil surface, easily dislodging the grains and sweeping them downslope.

We can get an appreciation of the contrast between normal and accelerated erosion rates by comparing the quantity of sediment derived from cultivated surfaces with that derived from naturally forested or reforested surfaces within a given region in which climate, soil, and topography are fairly uniform. *Sediment yield* is a technical term for the quantity of sediment removed by overland flow from a unit area of ground surface in a given unit of time. Yearly sediment yield is stated in tons per acre, or metric tons per hectare.

Figure 20.4 gives data of annual average sediment yield and runoff by overland flow from several types of upland surface in northern Mississippi. Notice that both surface runoff and sediment yield decrease greatly with increased effectiveness of the protective vegetative cover. Sediment yield from cultivated land undergoing accelerated erosion is over twenty times greater than from pine plantation land. The reforested land has a sediment yield rate representing the geologic norm of soil erosion for this region; it is about

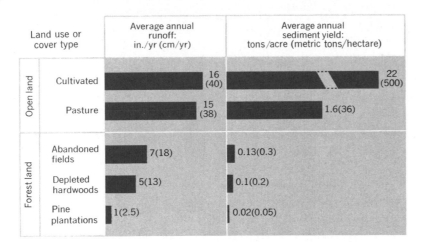

Land use or cover type		Average annual runoff: in./yr (cm/yr)	Average annual sediment yield: tons/acre (metric tons/hectare)
Open land	Cultivated	16 (40)	22 (500)
	Pasture	15 (38)	1.6(36)
Forest land	Abandoned fields	7(18)	0.13(0.3)
	Depleted hardwoods	5(13)	0.1(0.2)
	Pine plantations	1(2.5)	0.02(0.05)

figure 20.4
The bar graph shows that both runoff and sediment yield are much greater for open land than for land covered by shrubs and forest. Cultivated land has an enormous sediment yield, as compared with any of the other types. (Data of S. J. Ursic, 1965, Dept. of Agriculture.)

the same as for mature pine and hardwood forests that have not experienced cultivation.

The distinction between normal and accelerated slope erosion applies to regions in which the water balance shows an annual surplus. Under a semiarid climate with summer drought the natural plant cover consists of grasses and shrubs. Although sparse and providing rather poor ground cover of plant litter, it is strong enough that the geologic norm of erosion can be sustained. However, in these semiarid environments, the sparse plants provide only a minimum of protection to the soil surface. Consequently, the natural equilibrium is highly sensitive to upset. Depletion of plant cover by fires or the grazing of herds of domesticated animals can easily set off rapid erosion. These sensitive, marginal environments require cautious use, because they lack the potential to recover rapidly from accelerated erosion, once it has set in.

Erosion at a very high rate by overland flow is actually a natural geological process in certain favorable localities in semiarid and arid lands; it takes the form of *badlands*. One well-known area of badlands is the Big Badlands of South Dakota, along the White River. Badlands are underlain by clay formations, easily eroded by overland flow. Erosion rates are too fast to permit plants to take hold, and no soil can

develop. A maze of small stream channels is developed; ground slopes are very steep (Figure 20.5). Badlands such as these are self-sustaining and have been in existence at one place or another on continents throughout much of geologic time.

figure 20.5
Badlands, such as these in the Petrified Forest National Monument, Arizona, are like miniature mountains on bare clay formations. (B. Mears, Jr.)

Forms of accelerated soil erosion

Humid regions of substantial water surplus, for which the natural plant cover is forest or prairie grasslands, experience accelerated soil erosion when Man expends enough energy to remove the plant cover and keep the land barren by annual cultivation. With fossil fuels to power machines of plant and soil destruction, Man has easily overwhelmed the restorative forces of nature over vast expanses of continental surfaces. We now consider the consequences of these activities.

When a plot of ground is first cleared of forest and plowed for cultivation, little erosion will occur until the action of rain splash has broken down the soil aggregates and sealed the larger openings. Overland flow then begins to remove the soil in rather uniform thin layers, a process termed _sheet erosion_. Because of seasonal cultivation, the effects of sheet erosion are often little noticed until the upper horizons of the soil are removed or greatly thinned. Reaching the base of the slope, where the surface slope is sharply reduced to meet the valley bottom, soil particles come to rest and accumulate in a thickening layer termed _slope wash_. This deposit, too, has a sheetlike distribution and may be little noticed, except where it can be seen that fence posts or tree trunks are being slowly buried.

Material that continues to be carried by overland flow to reach a stream in the valley axis is then carried farther down valley and may accumulate as alluvium in layers on the valley floor. _Alluvium_ is a word applied generally to any stream-laid deposit. Deposition of alluvium results in burial of fertile soil under infertile, sandy layers. The alluvium chokes the channels of small streams, causing the water to flood broadly over the valley bottoms (Figure 20.6).

Where slopes are steep, runoff from torrential rains produces a more intense activity, _rill erosion,_ in which innumerable, closely spaced channels are scored into the soil and regolith (Figure 20.7). If these rills are not destroyed by soil tillage, they may soon begin to integrate into still larger channels, called _gullies_. Gullies are steep-walled, canyonlike trenches whose upper ends grow progressively upslope (Figure 20.8). Ultimately, a rugged, barren topography, like the badland forms of the dry climates,

figure 20.6
This thick layer of alluvium has choked the floor of a small valley. The silt and sand were carried by a stream in flood, draining cultivated fields highly susceptible to soil erosion. (Soil Conservation Service, U. S. Department of Agriculture)

figure 20.7
Shoestring rills on a barren slope. (Soil Conservation Service.)

results from accelerated soil erosion that is allowed to proceed unchecked.

The natural soil with its well-developed horizons is a nonrenewable natural resource. The rate of soil formation is extremely slow in

figure 20.8
This great gully, eroded into deeply weathered regolith, was typical of certain parts of the Piedmont region of South Carolina and Georgia before remedial measures were applied. (Soil Conservation Service.)

comparison with the rate of its destruction once accelerated erosion has begun and is allowed to go unchecked. Soil erosion as a potentially disastrous form of environmental degradation was brought to public attention decades ago in the United States.

Curative measures developed by the Soil Conservation Service have proved effective in stopping accelerated soil erosion and permitting the return to slow erosion rates approaching the geologic norm. These measures include construction of terraces to eliminate steep slopes, permanent restoration of overly steep slope belts to dense vegetative cover, and the healing of gullies by placing check dams in the gully floors.

Geologic work of streams

The geologic work of streams consists of three closely interrelated activities: erosion, transportation, and deposition. _Erosion_ by a stream is the progressive removal of mineral material from the floor and sides of the channel, whether this be carved in bedrock or regolith. _Transportation_ consists of movement of the eroded particles by dragging along the bed, by suspension in the body of the stream, or in solution. _Deposition_ is the accumulation of transported particles on the stream bed and floodplain, or on the floor of a standing body of water into which the stream empties. Obviously, erosion cannot occur without some transportation taking place, and the transported particles must eventually come to rest. Erosion, transportation, and deposition are simply three phases of a single activity.

Stream erosion

Streams erode in various ways, depending on the nature of the channel materials and the tools with which the current is armed. The force of the flowing water alone, exerting impact and a dragging action on the bed and banks, can erode poorly consolidated alluvial materials such as gravel, sand, silt, and clay. This erosion process, called *hydraulic action,* is capable of excavating enormous quantities of unconsolidated materials in a short time (Figure 20.9). The undermining of the banks causes large masses of alluvium to slump into the river where the particles are quickly separated and become a part of the stream's load. This process of bank caving is an important source of sediment during high river stages.

Where rock particles carried by the swift current strike against bedrock channel walls, chips of rock are detached. The rolling of cobbles and boulders over the stream bed will further crush and grind smaller grains to produce an assortment of grain sizes. These processes of mechanical wear constitute *abrasion;* it is the principal means of erosion in bedrock too strong to be affected by simple hydraulic action (Figure 20.10).

Finally, the chemical processes of rock weathering — acid reactions and solution — are effective in removal of rock from the stream channel and may be referred to as *corrosion.* Effects of corrosion are conspicuous in limestone, which develops cupped and fluted surfaces.

Stream transportation

The solid matter carried by a stream is its *load,* carried in three forms. Dissolved matter is transported invisibly in the form of chemical ions. All streams carry some dissolved salts resulting from mineral alteration. Clay and silt are carried in *suspension;* that is, they are held up in the water by the upward elements of flow in turbulent eddies in the stream. This fraction of the transported matter is the *suspended load.* Sand, gravel, and cobbles move as *bed load* close to the channel floor by rolling or sliding and an occasional low leap.

The load carried by a stream varies enormously in the total quantity present and the size of the fragments, depending on the discharge and stage of the river. In flood, when velocities are high, the water is turbid with suspended load.

Suspended loads of large rivers

A large river such as the Mississippi carries most of its load — about 90 percent — in suspension. Most of the suspended load comes from its great western tributary, the Missouri River, which is fed from semiarid lands, including the Dakota Badlands.

The Yellow River (Hwang Ho) of China heads the world list in annual suspended sediment load, while its watershed sediment yield is one

figure 20.9

A river in flood eroded this huge trench at Cavendish, Vermont in November 1927. An area 1 mi (1.6 km) wide and 3 mi (5 km) long, once occupied by eight farms, was cut away by the flood waters. Damage was great because the material consisted of sand and gravel, which offered little resistance. (Wide World Photos.)

(right) **Falls of the Yellowstone River, Yellowstone National Park, Wyoming.** The gorge is carved into volcanic rocks, colored through mineral alteration by rising hydrothermal solutions. (Orlo E. Childs)

(below) **The American Falls of Niagara Falls, New York.** A great mass of broken limestone lies at the base of the falls. (Orlo E. Childs)

(left) **Potholes worn in granite, Sand Creek, Wyoming.** (Arthur N. Strahler)

Gorges and Falls

(below) **Grand Canyon of the Colorado River at Toroweap, Arizona.** The view is downstream, toward the west. Basaltic lavas have poured down the canyon walls at the right; a cinder cone lies at the left. The narrow gorge is about 3000 ft (900 m) deep at this point. (Donald L. Babenroth)

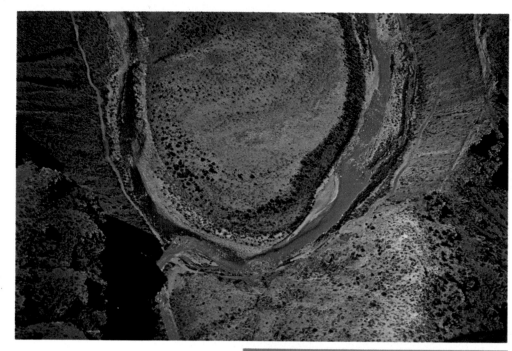

(left) **In this vertical air view of the gorge of the Salt River, Arizona, the turbid storm runoff of a small tributary contrasts sharply with the clear water of the main stream. (Mark A. Melton)**

Stream Channels

(left) **Green irrigated areas of floodplain (right) and low alluvial terrace (left) border a small meandering river, fed from mountain snowmelt, near Greybull, Montana. (Arthur N. Strahler)**

(above) **Rounded cobbles of volcanic rock comprise the bed load of this ephemeral desert stream in Arizona. (Arthur N. Strahler)**

(right) **Deep branching gullies have carved up an overgrazed pasture near Shawnee, Oklahoma. Contour terracing and check dams have halted the headward growth of the gullies. (Mark A. Melton)**

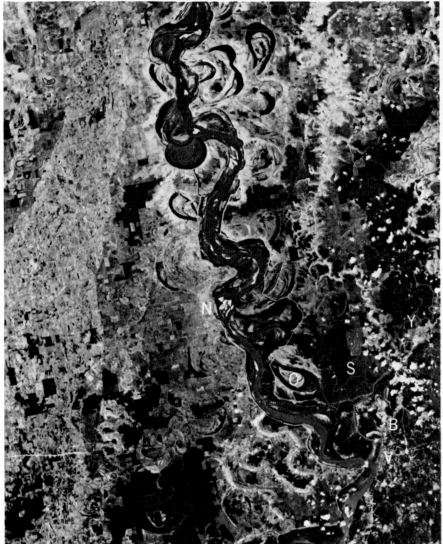

(above) Part of an oxbow lake from an abandoned section of the Mississippi River near Mound Bayou, Louisiana. Soil color differences in the cultivated area reveal old bars and swales. (Arthur N. Strahler)

Mississippi River

(below) Cultivated natural levees (right) and forested backswamp (background) of the Mississippi River, just south of New Orleans, Louisiana. An artificial levee runs close to the river channel. (Orlo E. Childs)

(above) Apollo 9 infrared photograph of the Mississippi River north of Vicksburg, Mississippi. The area shown is about 60 mi (95 km) wide. Oxbow lakes and marshes mark old cutoffs of meander bends. Lighter areas close to the river are higher natural levees, well-drained and cultivated. Dark patches are heavily forested backswamp areas and abandoned channels, lower in elevation than the levees. Vicksburg, at the lower right, is situated where the river impinges upon the higher eastern bluffs, formed of loess. Here the Yazoo River enters from the north. Older floodplain covers the left side of the area. (NASA —
No. AS9-26A-3741) V—Vicksburg
O—oxbow lake Y—Yazoo River
S—backswamp B—bluffs
N—natural levee

Desert Fluvial Processes

(right) **This barren desert landscape near Death Valley, California, is dominated by knife-edged divides and V-shaped ravines carved by running water. (Mark A. Melton)**

(below) **Bedrock is exposed in places on this pediment, carved upon ancient schist. Fragments of schist form a desert pavement. Cabeza Prieta Range, Arizona. (Mark A. Melton)**

(left) **A flash flood fills a desert stream channel with raging, turbid waters. A distant thunderstorm produced the runoff. Coconino Plateau, Arizona. (Arthur N. Strahler)**

(left) **Badlands are sustained on weak clay formations where no protective plant cover can gain hold. Mancos formation, Henry Mountains region, Utah. (Alan H. Strahler)**

(above) **In this air view across the Harquehala Mountains, Arizona, broad rock pediments slope gently away from the mountains toward a central zone of alluvium. (Mark A. Melton)**

(left) **Great alluvial fans extend out upon the floor of Death Valley, California. (Mark A. Melton)**

figure 20.10
Potholes carved in granite bear witness to strong abrasion in the channel of the James River, Virginia. (U.S. Geological Survey.)

summer rains sweeps up a large amount of sediment.

Most investigators generally agree that land cultivation has greatly increased the sediment load of rivers of Asia, Europe, and North America. The increase because of Man's activities is thought to be greater by a factor of 2½ than the geologic norm for the entire world land area. For the more strongly affected river basins, the factor may be 10 or more times larger than the geologic norm. This form of environmental impact has attracted little attention because it has accumulated over many centuries of agricultural development.

of the highest known for a large river basin. The explanation lies in a high soil-erosion rate on intensively cultivated upland surfaces of wind-deposited silt (loess) in Shensi and Shansi provinces (Figure 20.11). Much of the drainage area is in a semiarid climate with dry winters; vegetation is sparse, and the runoff from heavy

Capacity of a stream to transport load

The maximum solid load of debris that can be carried by a stream at a given discharge is a measure of the stream's *capacity*. Load is stated as the weight of material moved through the stream cross section in a given unit of time,

figure 20.11
Deeply eroded windblown silt (loess) in Shansi Province, China. Steep slopes have been terraced for cultivation, arresting the extension of gully heads. Nevertheless, this region leads the world in sediment yield. (U.S. Geological Survey.)

commonly in units of tons per day. Total solid load includes both the bed load and the suspended load.

Where a stream is flowing in a channel of hard bedrock, it may not be able to pick up enough alluvial material to supply its full capacity for bed load. When a flood occurs, the channel cannot be quickly deepened in response. Such conditions exist in streams occupying deep gorges and having steep gradients. Conditions are different where thick layers of silt, sand, and gravel underlie the channel. As the speed of flow increases, the stream easily picks up and sets in motion all of the material that it is capable of moving. In other words, the increasing capacity of the stream for load is easily satisfied.

Capacity for load increases sharply with an increase in the stream's velocity, because the swifter the current the more intense is the turbulence and the stronger is the dragging force against the bed. Capacity to move bed load goes up about as the third to fourth power of the velocity. This means that when a stream's velocity is doubled in flood, its ability to transport bed load is increased from eight to sixteen times. It is no wonder, then, that most of the conspicuous changes in the channel of a stream occur in flood stage, with few important changes occurring in low stages.

Stream gradation

A trunk stream, fully developed within its drainage basin, has undergone a long period of adjustment of its channel so that it can discharge not only the surplus water produced by the basin but also the solid load that the tributary channels supply. The transport of bed load requires a gradient. A stream channel system adjusts its gradient to achieve an average balanced state of operation, year in and year out and from decade to decade. In this equilibrium condition the stream is said to be *graded*.

It will be helpful in developing the concept of a graded stream system to investigate the changes that will take place along a stretch of stream that is initially poorly adjusted to the transport of its load. Such an ungraded channel is illustrated in Figure 20.12. The starting profile is imagined to be brought about by crustal uplift in a series of fault steps, bringing to view a surface that was formerly beneath the ocean and exposing it to fluvial processes for the first time. Overland flow collects in shallow depressions, which fill and overflow from higher to lower levels. In this way a through-going channel originates and begins to conduct runoff to the sea.

Figure 20.13 illustrates the gradation process in a series of block diagrams. Over *waterfalls* and

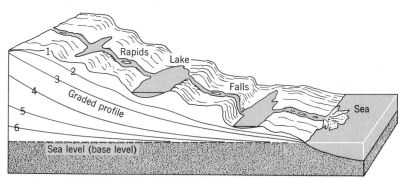

figure 20.12
Schematic diagram of gradation of a stream. Originally the channel consists of a succession of lakes, falls, and rapids. (From A. N. Strahler, 1972, _Planet Earth_, Harper & Row, New York.)

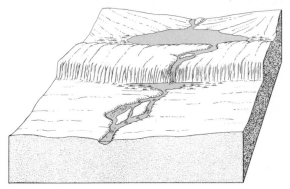

A. Stream established on a land surface dominated by landforms of recent tectonic activity.

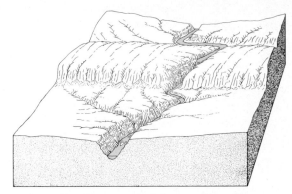

B. Gradation in progress; lakes and marshes drained; deepening gorge; tributary valleys extending.

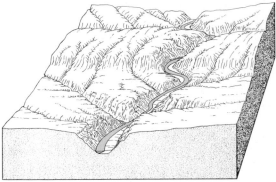

C. Graded profile attained; beginning of floodplain development; valley widening in progress.

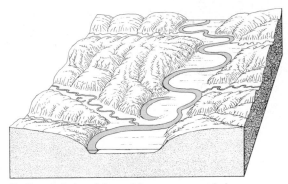

D. Floodplain widened to accommodate meanders; floodplains extended up tributary valleys.

figure 20.13
Evolution of a stream and its valley. (After E. Raisz.)

rapids, which are simply steep-gradient portions of the channel, flow velocity is greatly increased and abrasion of bedrock is most intense (Figure 20.13*A*). As a result, the falls are cut back and the rapids trenched, while the ponded reaches are filled by sediment and are lowered in level as the outlets are cut down. In time the lakes disappear and the falls are transformed into rapids. Erosion of rapids reduces the gradient to one more closely approximating the average gradient of the entire stream (Figure 20.13*B*). At the same time, branches of the stream system are being extended into the landmass, carving out a drainage basin and transforming the original landscape into a fluvial landform system.

In the early stages of gradation and extension, the capacity of the stream exceeds the load supplied to it, so that little or no alluvium accumulates in the channels. Abrasion continues to deepen the channels, with the result that they come to occupy steep-walled *gorges* or *canyons* (Figure 20.14). Weathering and mass wasting of these rock walls contributes an increasing supply of rock debris

figure 20.14
The Royal Gorge of the Arkansas River in the Colorado Rockies illustrates the canyon of a river with a steep gradient. Seen above is a suspension bridge, 1053 ft (321 m) above river level. (Josef Muench.)

to the channels. Debris shed from land surfaces contributing overland flow to the newly developed branches is also on the increase.

We can anticipate a gradual decrease in the stream's capacity for bed load, resulting from the gradual reduction in the channel gradient. This decrease will be converging on the increasing load with which it is being supplied. There will come a point in time at which the supply of load exactly matches the stream's capacity to transport it. At this point the stream has achieved the graded condition and possesses a *graded profile,* descending smoothly and uniformly in the downstream direction (Figure 20.12).

It is important to understand that the balance between load and a stream's capacity exists only as an average condition over periods of

many years. We are aware that streams scour their channels in flood and deposit load when in falling stage. Perhaps in terms of conditions of the moment, a stream is rarely in equilibrium; but, over long periods of time, the graded stream maintains its level. After attaining this state of balance, the stream continues to cut sidewise into its banks on the outsides of bends. This *lateral cutting* does not appreciably alter the gradient and therefore does not materially affect the equilibrium.

The first indication that a stream has attained a graded condition is the beginning of floodplain development. On the outside of a bend the channel shifts laterally into a curve of larger radius and thus undercuts the valley wall. On the inside of the bend alluvium accumulates in the form of a sand bar. Widening of the bar deposit produces a crescentic piece of low ground, which is the first stage in flood plain development. This stage is illustrated in Figure 20.13C. As lateral cutting continues, the floodplain strips are widened, and the channel develops sweeping bends, called *meanders* (Figure 20.13D). The floodplain is then widened into a continuous belt of flat land between steep valley walls.

Floodplain development reduces the frequency with which channel scour attacks and

figure 20.15
Following stream gradation the valley walls become more gentle in slope and the bedrock is covered by soil and weathered rock. (After W. M. Davis.)

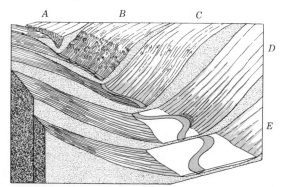

undermines the adjacent valley wall. Weathering, mass wasting, and overland flow then act to reduce the steepness of the valley-side slopes (Figure 20.15). As a result, in a humid climate, the gorgelike aspect of the valley gradually disappears and eventually gives way to an open valley with soil-covered slopes protected by a dense plant cover.

Environmental significance of gorges and waterfalls

Deep gorges and canyons of major rivers exert environmental controls in a variety of ways. There is little or no room for roads or railroads between the stream and the valley sides, so that road beds must be cut or blasted at great expense and hazard from the sheer rock walls. Maintenance is expensive because of flooding or undercutting by the stream and the sliding and falling of rock, which can damage or bury the road bed. Yet a gorge may afford the only feasible passage through a mountain range. The Royal Gorge of the Arkansas River, in the Rocky Mountain Front Range of southern Colorado, is a striking example (Figure 20.14). This illustration also brings out the point that a deep canyon is a barrier to movement across it and may require construction of expensive bridges.

Although small waterfalls are common features of alpine mountains carved by glacial erosion (Chapter 22), large waterfalls on major rivers are comparatively rare the world over. New river channels resulting from flow diversions caused by ice sheets of the Pleistocene Epoch provide one class of falls and rapids of large discharge. Certainly the preeminent example is Niagara Falls (Figure 20.16). Overflow of Lake Erie into Lake Ontario happened to be situated over a gently inclined layer of limestone, beneath which lies easily eroded shale. As the detailed diagram shows (Figure 20.17), the fall is maintained by continual undermining of the limestone by erosion in the plunge-pool at the

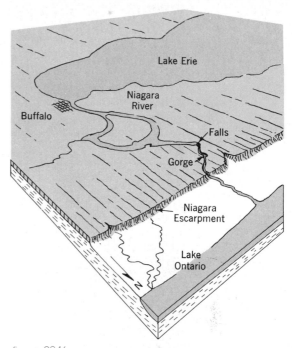

figure 20.16

A bird's-eye view of the Niagara River with its falls and gorge carved in strata of the Niagara Escarpment. View is toward the southwest from a point over Lake Ontario. Redrawn from a sketch by G. K. Gilbert, 1896. (From A. N. Strahler, 1971, The Earth Sciences, 2nd ed., Harper & Row, New York.)

base of the fall. The height of the falls is now 170 ft (52 m) and its discharge about 200,000 cfs (17,000 cms). The drop of Niagara Falls is utilized for the production of hydroelectric power by the Niagara Power Project, in which water is withdrawn upstream from the falls and carried in tunnels to generating plants located 4 mi (6.4 km) downstream from the falls. Capability of this project is 2400 megawatts of power, making it the largest single producer in the western hemisphere.

Most large rivers of steep gradient do not possess falls, and it is necessary to build dams to create artificially the vertical drop necessary for turbine operation. An example is the Hoover Dam, behind which lies Lake Hoover (formerly Lake Mead), occupying the

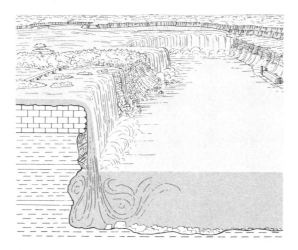

figure 20.17
Niagara Falls is formed where the river passes over the eroded edge of a massive limestone layer. Continual undermining of weak shales at the base keeps the fall steep. (After G. K. Gilbert and E. Raisz.)

canyon of the Colorado River. With a dam height of 726 ft (220 m) the generating plant of Hoover Dam is capable of producing 1345 megawatts of power, about half as much as the Niagara Project.

We should not lose sight of the esthetic and recreational values of gorges, rapids, and waterfalls of major rivers. The Grand Canyon of the Colorado River, probably more than any single product of fluvial processes, highlights the scenic value of a great river gorge. Against the advantages in obtaining hydroelectric power and fresh water for urban supplies and irrigation by construction of large dams, we must weigh the permanent loss of some large segments of our finest natural scenery, along with the destruction of ecosystems adapted to the river environment. It is small wonder, then, that new dam projects are meeting with stiff opposition from concerned citizen groups who fear that the harmful environmental impacts of such structures far outweigh their future benefits.

Goal of fluvial denudation

With the concept of a graded stream in mind, this is a good place to expand our thinking to take in some of the broader aspects of fluvial denudation. Consider a landscape made up of many drainage basins and their branching stream networks. The region is rugged, with steep mountainsides and high, narrow crests (Figure 20.18A). Rock debris is being transported out of each drainage basin by the graded trunk streams at the same average rate as the debris is being contributed from the land surfaces within the basin. Obviously, this

figure 20.18
Reduction of a landmass by fluvial erosion. (A) In early stages, relief is great, slopes are steep, and the rate of erosion is rapid. (B) In an advanced stage, relief is greatly reduced, slopes are gentle, and rate of erosion is slow. Soils are thick over the broadly rounded hill summits. (C) After many millions of years of fluvial denudation, a peneplain is formed. Slopes are very gentle and the landscape is an undulating plain. Floodplains are broad, and stream gradients are extremely low. All of the land surface lies close to base level.

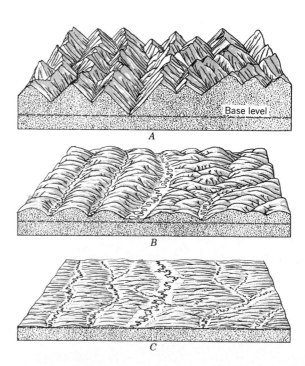

equilibrium cannot be sustained, since the export of debris must lower the land surface generally. This being the case, the average altitude of the land surface is steadily declining, and this decline must be accompanied by a reduction in the average gradients of all streams (Figure 20.18*B*). In essence, the fluvial system is consuming its own mass. The mass of rock lying above sea level and available for removal by fluvial action is called the *landmass*.

To describe the ensuing events briefly, as time passes, the streams and valley-side slopes of the drainage basins must undergo gradual change to lower gradients. In theory, the ultimate goal of the denudation process is a reduction of the landmass to a featureless plain at sea level. In this process, sea level projected beneath the landmass represents the lower limiting level, or *base level*, of the fluvial denudation (labeled on Figures 20.12 and 20.18). But, because the rate of denudation becomes progressively slower, the land surface approaches the base level surface of zero elevation at a slower and slower pace. Under this program, the ultimate goal can never be

reached. Instead, after the passage of some millions of years, the land surface is reduced to an undulating surface of low elevation, called a *peneplain* (Figure 20.18*C*). This term was coined by an American geographer, W. M. Davis, by combining the words ''penultimate'' and ''plain.''

Production of a peneplain requires a high degree of crustal and sea-level stability for a period of many millions of years. One region that has been cited as a possible example of contemporary peneplain is the Amazon-Orinoco basin of South America. Peneplains that have been uplifted after their development and are now high-standing land surfaces are numerous within the continental shields. These regions are characterized by an upland surface of uniform elevation. The upland is trenched by stream valleys graded with respect to a lower base level. Figure 20.19 shows a landscape interpreted as an uplifted peneplain deeply trenched by a stream valley. Notice how the gently undulating upland surface, used for agriculture, contrasts with the steep canyon walls clothed in rainforest.

figure 20.19
The gently rolling upland surface in the distance is part of the St. John peneplain; its elevation is about 2000 ft (600 m). The peneplain is deeply trenched by Canyon de San Cristobal (foreground), carved by the Rio Usabon. (U.S. Geological Survey.)

Aggradation and alluvial terraces

A graded stream, delicately adjusted to its supply of water and rock waste from upstream sources, is highly sensitive to changes in those inputs. Changes in climate and in surface characteristics of the watershed bring changes in discharge and load at downstream points, and these changes in turn require channel readjustments.

Consider first the effect of an increase in bed load beyond the capacity of the stream. At the point on a channel where the excess load is introduced, the coarse sediment accumulates on the stream bed in the form of bars of sand, gravel, and pebbles. These deposits raise the elevation of the stream bed, a process called *aggradation*. As more bed materials accumulate the stream channel gradient is increased; the increased flow velocity enables bed materials to be dragged downstream and spread over the channel floor at progressively more distant downstream reaches.

Aggradation typically changes the channel cross section from one of narrow and deep form to a wide, shallow cross section. Because bars are continually being formed, the flow is divided into multiple threads, and these rejoin and subdivide repeatedly to give a typical *braided channel* (Figure 20.20). The coarse channel deposits spread across the former floodplain, burying fine-textured alluvium with coarse material. We have already seen that alluvium accumulates as a result of accelerated soil erosion.

How is aggradation induced in a stream system by natural processes? Figure 20.20 illustrates one natural cause of aggradation that has been of major importance in stream systems of North America and Eurasia during the Pleistocene Epoch. Advance of a valley glacier has resulted in the input of a large quantity of coarse rock debris at the head of the valley. Valley aggradation was widespread in a broad

figure 20.20

The braided stream in the foreground is aggrading the floor of a glacial trough. A shrunken glacier in the distance provides the meltwater and debris, Peters Creek, Chugacn Mountains, Alaska. (Steve McCutcheon, Alaska Pictorial Service.)

zone marginal to the great ice sheets of the Pleistocene Epoch; the accumulated alluvium filled most valleys to depths of many tens of feet. Figure 20.21 shows a valley filled in this manner by an aggrading stream. The case could represent any one of a large number of valleys in New England or the Middle West. Suppose, next, that the source of bed load is cut off or greatly diminished. In the case illustrated in Figure 20.21, the ice sheets have disappeared from the headwater areas and, with them, the supplies of coarse rock debris. Reforestation of the landscape has restored a protective cover to valley-side and hill slopes of the region, holding back coarse mineral particles from entrainment in overland flow. Now the streams have copious water discharges but little bed load. In other words, they are operating below capacity. The result is channel scour and deepening. The channel form becomes deeper and narrower. Gradually, the stream profile level is lowered, a process of _degradation_. Because the stream is very close to being in the graded condition at all times, its dominant activity is lateral (sidewise) cutting by growth of meander bends, as shown in Figure 20.21B. The valley alluvium is gradually excavated and carried downstream, but it cannot all be removed because the channel encounters hard bedrock in many places. Consequently, as shown in Figure 20.21C, steplike surfaces remain on both sides of the valley. The treads of these steps are _alluvial terraces_.

Alluvial terraces have always attracted occupation by Man because of their advantages over both the valley-bottom flood-plain, which is subject to annual flooding, and the hill slopes beyond, which may be too steep and rocky to cultivate. Besides, terraces were easily tilled and made prime agricultural land. Towns were easily laid out on the flat ground of a terrace, and roads and railroads were easily run along the terrace surfaces parallel with the river.

Aggradation induced by Man

Channel aggradation is a common form of environmental degradation brought on by Man's activities. Accelerated soil erosion following cultivation, lumbering, and forest fires is the most widespread source of sediment for valley aggradation. Aggradation of channels has also been a serious form of environmental

degradation in coal-mining regions, along with water pollution (acid mine drainage) mentioned in Chapter 19. Throughout the Appalachian coal fields, channel aggradation is widespread because of the huge supplies of coarse sediment from mine wastes (Figure 20.22). Strip mining has enormously increased the aggradation of valley bottoms because of the vast expanses of broken rock available to entrainment by runoff.

Urbanization and highway construction are also major sources of excessive sediment,

figure 20.21
Alluvial terraces form when a graded stream slowly cuts away the alluvial fill in its valley. The letter "R" marks points where bedrock lies at the surface, protecting the terrace above it from being undercut.

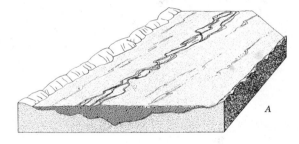

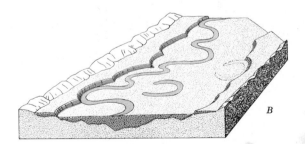

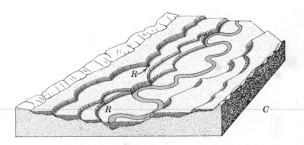

figure 20.22

This valley bottom in Kentucky is choked with coarse debris from a nearby strip mine area. The natural channel has been completely buried under the rising alluvium. (U.S. Dept. of the Interior.)

causing channel aggradation. Large earth-moving projects are involved in creating highway grades and preparing sites for industrial plants and housing developments. While these surfaces are eventually stabilized, they are vulnerable to erosion for periods ranging from months to years. The regrading involved in these projects often diverts overland flow into different flow paths, further upsetting the activity of streams in the area.

Mining, urbanization, and highway construction not only cause drastic increases in bed load, which cause channel aggradation close to the source, but also increase the suspended load of the same streams. Suspended load travels downstream and is eventually deposited in lakes, reservoirs, and estuaries far from the source areas. This sediment is particularly damaging to the bottom environments of aquatic life. The accumulation of fine sediment reduces the capacity of reservoirs, limiting the useful life of a large reservoir to perhaps a century or so. Excess sediment also results in rapid filling of tidal estuaries, requiring increased dredging of channels.

Alluvial rivers and their floodplains

An *alluvial river* is one that flows on a thick accumulation of alluvial deposits constructed by the river itself in earlier stages of its activity. A characteristic of an alluvial river is that it experiences overbank floods with a frequency ranging between annual and biennial occurrence during the season of large water surplus over the watershed. Overbank flooding of an alluvial river normally inundates part or all of a floodplain that is bounded on either side by rising slopes, called *bluffs*.

Typical landforms of an alluvial river and its floodplain are illustrated in Figure 20.23. Dominating the floodplain is the meandering river channel itself, and also abandoned reaches of former channel (Figure 20.24). Meanders develop narrow necks, which are cut through, shortening the river course and leaving a meander loop abandoned. This event is called a *cutoff*. It is quickly followed by deposition of silt and sand across the ends of the abandoned channel, producing an *oxbow lake*. The oxbow lake is gradually filled in with fine sediment brought in during high floods and with organic matter produced by aquatic plants. Eventually the oxbows are converted into swamps, but their identity is retained indefinitely.

During periods of overbank flooding, when the entire floodplain is inundated, water spreads from the main channel over adjacent floodplain deposits. As the current rapidly slackens, sand and silt are deposited in a zone adjacent to the channel. The result is an accumulation known as a *natural levee*. Because deposition is heavier closest to the channel and decreases away from the channel, the levee surface slopes away from the channel (Figure 20.23). In Figure 20.25, notice that the higher ground of the natural levees is revealed by a line of trees on either side of the channel. Between the levees and the bluffs is lower ground, the *back-swamp* (Figure 20.23).

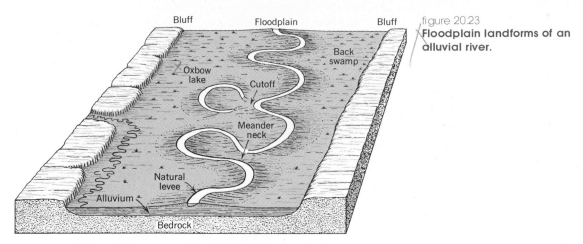

figure 20.23
Floodplain landforms of an alluvial river.

Overbank flooding results not only in the deposition of a thin layer of silt on the floodplain, but brings an infusion of dissolved mineral substances that enter the soil. As a result of the resupply of nutrient bases, floodplain soils retain their remarkable fertility in regions of soil-water surplus from which these bases are normally leached away.

figure 20.24
This vertical air photograph, taken from an altitude of about 20,000 ft (6100 m), shows meanders, cutoffs, oxbow lakes and swamps, and floodplain of the Hay River, Alberta. (Department of Energy, Mines and Resources, Canada.)

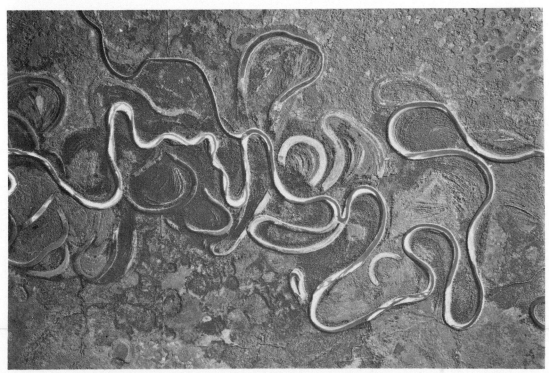

Alluvial rivers and Man

Alluvial rivers attracted Man's habitations long before the dawn of recorded history. Early civilizations arose in the period 4000 to 2000 B.C. in alluvial valleys of the Nile River in Egypt, the Tigris and Euphrates Rivers of Mesopotamia, the Indus River of what is now Pakistan, and the Yellow River of China. Fertile and easily cultivated alluvial soils, situated close to rivers of reliable flow, from which irrigation water was easily lifted or diverted, led to intense utilization of these fluvial zones. With respect to human culture, the alluvial river and its floodplain are sometimes referred to as the riverine environment. Today, about half of the world's population lives in southern and southeastern Asia; the bulk of these persons are small farmers cultivating alluvial soils of seven great river floodplains.

To the agricultural advantages of the alluvial zone are added the value of the river itself as an artery of transportation. Navigability of large alluvial rivers led to growth of towns and cities, many situated at the outsides of meander bends, where deep water lies close to the bank, or at points on the floodplain bluffs where the river is close by (e.g., Memphis, Tennessee, and Vicksburg, Mississippi).

Flood abatement measures

In the face of repeated disastrous floods, vast sums of money have been spent on a wide variety of measures to reduce flood hazards on the floodplains of alluvial rivers. The economic, social, and political aspects of flood abatement are beyond the scope of physical geography; we shall review only the engineering principles applied to the problem. Two basic forms of regulation are: first, to detain and delay runoff by various means on the ground surfaces and in smaller tributaries of the watershed; and second, to modify the lower reaches of the river where floodplain inundation is expected.

The first form of regulation aims at treatment of watershed slopes, usually by reforestation or planting of other vegetative cover to increase the amount of infiltration and reduce the rate of overland flow. This type of treatment, together with construction of many small flood-storage dams in the valley bottoms, can greatly reduce the flood crests and allow the discharge to pass into the main stream over a longer period of time.

Under the second type of flood control, designed to protect the floodplain areas directly, two quite different theories can be practiced. First, the building of *artificial levees*, parallel with the river channel on both sides, can function to contain the overbank flow and prevent inundation of the adjacent floodplain (Figure 20.26). Artificial levees are broad embankments built of earth and must be high enough to contain the greatest floods; otherwise they will be breached rapidly by great gaps at the points where water spills over (Figure 20.26). Under the control of the Mississippi River Commission, which began in 1879, a vast system of levees was built along the Mississippi River in the expectation of containing all floods. Levees have been continuously improved and now total more than 2500 mi (400 km) in length and in places are as high as 30 ft (10 m).

Because of natural and artificial levees, the river channel is slowly built up to a higher level than the floodplain, making these "bottom lands," as they are called, subject to repeated inundations and consequent heavy loss of life and property. Floodplains of many of the world's great rivers present serious problems of this type. In China, in 1887, the Hwang Ho inundated an area of 50,000 sq mi (130,000 sq km), causing the direct death of a million people and the indirect death of a still greater number through ensuing famine.

The second theory, practiced in more recent years on the Mississippi River by the U.S. Army

figure 20.25
The Wabash River in flood near Delphi, Indiana, February 1954. An ice dam clogs the river channel, while lines of trees mark the crest of the bordering natural levees. The floodplain itself is inundated on both sides of the channel and reaches to the base of the bluff, at left. (U.P.I. Telephoto.)

figure 20.26
This air view, taken in April 1952, shows a break in the artificial levee adjacent to the Missouri River in western Iowa. Water is spilling from the high river level at right to the lower floodplain level at left. (U.S. Department of Agriculture.)

Corps of Engineers, is to shorten the river course by cutting channels directly across the great meander loops to provide a more direct river flow. Shortening has the effect of increasing the river slope which, in turn, increases the velocity. Greater velocity enables a given flood discharge to be moved through a channel of smaller cross-sectional area; the flood stage is correspondingly reduced. Channel improvement, begun in the 1930s, initially had a measureable effect in reducing flood crests along the lower Mississippi, and the levees were not in such great danger of being overtopped. However, new meander bends grew rapidly and proved difficult to control.

Certain parts of the floodplain are also set aside as temporary basins into which the river is to be diverted according to plan to reduce the flood crest. In the delta region, floodways are designed to conduct flow from river channels to the ocean by alternate routes.

Fluvial processes in an arid climate

The general appearance of desert regions is strikingly different from that of humid regions, reflecting differences in both vegetation and landforms. Rain falls in dry climates as well as in moist, and most landforms of desert regions are formed by running water. A particular locality in a dry desert may experience heavy rain only once in several years but, when it does fall, stream channels carry water and perform important work as agents of erosion, transportation, and deposition. Although running water is a rather rare phenomenon in dry deserts, it works with more spectacular effectiveness on the fewer occasions when it does act. This is explained by the meagerness of vegetation in dry deserts. The few small plants that survive offer little or no protection to soil or bedrock. Without a thick vegetative cover to protect the ground and hold back the swift downslope flow of water, large quantities of coarse rock debris are swept into the streams. A dry channel is transformed in a few minutes into a raging flood of muddy water, heavily charged with rock fragments.

An important contrast between regions of arid and humid climates lies in the way in which the water enters and leaves a stream channel. In a humid region with a high water table sloping toward the stream channels, ground water moves steadily toward the channels, into which it seeps, producing permanent, or perennial streams. Such streams are described as *effluent* (Figure 20.27A).

In arid regions, where streams flow across plains of gravel and sand, water is lost from the channels by seepage to a water table that lies

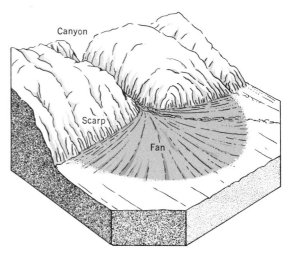

figure 20.28
A simple alluvial fan.

below the level of the streams. Such streams are designated as *influent* (Figure 20.27B). The water table beneath the channel tends to rise, forming a water-table mound.

Loss of discharge by influent seepage and evaporation strongly affects streams in alluvium-filled valleys of arid regions. Aggradation occurs and braided channels are conspicuous. Streams of desert regions are often short and terminate in alluvial deposits or on shallow, dry lake floors.

Alluvial fans

One very common landform built by braided, aggrading streams is the *alluvial fan*, a low cone of alluvial sands and gravels resembling in outline an open Japanese fan (Figure 20.28). The apex, or central point of the fan, lies at the mouth of a canyon or ravine. The fan is built out on an adjacent plain. Alluvial fans are of many sizes; some desert fans are many miles across (Figure 20.29).

Fans are built by streams carrying heavy loads of coarse rock waste from a mountain or upland region. The braided channel shifts

figure 20.27
Effluent and influent streams. (From A. N. Strahler, 1971, The Earth Sciences, 2nd ed., Harper & Row, New York.)

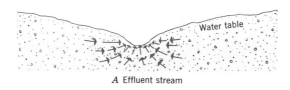

A Effluent stream

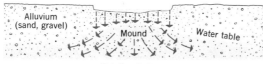

B Influent stream

figure 20.29
A great alluvial fan in Death Valley, built of debris swept out of a large canyon. Observe the braided stream channels. (Spence Air Photos.)

constantly, but its position is firmly fixed at the canyon mouth. The lower part of the channel, below the apex, sweeps back and forth. This activity accounts for the semicircular fan form and the downward slope in all radial directions from the apex.

Large, complex alluvial fans also include mudflows (Figure 18.16). As shown in Figure 20.30, mud layers are interbedded with sand and gravel layers. Water infiltrates the fan at its head, making its way to lower levels along sand layers that serve as aquifers. The mudflow

layers serve as barriers to ground water movement, or aquicludes. Ground water trapped beneath an aquiclude is under hydraulic pressure from water higher in the fan apex. When a well is drilled into the lower slopes of the fan, water rises spontaneously as artesian flow.

Alluvial fans are the dominant class of ground water reservoirs in the southwestern United States. Sustained heavy pumping of these reserves for irrigation has lowered the water table severely in many fan areas. Rate of

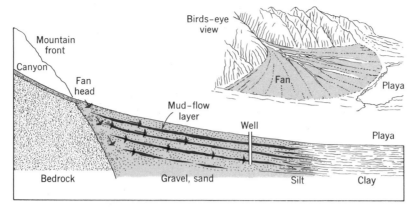

figure 20.30
An idealized cross section of an alluvial fan showing mudflow layers (aquicludes) interbedded with sand layers (aquifers). (From A. N. Strahler, 1972, Planet Earth, Harper & Row, New York.)

recharge is extremely slow in comparison. However, efforts are made to increase this recharge by means of water-spreading structures and infiltrating basins on the fan surfaces.

A serious side effect of excessive ground water withdrawal is that of subsidence of the ground surface. Important examples of subsidence are found in alluvial valleys filled to great depth with alluvial fan and lake-deposited sediments. For example, in one locality in the San Joaquin Valley of California, the water table has been drawn down over 100 ft (30 m). The resulting ground subsidence has amounted to about 10 ft (3 m) over a 35-year period.

The landscape of mountainous deserts

Where tectonic activity in the form of block faulting has been active in an area of continental desert, the assemblage of fluvial landforms is particularly diverse. The basin-and-range region of the western United States is such an area; it includes large parts of Nevada and Utah, southeastern California, southern Arizona and New Mexico, and adjacent parts of Mexico. Between uplifted and tilted blocks are down-dropped tectonic basins.

Fluvial denudation has carved up the mountain blocks into a ragged landscape of deep canyons and high divides, as shown in Figure 20.31A. Rock waste furnished by these steep mountain slopes is carried from the mouths of canyons to form numerous large alluvial fans. The fan deposits form a continuous apron extending far out into the basins.

As the mountain masses are lowered in height and reduced in extent, the fan slopes encroach on the mountain bases. Close to the receding mountain front there develops a sloping rock floor, called a *pediment*. As the remaining mountains shrink further in size, the pediment expands to form a rock platform thinly veneered with alluvium (Figure 20.31B).

In the centers of desert basins lie the saline lakes and dry lake basins referred to in Chapter 10. Accumulation of fine sediment and precipitated salts produces an extremely flat land surface, referred to in the southwestern United States and in Mexico as a *playa*. Salt flats are found where an evaporite layer forms the surface. In some playas shallow water forms a salt lake.

Figure 20.32 is an air photograph of a mountainous desert landscape — the Death Valley region of southeastern California. The three environmental zones are: (1) rugged mountain masses dissected into branching canyons with steep, rocky walls; (2) a piedmont zone of alluvial fans or a pediment; and (3) the playa occupying the central part of the basin.

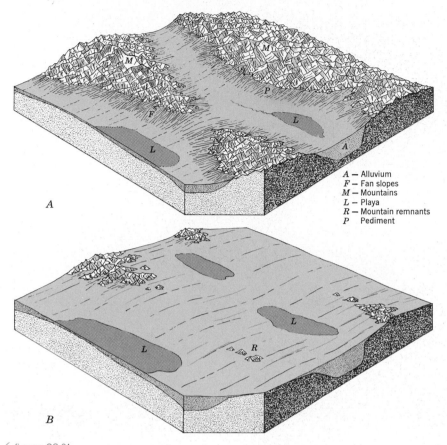

A — Alluvium
F — Fan slopes
M — Mountains
L — Playa
R — Mountain remnants
P   Pediment

figure 20.31
Idealized diagrams of landforms of the mountainous deserts of the southwestern United States. (A) Stage of rapid filling of tectonic basins with debris from high, rugged mountain blocks. (B) Advanced stage with small mountain remnants and broad playa and fan slopes.

It should be obvious that fluvial processes in such a region are limited to local transport of rock particles from a mountain range to the nearest adjacent basin, which receives all of the sediment and must be gradually filling as the mountains are diminishing in elevation. Since there is no outlfow to the sea, the concept of a base level of denudation has no meaning, and a peneplain does not represent the penultimate stage of denudation, as it does in the humid environment. Each arid basin becomes a closed system so far as mass transport is involved. Only the hydrologic system is open, with water entering as precipitation and leaving as evaporation. The desert land surface produced in an advanced stage of fluvial activity is an undulating plain, called a *pediplain*. It consists of areas of pediment surrounded by areas of alluvial fan and playa surfaces.

Fluvial processes in review

Running water as a fluid agent of denudation dominates the sculpturing of the continents. We have examined fluvial systems in which overland flow becomes organized into channel systems, converging the flow into larger and

figure 20.32
This air view of Death Valley, California, shows a mountainous desert landscape with deeply eroded mountain blocks and extensive fan surfaces. (Spence Air Photos.)

larger streams. As streams erode, transport, and deposit rock material, they shape a wide variety of secondary landforms. Gorges and canyons dominate the earlier stages in carving up of a landmass. These transient forms finally give way to broad floodplains on which meandering rivers move on low gradients. More subtle landforms, such as the natural levees, now dominate the fluvial landscape. Overbank flooding and widespread inundation bring major environmental impact to these flat alluvial lands, where they are intensively occupied by Man. Efforts to control flooding by engineering works often fail, despite enormous inputs of money. Doubts are now raised as to the wisdom of continuing to protect lands naturally subject to river inundations.

The impact of Man on the action of running water is severe in agricultural and urban areas. Soil erosion is easily induced, but controlled only with great difficulty. Channels are damaged by aggradation from upstream sources of sediment. Reservoirs impounded by great dams are being steadily filled with sediment, and we have good reason to question the desirability of building more of these expensive works.

In this chapter we have also compared the action of running water in deserts with that in humid climates. While the basic principles are the same everywhere, aridity emphasizes the transport and accumulation of bed load with only short distances separating the supply region from the accumulation region.

Landforms and Rock Structure

Chapter 21

FLUVIAL DENUDATION ACTS on any landmass exposed to the atmosphere, but landmasses vary greatly in their formative processes. Consequently, landmasses differ greatly from place to place in rock composition and altitude above sea level; they also differ greatly from place to place in rock composition and rock structure. As you might expect, landforms of fluvial denudation can take on a wide variety of shapes and patterns, expressing the complexity of the crustal rock from which they are carved.

Recall from the previous chapter that the mass or rock lying above base level constitutes the landmass. In theory, denudation cannot reduce the land surface below the base level, which is the inland extension of sea level. But the earth's crust can be lifted higher above sea level by means of tectonic movements. Crustal rise thus increases the available landmass. Through repeated crustal uplifts and repeated episodes of denudation, deep-seated rocks and structures appear at the surface. In this way the root structures of mountain belts come to be exposed, along with batholiths of plutonic rocks.

In Chapter 20 we classified all landforms as being either initial or sequential in origin. As a landmass is first brought into existence by tectonic activity or volcanic activity, its surface configuration is made up of initial landforms. Initial landforms produced directly by volcanic activity and by faulting were described in Chapter 17. Denudation soon converts the initial landforms into sequential landforms. In this chapter our interest is in the erosional class of sequential landforms, controlled in shape, size, and arrangement by the underlying rock structure.

Denudation acts more rapidly on the weaker rock types, lowering them to produce valley floors and leaving the stronger rocks to stand out in bold relief as ridges and uplands. This is a subject we shall pursue further in the present chapter. Rock structure controls the placement of streams and the shapes and heights of the invervening divides. A distinctive assemblage of landforms and stream patterns is developed for each of the major types of crustal structures. The habitats of plants and animals are varied by these landforms. Man occupies these habitats and exploits them for agricultural lands, communication lines, and urban development. In all of these activities a distinctive set of constraints and opportunities is imposed by each of the types of structurally controlled landforms.

Structural groups of landmasses

Landmasses fall into three major groups, and each of these in turn contains two or more distinctive types. As major groups, we can recognize the following.

Group A. Undisturbed sedimentary strata. These are thick covers of sedimentary rocks overlying ancient shield rocks. The strata are most commonly of marine origin, deposited on the floors of shallow continental shelves and inland seas; they have been brought above sea level by crustal rise and are now in the process of undergoing fluvial denudation. Because no significant amount of bending or faulting has affected these strata, their attitude is nearly horizontal over very large areas.

Group B. Disturbed structures of tectonic activity. These landmasses show strongly the effects of bending and breaking of the crust by mountain-building processes. Sedimentary strata showing folding and faulting are included in this group. Metamorphic rocks produced in root zones of ancient mountain belts are another type.

Group C. Eroded igneous masses. Exposed plutons — large bodies of intruded igneous rock — are one important type of igneous mass. A quite different type includes extinct composite and shield volcanoes undergoing erosion.

Figure 21.1 illustrates eight landmass types within the three major groups. Within Group A, two important types are coastal plains and horizontal strata. Group B includes folds, domes, fault blocks, and metamorphic belts.

Group C includes plutons and eroded volcanoes.

A brief overview of each landmass type will give you some insight into the meaning of diverse landforms of the continents and the ways in which these landforms affect Man's environment and provide a variety of mineral

figure 21.1
Several distinctive types of landmasses can be recognized on the basis of rock composition and structure.

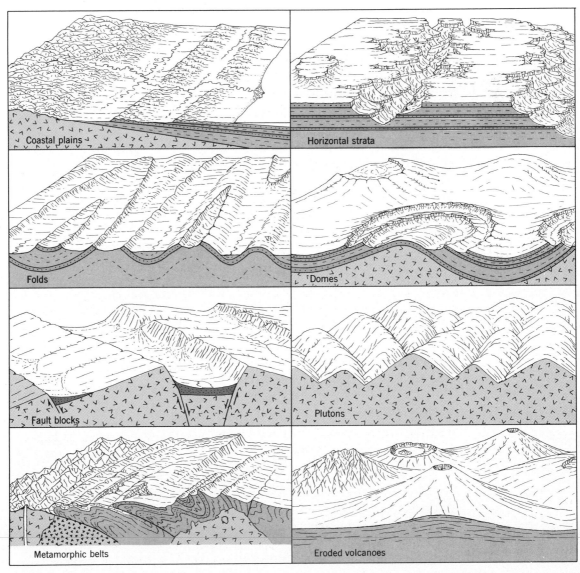

resources. Each of our national parks and wilderness areas owes its distinctive scenery to the influence of a particular landmass type. These protected natural areas are a national resource on which no dollar value can be placed; their diversity depends largely on the kinds of rocks and structures out of which their landforms have been carved by fluvial denudation.

Rock structure as a landform control

As denudation takes place, landscape features develop in close conformity with patterns of bedrock composition and structure. Figure 21.2 shows five types of sedimentary rock, together with a mass of much older igneous rock on which the sediments were deposited. The diagram shows the usual landform habit of each rock type, whether to form valleys or mountains. The cross section shows conventional rock symbols used by geologists. These rock strata have been strongly tilted and deeply eroded.

Shale is a weak rock and is reduced to the lowest valley floors of the region. Limestone, easily subject to carbonic acid action, is also a valley former in humid climates. On the other hand, limestone is highly resistant and usually stands high in arid climates. Sandstone and

figure 21.2
Landforms evolve through the slow erosional removal of weaker rock, leaving the more resistant rock standing as ridges or mountains.

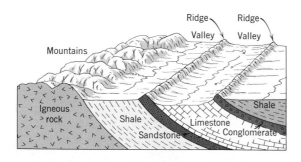

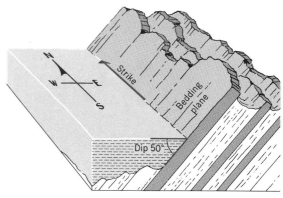

figure 21.3
Strike and dip.

conglomerate are typically resistant rocks; they form ridges and uplands. As a group, the igneous rocks are resistant to denudation; they typically form uplands rising above sedimentary strata.

The metamorphic rocks are, as a group, more resistant to denudation than their sedimentary parent types. However, as shown in Figure 21.25, these are conspicuous differences among the types of metamorphic rocks.

Natural layers and planes of weakness are characteristic of the structure of each type of rock. We need a system of geometry to enable us to measure and describe the attitude of these natural planes and to indicate them on maps. Examples of such planes are the bedding layers of sedimentary strata, the sides of a dike, and the joints in granite. Rarely are these planes truly horizontal or vertical.

The acute angle formed between a natural rock plane and an imaginary horizontal plane is termed the _dip_. The amount of dip is stated in degrees ranging from 0° for a horizontal plane to 90° for a vertical plane. Figure 21.3 shows the dip angle for an outcropping layer of sandstone, against which a horizontal water surface rests. The compass direction of the line of intersection between the inclined rock plane and an imaginary horizontal plane is the _strike_. In Figure 21.3 the strike is north.

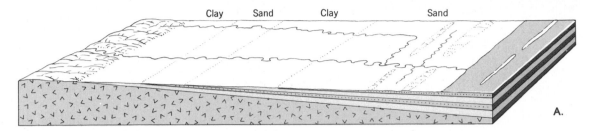

Clay   Sand   Clay   Sand

A.

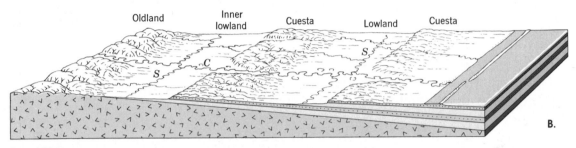

Oldland   Inner lowland   Cuesta   Lowland   Cuesta

B.

figure 21.4
Development of a broad coastal plain. (A) Early stage; plain recently emerged. (B) Advanced stage; cuestas and lowlands developed. S = subsequent stream; C = consequent stream. (After A. K. Lobeck.)

Coastal plains

Coastal plains are found along stable continental margins largely free of tectonic activity. Figure 21.4A shows a coastal zone that has recently emerged from beneath the sea. Formerly this zone was a shallow continental shelf accumulating successive layers of sediment brought from the land and distributed by currents. On the newly formed land surface streams flow directly seaward, down the gentle slope. A stream of this origin is a *consequent stream,* defined as any stream whose course is controlled by the initial slope of a land surface. Consequent streams occur on many landforms, such as volcanoes, fault blocks, or beds of drained lakes.

In an advanced stage of coastal-plain development a new series of streams and topographic features has developed (Figure 21.4B). Where more easily eroded strata (usually clay or shale) are exposed, denudation is rapid, making *lowlands.* Between them rise broad belts of hills called *cuestas.* Cuestas are commonly underlain by sand, sandstone, limestone, or chalk. The lowland lying between the area of older rock and the first cuesta is called the *inner lowland.*

Streams that develop along the trend of the lowlands, parallel with the shoreline, are of a class known as *subsequent streams.* They take their position along any belt or zone of weak rock and therefore follow closely the pattern of rock exposure. Subsequent streams occur in many regions, and we shall mention them again in the discussion of folds, domes, and faults. The drainage lines on a fully dissected coastal plain combine to form a *trellis drainage pattern.* In this pattern the subsequent streams trend at about right angles to the consequent streams.

The coastal plain of the United States is a major geographical region, ranging in width from 100 to 300 mi (160 to 500 km) and extending for 2000 mi (3000 km) along the

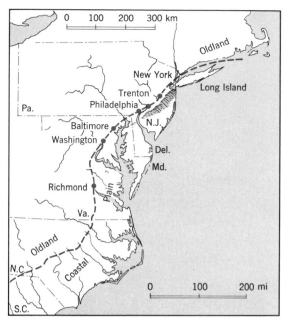

figure 21.5
The coastal plain of the Atlantic seaboard states shows little cuesta development except in New Jersey. The inner limit of the coastal plain is marked by a series of Fall Line cities. (After A. K. Lobeck.)

figure 21.6
The Alabama-Mississippi coastal plain is belted by a series of sandy cuestas and shale lowlands (After A. K. Lobeck.)

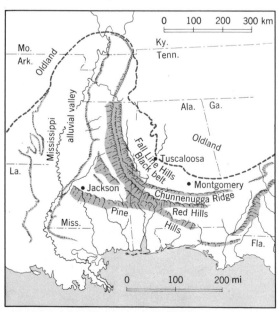

Atlantic and Gulf coasts. The coastal plain starts at Long Island, New York, as a partly submerged cuesta, and widens rapidly southward so as to include much of New Jersey, Delaware, Maryland, and Virginia (Figure 21.5). Throughout this portion the coastal plain has but one cuesta. The inner lowland is a continuous broad valley developed on weak clay strata.

In Alabama and Mississippi the coastal plain is fully dissected. Cuestas and lowlands run in belts roughly parallel with the coast (Figure 21.6). The cuestas are underlain by sandy formations and support pine forests. Limestone forms lowlands such as the Black Belt in Alabama, named for its dark, fertile soils.

Environmental and resource aspects of coastal plains

Broad coastal plains, such as those of the eastern United States, show intensive agricultural development because of the soil fertility and easy cultivation of broad lowlands. Cuestas provide valuable pine forests in the southern United States.

Transportation tends to follow the lowlands and to connect the larger cities located there. For example, important highways and railroads connect New York City with Trenton, Philadelphia, Baltimore, Washington, and Richmond, all of which are situated in the inner lowland (Figure 21.5). The inner limit of coastal plain strata is known as the Fall Line, because here the major rivers have developed falls and rapids on the resistant oldland rocks. Cities grew at the Fall Line, where the rivers become navigable tidal estuaries. Thus geology has played a part in the location of the nuclei of Megalopolis.

The seaward dip of sedimentary strata in a coastal plain provides a structure favorable to the development of artesian water wells (Figure

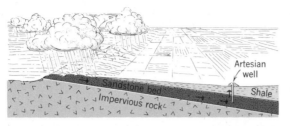

figure 21.7
An artesian well requires a dipping sandstone layer. (After E. Raisz.)

21.7). Water penetrates deeply into sandy cuesta strata, overlain by aquicludes of shale or clay. When a well is drilled into the sand formation considerably seaward of its surface exposure, water under hydraulic pressure reaches the surface. Artesian water in large quantities is available in many parts of the Atlantic and Gulf coastal plains, although it is no longer sufficient to supply the demands of densely populated and industrialized communities.

Of the economic mineral resources of broad coastal plains the most important are petroleum and natural gas. The Gulf Coastal Plain is a major oil-producing region. Here the strata beneath the coastal plain and offshore continental shelf are extremely thick. Lignite, a low grade of coal, occurs beneath a large area of Alabama, Mississippi, Louisiana, and Texas (see Figure 21.12).

Other mineral deposits of economic importance in coastal plains include: sulfur, occurring in the coastal plain of Louisiana and Texas; phosphate beds, found in Florida; and clays, used in manufacture of pottery, tile, and brick in New Jersey and the Carolinas.

Horizontal strata

Extensive areas of the continental shields are covered by thick sequences of horizontal sedimentary strata. At various times in the 600 million years following the end of Precambrian time, these strata were deposited in shallow inland seas. Following crustal uplift, with little disturbance other than minor warping or

figure 21.8
Horizontal strata exposed in the walls of Grand Canyon, Arizona. (A. N. Strahler.)

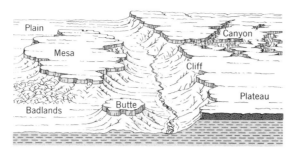

figure 21.9

In arid climates, distinctive erosional landforms develop in horizontal strata.

faulting, these areas became continental surfaces undergoing denudation.

In arid climates, where vegetation is scant and the action of overland flow especially effective, sharply defined landforms develop on horizontal sedimentary strata. The normal sequence of landforms is sheer rock wall, called a *cliff*. At the base of the cliff is a smooth inclined slope. This slope flattens out to make a bench, terminated at the outer edge by the cliff of the next lower set of forms. In the walls of the great canyons of the Colorado Plateau region, these forms are wonderfully displayed (Figure 21.8).

In the arid regions erosion strips away successive rock layers. Cliffs, capped by hard rock layers, retreat as near-perpendicular surfaces because the weak clay or shale formations exposed at the cliff base are rapidly washed away by storm runoff and channel erosion. When undermined, the rock in the upper cliff face repeatedly breaks away along vertical fractures.

Cliff retreat produces *mesas*, table-topped plateaus bordered on all sides by cliffs (Figure 21.9). Mesas represent the remnants of a formerly extensive layer of resistant rock. As a mesa is reduced in area by retreat of the rimming cliffs, it maintains its flat top. Before its complete consumption the final landform is a small steep-sided hill known as a *butte* (Figure 21.10).

Regions of horizontal strata have branching stream networks formed into a *dendritic drainage pattern* (Figure 21.11). The smaller

figure 21.11

A dendritic drainage pattern formed on horizontal strata.

figure 21.10

This early phograph shows a butte of horizontal red sandstones capped by a white gypsum layer, near Cambria, Wyoming. (U.S. Geological Survey.)

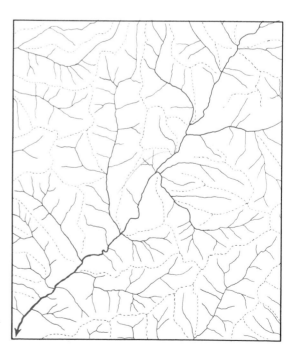

streams in this pattern take a great variety of directions. This treelike branching pattern is also found in areas of granite plutons (see Figure 21.28).

Resources of horizontal strata

Horizontal strata have only those minerals and rocks of economic value that are associated with sedimentary rocks. Building stone, such as the Bedford limestone in Indiana or the Berea sandstone in Ohio, is a valuable resource. Limestone may be quarried for use in manufacture of portland cement or as flux in iron smelting. Some important deposits of lead, zinc, and iron ores occur in sedimentary rocks.

For example, the lead and zinc mines of the Tristate district (Missouri, Kansas, Oklahoma) are in horizontal limestones. Uranium ores are important in strata of the Colorado Plateau.

Perhpas the greatest mineral resources occurring in sedimentary strata are coal and petroleum. In Chapter 16 we described the occurrence of these fossil fuels. Recall from Chapter 18 that strip mining of coal is an extreme form of environmental degradation, with various undesirable side effects besides the massive scarification of the landscape. As the map of North American coal fields shows (Figure 21.12), large reserves of coal lie close to the surface in areas of horizontal strata in the Great Plains region west of the Mississippi,

figure 21.12

Coal fields of the United States and southern Canada. The lignite area of the southern states lies within the coastal plain. The anthracite fields are in folded strata.

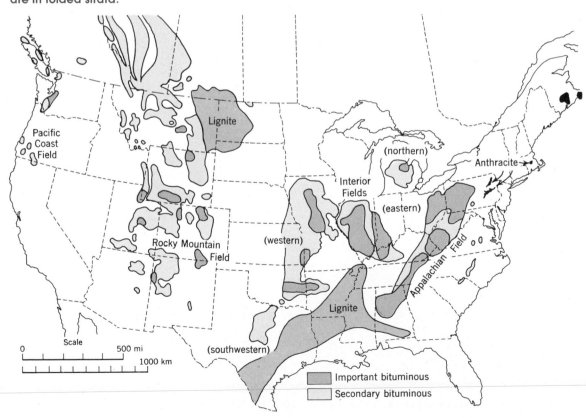

in the northern High Plains still further west, in sedimentary basins of the Rocky Mountains, and in the Colordo Plateau near the common corner of the states of Utah, Arizona, Colorado and Nex Mexico. The great demands for this coal in decades to come make it likely that large areas will be strip-mined, and that power generating plants of large capacity will be built close to the supplies of coal. Environmentalists are greatly concerned with the land scarification and air pollution that such developments can bring.

Fold belts

We turn next to landmasses of the second group, developed on structures of tectonic origin. According to the model of mountain making developed in Chapter 17, strata of a geosyncline are ultimately compressed into folds along narrow belts adjacent to the zones of active subduction (see Figure 17.27). Long after folding has ceased, denudation exposes the mountain roots, bringing the folded strata into relief, as illustrated in Figure 17.20. Alternating hard and soft rock layers give the crust a strong grain, much like the grain in a wooden plank. Denudation removes the weaker bands of the grain, allowing the harder bands to stand out in relief, much like the texture of surfwood, etched by sand and waves.

In the language of geology, a troughlike downbend of strata is called a *syncline;* the archlike upbend next to it is called an *anticline*. Obviously, in a belt of folded strata, synclines alternate with anticlines, like a succession of wave troughs and wave crests on the ocean surface.

A series of three block diagrams (Figure 21.13) shows some of the distinctive landforms resulting from fluvial denudation of a belt of folded strata. Erosion of these simple, open folds produces a *ridge-and-valley landscape,* as weaker formations of shale and limestone are

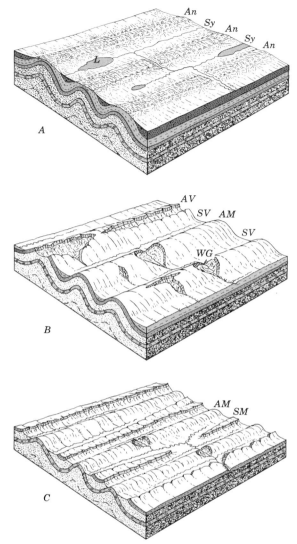

figure 21.13
Stages in the erosional development of folded strata. (**A**) While folding is still in progress, erosion cuts down the anticlines; alluvium fills the synclines, keeping relief low. **An** = anticline; **Sy** = syncline; **L** = lake. (**B**) Long after folding has ceased, erosion exposes a highly resistant layer of sandstone or quartzite. **AV** = anticlinal valley; **SV** = synclinal valley; **WG** = watergap. (**C**) Continued erosion partly removes the resistant formation but reveals another below it. **AM** = anticlinal mountain; **SM** = synclinal mountain.

eroded away, leaving hard sandstone strata to stand in bold relief as long, narrow ridges.

On eroded folds, the stream network is distorted into a trellis drainage pattern (Figure

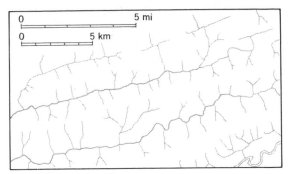

figure 21.14
A trellis drainage pattern on folds.

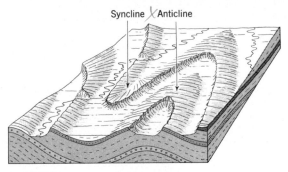

figure 21.15
Folds with crests that plunge downward give rise to zigzag ridges following erosion.

21.14). The principal elements of this pattern are long, parallel subsequent streams occupying the narrow valleys of weak rock.

The folds illustrated in Figure 21.13 are continuous and even-crested; they produce ridges that are approximately parallel in trend and continue for great distances. In some fold regions, however, the folds are not continuous and level-crested. Instead, the fold crests rise or descend from place to place. A descending fold crest is said to plunge. Plunging folds give rise to a zigzag line of ridges (Figure 21.15).

Environmental and resource aspects of fold regions

Some of the environmental and resource aspects of dissected fold regions are illustrated by the Appalachians of south-central and eastern Pennsylvania (Figure 21.16). The ridges, of resistant sandstones and con-glomerates, rise boldly to heights of 2000 ft (600 m) above broad lowlands underlain by weak shales and limestones. Major highways run in the valleys, crossing from one valley to another through the watergaps of streams that have cut through the ridges. Important cities are situated near the watergaps of major streams. An example is Harrisburg, located where the Susquehanna River issues from a

series of watergaps. Where no watergaps are conveniently located, roads must climb in long, steep grades over the ridge crests. The ridges are heavily forested; the valleys are rich agricultural belts.

In various fold regions of the world (e.g., in the Pennsylvania Appalachians), an important resource is anthracite, or hard coal (Figure 21.17). This coal occurs in strata that have been tightly folded. Pressure has converted the coal from bituminous into anthracite. Because of extensive erosion, all coal has been removed except that which lies in the central parts of synclines. The coal seams dip steeply; workings penetrate deeply to reach the coal

figure 21.16
A great synclinal fold, involving three resistant quartzite formations and thick intervening shales, has been eroded to form bold ridges through which the Susquehanna River has cut a series of watergaps. (After A. K. Lobeck.)

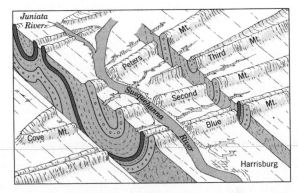

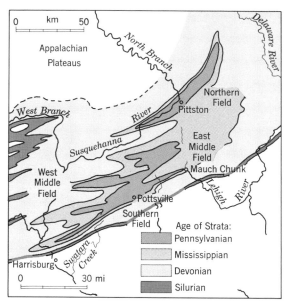

figure 21.17

Anthracite coal basins of central Pennsylvania correspond with areas of Pennsylvanian strata, downfolded into long synclinal troughs.

illustrated in the two block diagrams of Figure 21.18. Strata are first removed from the summit region of the dome, exposing older strata beneath. Eroded edges of steeply dipping strata form sharp-crested sawtooth ridges called _hogbacks_ (Figure 21.19). When the last of the strata have been removed, the ancient shield rock is exposed in the central core of the dome, which then develops a mountainous terrain.

The stream network on a deeply eroded dome shows dominant subsequent streams forming a circular system, called an _annular drainage pattern_ (Figure 21.20). The shorter tributaries

that lies in the bottoms of the synclines. Seams near the surface are worked by strip mining.

Broad, gentle anticlinal folds may form important traps for the accumulation of petroleum. The principle is shown in Figure 16.18. Oil migrates in permeable sandstone beds to the anticlinal crest, where it is trapped by an impervious cap rock of shale. Many of the oil and gas pools of western Pennsylvania, where petroleum production first succeeded, are on low anticlines.

Sedimentary domes

Another type of tectonic structure is the _sedimentary dome_, a circular structure in which strata have been raised into a domed shape. Sedimentary domes occur in various places within the covered shield areas of the continents. Igneous intrusions at great depth may have been responsible for these uplifts.

Erosion features of a sedimentary dome are

figure 21.18

Erosion of sedimentary strata from the summit of a dome structure. Above: Strata partially removed, forming an encircling hogback ridge. Below: Strata removed from center of dome, revealing a core of older igneous or metamorphic rock.

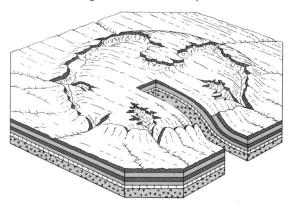

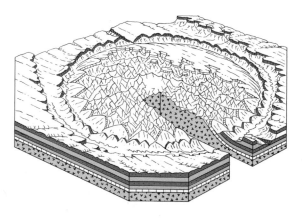

figure 21.19
The Virgin River cuts across a hogback of steeply dipping strata on the flank of the Virgin anticline, southwestern Utah. (Frank Jensen.)

make a radial arrangement. The total pattern resembles a trellis pattern bent into a circular form.

The Black Hills dome

A classic example of a large and rather complex sedimentary dome is the Black Hills dome of western South Dakota and eastern Wyoming (Figure 21.21). Valleys that encircle this dome are splendid locations for railroads and highways. So, it is natural that towns and cities should have grown in these valleys. One valley in particular, the Red Valley, is continuously developed around the entire dome and has been called the Race Track because of its shape. It is underlain by a weak shale, which is easily washed away. The cities of Rapid City, Spearfish, and Sturgis are located in the Red Valley. On the outer side of the Red

Valley is a high, sharp hogback of Dakota sandstone, known simply as Hogback Ridge. It rises some 500 ft (150 m) above the level of the Red Valley. Farther out toward the margins of the dome the strata are less steeply inclined and form a series of cuestas. Artesian water is obtained from wells drilled in the surrounding plain.

The eastern central part of the Black Hills consists of a mountainous core of intrusive and metamorphic rocks. These mountains are richly forested, whereas the intervening valleys are beautiful open parks. Thus the region is attractive as a summer resort area. Harney Park, elevation 7242 ft (2207 m), is highest of the peaks of the core. In the northern part of the central core, in the vicinity of Lead and Deadwood, there are valuable ore deposits. At Lead is the fabulous Homestake Mine, one of the world's richest gold-producing mines.

The western central part of the Black Hills consists of a limestone plateau deeply carved by streams. The original dome had a flattened summit. The limestone plateau represents one of the last remaining sedimentary rock layers to be stripped from the core of the dome.

figure 21.20
The drainage pattern on an eroded dome combines annular and radial elements.

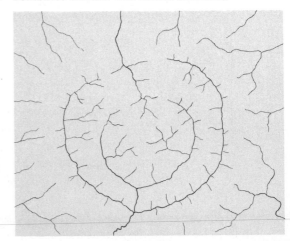

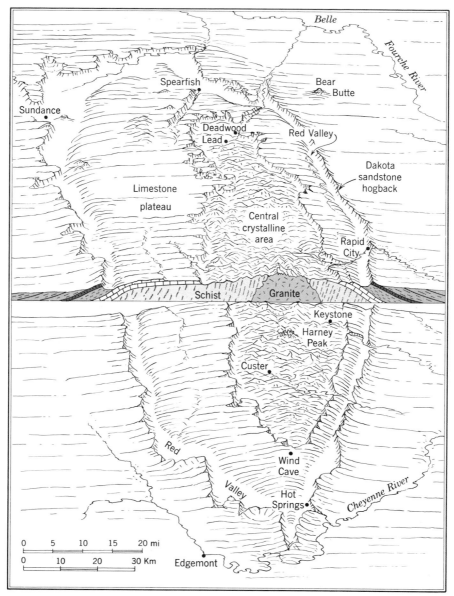

figure 21.21

The Black Hills consist of a broad, flat-topped dome, deeply eroded to expose a core of igneous and metamorphic rocks.

Erosion forms on fault structures

In Chapter 17 we found that active normal faulting produces a sharp surface break called a fault scarp (Figure 17.14). Erosion quickly modifies and obliterates this scarp but, because the fault plane extends hundreds of feet down into bedrock, its effects on erosional landforms persist for long spans of geologic time. Figure 21.22 shows both the original fault scarp (above), and its later landform expression (below). Even though the cover of sedimentary

(right) Intricately carved strata of Eocene age, Bryce Canyon National Park, Utah. These limestone beds were laid down in a former freshwater lake, then uplifted to form a high plateau. (Orlo E. Childs)

(above) A butte of sandstone (Mesa Verde formation) not far from Ship Rock, northwestern New Mexico. Weak shales (Mancos formation) form the base of the butte. (Arthur N. Strahler)

Horizontal Strata

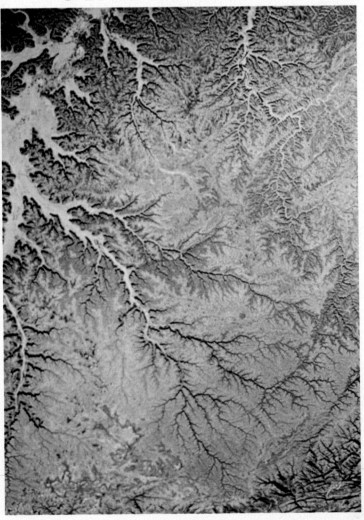

(one above) Sandstones and shales carved into deep canyons in a semiarid climate. View from Dead Horse Point, Canyon Lands National Park, Utah. (Alan H. Strahler)

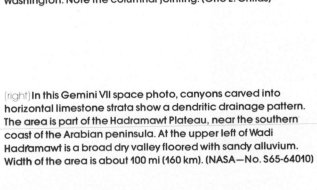

(above) Horizontal flows of basalt lava, piled one upon the other, form these cliffs in the south wall of the Columbia River gorge, Washington. Note the columnar jointing. (Orlo E. Childs)

(right) In this Gemini VII space photo, canyons carved into horizontal limestone strata show a dendritic drainage pattern. The area is part of the Hadramawt Plateau, near the southern coast of the Arabian peninsula. At the upper left of Wadi Hadramawt is a broad dry valley floored with sandy alluvium. Width of the area is about 100 mi (160 km). (NASA—No. S65-64010) (below) A broad anticlinal fold in Death Valley, California. The soft, easily eroded strata show sharp upturning around the sides and end of the anticline. (Mark A. Melton)

(below) Flatirons of massive Jurassic sandstone have been carved from upturned strata near Marsh Pass, Arizona. (Alan H. Strahler)

(above) Hogbacks of steeply dipping sandstone, Paria, Utah. Weak shale underlies the valleys between hogbacks. (Donald L. Babenroth)

(above) A near-vertical bed of massive sandstone gives the Garden of the Gods a distinctive appearance. This locality is near Colorado Springs, Colorado. (Arthur N. Strahler)

Domes and Folds

(left) Zigzag ridges, formed on deeply eroded plunging folds of Paleozoic strata, characterize the Newer Appalachians of southcentral Pennsylvania. In this image, obtained by ERTS-1 orbiting satellite, the Susquehanna River cuts across the narrow ridges in a series of watergaps north of Harrisburg. The area shown is about 100 mi (160 km) across. (NASA — EROS Data Center, No. 81171152455N)

(above) **This sheer rock wall 1000 ft (300 m) high is a fault scarp. Uplift along the fault raised a plateau of massive limestone, Big Bend National Park, Texas. (Mark A. Melton)**

(right) **Photographed by astronauts aboard Gemini XII space vehicle, the Red Sea stretches southeastward toward the horizon. The Red Sea is about 125 mi (200 km) wide; its straight, parallel shores reveal its origin as a widening belt of ocean floor between two separating lithospheric plates. The Arabian Peninsula lies to the left; Egypt to the right. The Gulf of Suez (bottom, center) and Gulf of Aqaba (left) are narrower rift depressions — grabens — bounded by normal faults. The triangular Sinai Peninsula lies between them. (NASA — No. S66-63481)**

(left) **Small-scale normal faulting in red shale beds, Kaibab Plateau, Arizona. (Arthur N. Strahler)**

Landforms of Faulting

(above) **A great block mountain forms the east wall of Death Valley, California. At the base of the mountain is an active fault scarp, partly buried by alluvial fans. (Mark A. Melton)**

(right) **A small graben. The valley represents a block faulted down between two normal faults. Coconino Plateau, Arizona. (Donald L. Babenroth)**

(right) **Mt. Hood**, a recently formed composite volcano of the Cascade Range, Oregon, with a summit elevation just over 11,000 ft (3350 m). Glaciers have carved the sides of the cone. (Orlo E. Childs)

(below) A fresh cinder cone near Bend, Oregon. The rough surface at the base of the cone is a recent basaltic lava flow. (Orlo E. Childs)

(above) **Ship Rock**, New Mexico, is a volcanic neck enclosed by a weak shale formation. The peak rises 1700 ft (520 m) above the surrounding plain. (Mark A. Melton)

Volcanic Landforms

(left) Basaltic shield volcanoes of the Island of Hawaii. Closest is snow-capped, extinct Mauna Kea. Distant and left is Mauna Loa, an active shield volcano. Both have summit elevations over 13,000 ft (4000 m). (Arthur N. Strahler)

(below) **Waimaea Canyon**, over 2500 ft (760 m) deep, has been eroded into the flank of an extinct shield volcano on Kauai, Hawaii. Basaltic lava flows are exposed. (Arthur N. Strahler)

(above) **Halemaumau** fire pit, a steep-walled crater within the central depression (caldera) of Kilauea volcano, Hawaii. (Arthur N. Strahler)

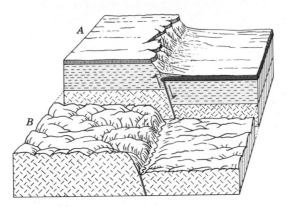

figure 21.22
(A) Fault scarp. (B) Fault-line scarp. (From
A. N. Strahler, 1971, The Earth Sciences, 2nd ed.,
Harper & Row, New York.)

strata has been completely removed, exposing the ancient shield rock, the fault continues to produce a landform. Because the fault plane is a zone of weak rock, crushed during faulting, it is occupied by a subsequent stream. The scarp persisting along the upthrown side is called a *fault-line scarp*. Figure 21.23 shows a fault-line scarp within the Canadian Shield.

Figure 21.24 shows some erosional features of a large, tilted fault block. The freshly uplifted block has a steep mountain face, but this is rapidly dissected into deep canyons. The upper mountain face reclines in angle as it is removed, while rock debris accumulates in the form of alluvial fans adjacent to the fault block. Vestiges of the fault plane are preserved as triangular facets, the snubbed ends of ridges between canyon mouths.

Environmental and resource aspects of faults and block mountains

Faults are of environmental and economic significance in several ways. Fault planes are usually zones along which the rock has been pulverized, or at least considerably fractured.

This breakage has the effect of permitting ore-forming solutions to rise along fault planes. Many important ore deposits lie in fault planes or in rocks that faults have broken across.

Another related phenomenon is the easy rise of ground water along fault planes. Springs, both cold and hot, are commonly situated along fault lines. They occur along the base of a fault-block mountain. Examples are Arrowhead Springs along the base of the San Bernardino Range and Palm Springs along the foot of the San Jacinto Mountains, both in southern California.

Petroleum, too, finds its way along fault planes where the rocks have been rendered permeable by crushing, or it becomes trapped in porous beds that have been faulted against impervious shale beds. Some of the most intensive searches for oil center about areas of

figure 21.23
A fault-line scarp near Great Slave Lake,
Northwest Territories, Canada. (Canadian Forces
Photograph No. 5120-105R.)

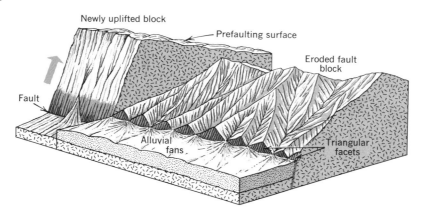

figure 21.24
Erosion of a tilted fault block produces a rugged mountain range. A line of triangular facets at the mountain base marks the position of the original fault plane.

faulted sedimentary strata because of the great production that has come from pools of this type.

Fault scarps and fault-line scarps can form imposing topographic barriers across which it is difficult to build roads and railroads. The great Hurricane Ledge of southern Utah is a feature of this type, in places a steep wall 2500 ft (760 m) high.

Grabens may be so large as to form broad lowlands. An illustration is the Rhine graben of Western Germany. Here a belt of rich agricultural land, 20 mi (32 km) wide and 150 mi (240 km) long, lies between the Vosges and Black Forest ranges, both of which are block mountains faulted up in contrast with the down-dropped Rhine graben block.

Landforms on eroded metamorphic rocks

Regions of metamorphic rock usually show a strong grain in the topography. Long ridges are separated by long, roughly parallel valleys (Figure 21.25). Neither ridges nor valleys have the sharpness of folded sedimentary strata, but the drainage pattern often takes the trellis form. Ridges and valleys reflect different rates of denudation of parallel belts of metamorphic rocks, such as schist, slate, quartzite, and marble. Marble forms valleys; slate and schist make hill belts; quartzite stands out boldly and may produce conspicuous narrow hogbacks.

Much of New England, particularly the Taconic and Green Mountains, illustrates the landforms of a metamorphic belt. The larger

figure 21.25
Metamorphic rocks form long, narrow parallel belts of valleys and mountains.

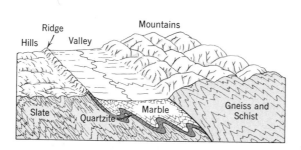

figure 21.26
Large plutons appear at the surface only after long-continued erosion has removed thousands of feet of overlying rocks. (After Longwell, Knopf, and Flint.)

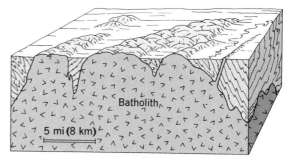

figure 21.27
The Idaho batholith, exposed in the Sawtooth Mountains of south central Idaho. View is northward. (From Geology Illustrated by John S. Shelton. W. H. Freeman and Company. Copyright © 1966.)

valleys trend north and south and are underlain by marble. These are flanked by ridges of gneiss, schist, slate, or quartzite.

Exposed plutons

Plutons, those huge masses of intrusive igneous rock in the form of batholiths, are eventually uncovered by erosion and gain surface expression (Figure 21.26). Plutons of granitic composition are a major ingredient in the mosaic of ancient rocks comprising the continental shields. A good example is the Idaho batholith, a granite mass exposed over an area of about 16,000 sq mi (40,000 sq km) — a region almost as large as New Hampshire and Vermont combined (Figure 21.27). Another example is the Sierra Nevada batholith of California; it makes up most of the high central part of that great mountain block.

The dendritic drainage pattern is often well developed on an eroded batholith (Figure 21.28). The pattern closely resembles the dendritic pattern of horizontal sedimentary rocks (Figure 21.11), and the two cannot easily be distinguished.

figure 21.28
This dendritic drainage pattern is developed on the deeply eroded Idaho batholith shown in Figure 21.27.

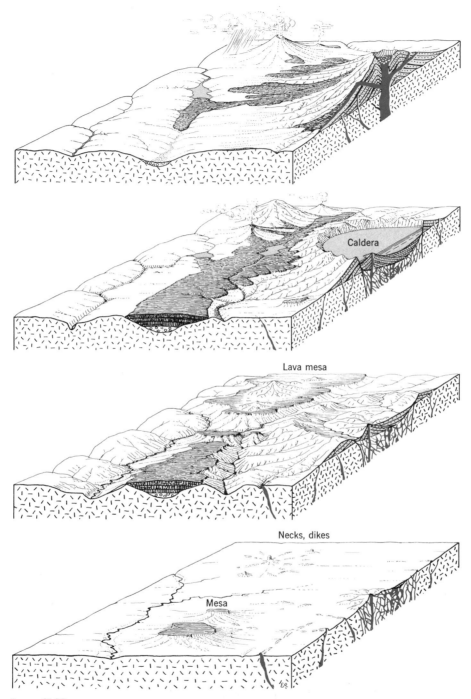

figure 21.29
Stages in the erosional development of volcanoes and lava flows.
(Drawn by E. Raisz.)

Erosional landforms of volcanoes and lava flows

Figure 21.29 shows successive stages in the erosion of volcanoes, lava flows, and a caldera. In the first block are active composite volcanoes in the process of building. Lava flows issuing from the volcanoes have spread down into a stream valley, following the downward grade of the valley and forming a lake behind the lava dam.

In the next block some changes have taken place. The most conspicuous change is the destruction of the largest volcano to produce a caldera. A lake occupies the caldera, and a small cone has been built inside. One of the other volcanoes, formed earlier, has become extinct. It has been dissected by streams, losing the smooth conical form. Smaller, neighboring volcanoes are still active, and the contrast in form is marked. A good example of a dissected extinct volcano is Mount Shasta in California

The system of streams on a dissected volcano cone is a *radial drainage pattern*. Because these streams take their positions on a slope of an initial land surface, they are of the consequent variety. It is often possible to recognize volcanoes from a drainage map alone (Figure 21.31) because of the perfection of the radial pattern.

In the third block of Figure 21.29, all volcanoes are extinct and have been deeply eroded. The caldera lake has been drained and the rim worn to a low, circular ridge. The lava flows have been able to resist erosion far better than the rock of the surrounding area and have come to stand as mesas high above the general level of the region.

The fourth block of Figure 21.29 shows an advanced stage of erosion of volcanoes. There remains now only a small sharp peak, called a *volcanic neck;* it is the solidified lava in the pipe of the volcano. Radiating from this are wall-like dikes, formed of magma, which filled radial fractures around the base of the volcano. Perhaps the finest illustration of a volcanic neck with radial dikes is Shiprock, New Mexico (Figure 21.32).

figure 21.30
Mt. Shasta in the Cascade Range is a dissected extinct volcano. The bulge on the left-hand slope is a more recent subsidiary cone, Shastina. (Eliot Blackwelder.)

figure 21.31
Radial drainage patterns of volcanoes in the East Indies. The letter C shows the location of a crater.

Shield volcanoes show erosion features quite different from those of the composite volcanoes. Figure 21.33 shows stages of erosion of Hawaiian shield volcanoes, beginning in the upper diagram with the active volcano and its central depression. Radial consequent streams cut deep canyons into the flanks of the extinct shield volcano, and these canyons are opened out into deep, steep-walled amphitheaters. Eventually, as the third diagram shows, the original surface of the shield volcano has been entirely obliterated, and a rugged mountain mass made up of sharp-crested divides and deep canyons remains.

The surfaces of volcanoes and lava flows remain barren and sterile for long periods after their formation. Certain types of lava surfaces are extremely rough and difficult to traverse; the Spaniards who encountered such terrain in

figure 21.32
Shiprock, New Mexico, is a volcanic neck. Radiating from it are dikes. (Spence Air Photos.)

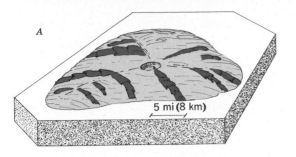

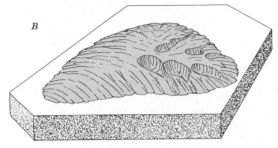

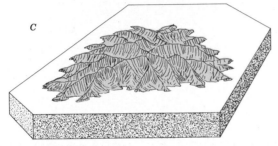

figure 21.33
Shield volcanoes in various stages of erosion make up the Hawaiian Islands. (A) Newly formed dome with central depression. (B) Early stage of erosion with deeply eroded valley heads. (C) Advanced erosion stage with steep slopes and mountainous relief.

tainous landscapes of volcanic origin. National parks have been made of Mt. Rainier, Mt. Lassen, and Crater Lake in the Cascade Range, a mountain mass largely of volcanic construction. Hawaii Volcanoes National Park recognizes the natural beauty of Mauna Loa and Kilauea; their displays of molten lava are a living textbook of igneous processes.

Landforms and rock structure in review

We have now added a new dimension to the realm of landforms produced by fluvial denudation. This dimension is the variety and complexity introduced by differences in rock composition and crustal structure. Because the various kinds of rocks offer different degrees of resistance to the forces of weathering, mass wasting, and running water, they exert a controlling influence on the shapes and sizes of landforms. We have seen, too, that streams carving up a landmass are controlled to a high degree by the structure of the rock on which they act. In this way distinctive drainage patterns evolve.

Diversity of landforms caused by variations in rock structure and composition have a profound influence on the environment. Each landform type is a distinctive habitat for plants; each landform type adds an element of uniqueness to the quality of the soil. Plant formations respond to these place-to-place differences in slope, drainage, and soil texture. Agriculture follows these same patterns, exploiting the most favorable habitats and rejecting those that are inhospitable. In this way the patterns of Man's cultural features have come to reflect the patterns of rock structure.

the southwestern United States named it "malpais" (bad ground). Most volcanic rocks in time produce highly fertile soils that are intensively cultivated.

Volcanoes are a valuable natural resource in terms of recreation and tourism. Few landscapes can rival in beauty the moun-

Landforms Made by Glaciers

Chapter 22

GLACIAL ICE HAS played a dominant role in shaping landforms of large areas in mid-latitude and subarctic zones. Glacial ice also exists today in two great accumulations of continental dimensions and in many smaller masses in high mountains. In this sense, glacial ice is an environmental agent of the present, as well as of the past, and is itself a landform. Glacial ice of Greenland and Antarctica strongly influences the radiation and heat balances of the globe. Moreover, these enormous ice accumulations represent water in storage in the solid state; they constitute a major component of the global water balance. Changes in ice storage can have profound effects on the position of sea level with respect to the continents.

Coastal environments of today have evolved with a rising sea level following melting of ice sheets of the last ice advance in the Pleistocene Epoch, or Ice Age. When we examine the evidence of former extent of those great ice sheets, we need to keep in mind that the evolution of modern Man as an animal species occurred during a series of climatic changes that placed many forms of environmental stress on all terrestrial plants and animals in the midlatitude zone. There were also important climatic effects extending into the tropical zone and even to the equatorial zone.

Glaciers

Most of us know ice only as a brittle, crystalline solid because we are accustomed to seeing it only in small quantities. Where a great thickness of ice exists — 300 ft (90 m) or more — the ice at the bottom behaves as a plastic material. The ice will then slowly flow in such a way as to spread out the mass over a larger area, or to cause it to move downhill, as the case may be. This behavior characterizes *glaciers,* defined as any large natural accumulations of land ice affected by present or past motion.

Conditions necessary for the accumulation of glacial ice are simply that snowfall of the winter shall, on the average, exceed the amount of ablation of snow that occurs in summer. The term *ablation* includes both evaporation and melting of snow and ice. Thus each year a layer of snow is added to what has already accumulated. As the snow compacts by surface melting and refreezing, it turns into a granular ice, then is compressed by overlying layers into hard crystalline ice. When the ice becomes so thick that the lower layers become plastic, outward or downhill flow commences, and an active glacier has come into being.

At sufficiently high altitudes, even at low latitudes, glaciers develop both because air temperature is low and mountains receive heavy orographic precipitation. Glaciers that form in high mountains are characteristically long and narrow because they occupy former stream valleys. These mountain glaciers bring the plastic ice from small collecting grounds high on the range down to lower altitudes, and consequently to warmer temperatures. Here the ice disappears by ablation. Such *alpine glaciers* are a distinctive type (Figure 22.1).

In arctic and polar regions, prevailing temperatures are low enough that ice can accumulate over broad areas, wherever uplands exist to intercept heavy snowfall. As a result, the uplands become buried under enormous plates of ice whose thickness may reach several thousand feet. During glacial periods, the ice spreads over surrounding lowlands, enveloping all landforms it encounters. This extensive type of ice mass is called an *ice sheet*.

Figure 22.2 illustrates a number of features of alpine glaciers. The center illustration is of a simple glacier occupying a sloping valley between steep rock walls. Snow is collected at the upper end in a bowl-shaped depression, the *cirque*. The upper end lies in the zone of

figure 22.1
The Eklutna Glacier, Chugach Mountains, Alaska, seen from the air. A deeply crevassed ablation zone (foreground) contrasts with the smooth-surfaced firn zone (background). (Steve McCutcheon, Alaska Pictorial Service.)

accumulation. Layers of snow in the process of compaction and recrystallization constitute the *firn*. The smooth firn field is slightly concave up in profile (upper right). Flowage in the glacial ice beneath carries the excess ice out of the cirque, downvalley. An abrupt steepening of the grade forms a rock step, over which the rate of ice flow is accelerated and produces deep crevasses (gaping fractures), which form an ice fall. The lower part of the glacier lies in the zone of ablation. Here the rate of ice wastage is rapid, and old ice is exposed at the glacier surface, which is extremely rough and deeply crevassed. The glacier terminus is heavily charged with rock debris.

The uppermost layer of a glacier is brittle and fractures readily into crevasses, whereas the ice beneath behaves as a plastic substance and moves by slow flowage (lower left). If one were to place a line of stakes across the glacier surface, the glacier flow would gradually deform that line into a parabolic curve, indicating that rate of movement is fastest in the center and diminishes toward the sides.

A simple glacier readily establishes a dynamic equilibrium in which the rate of accumulation at the upper end balances the rate of ablation at the lower end. Equilibrium is easily upset by changes in the average annual rates of nourishment or ablation.

Rate of flowage of glaciers is very slow, indeed, amounting to a few inches per day for large ice sheets and, for the more sluggish alpine glaciers, up to several feet per day for an active alpine glacier.

Glacial erosion

Glacial ice is usually heavily charged with rock fragments, ranging from pulverized rock flour to huge angular boulders of fresh rock. Some of this material is derived from the rock floor on which the ice moves. In alpine glaciers rock debris is also derived from material that slides or falls from valley walls. Glaciers are capable of great erosive work. *Glacial abrasion* is erosion caused by ice-held rock fragments that scrape and grind against the bedrock. By *plucking*, the moving ice lifts out blocks of bedrock that have been loosened by freezing of water in joint fractures.

Rock debris obtained by erosion must eventually be left stranded at the lower end of a glacier when the ice is dissipated. Both erosion and deposition result in distinctive landforms.

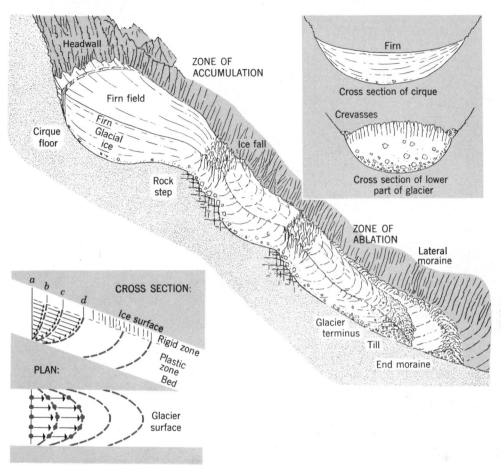

figure 22.2
Structure and flowage of a simple alpine glacier.

Landforms made by alpine glaciers

Landforms made by alpine glaciers are shown in a series of diagrams in Figure 22.3. Previously unglaciated mountains are attacked and modified by glaciers, after which the glaciers disappear and the remaining landforms are exposed to view.

Figure 22.3*A* shows a region sculptured entirely by weathering, mass wasting, and streams. The mountains have a smooth, full-bodied appearance, with rather rounded divides. Soil and regolith are thick. Imagine now that a climatic change results in the accumulation of snow in the heads of most of the valleys high on the mountain sides.

An early stage of glaciation is shown at the right side of Figure 22.3*B*, where snow is collecting and cirques are being carved by the outward motion of the ice and by intensive frost shattering of the rock near the masses of compacted snow.

In Figure 22.3*B*, glaciers have filled the valleys and are integrated into a system of tributaries that feed a trunk glacier. Tributary glaciers join the main glacier with smooth, accordant junctions. The cirques grow steadily larger. Their rough, steep walls soon replace the

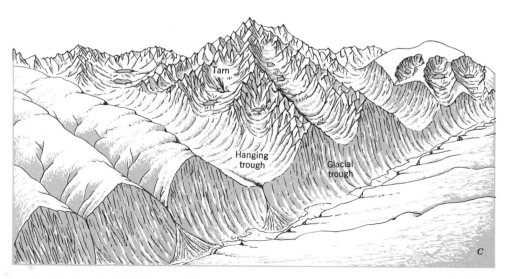

figure 22.4
The Swiss Alps appear from the air as a sea of sharp rock crests and toothlike horns. In the foreground is a cirque. (Swissair Photo.)

smooth, rounded slopes of the original mountain mass. Where two cirque walls intersect from opposite sides, a jagged, knifelike ridge is formed. Where three or more cirques grow together, a sharp-pointed peak is formed. The name *horn* is applied to such peaks in the Swiss Alps (Figure 22.4). One of the best known is the striking Matterhorn.

Glacier flow constantly deepens and widens its channel, so that after the ice has finally disappeared, a deep, steep-walled *glacial trough* remains (Figure 22.3C). The U-shape of its cross-profile is characteristic of a glacial trough. Tributary glaciers also carve U-shaped troughs, but they are smaller in cross section, with floors lying high above the floor level of the main trough; they are called *hanging troughs*. Streams, which later occupy the abandoned trough systems, form scenic

waterfalls and cascades where they pass down from the lip of a hanging trough to the floor of the main trough. High up in the smaller troughs the bedrock is unevenly excavated, so that the floors of troughs and cirques contain rock basins and rock steps. The rock basins are occupied by small lakes, called *tarns*. Major troughs sometimes hold large, elongated lakes.

Debris is carried by an alpine glacier within the ice and also dragged along between the ice and the valley wall. Where two ice streams join, this marginal debris is dragged along to form a narrow band riding on the ice in midstream (Figure 22.1). At the terminus of a glacier debris accumulates in a heap called a *terminal moraine*. This heap takes the form of a curved embankment lying across the valley floor and bending upvalley along each wall of the trough (Figure 22.5).

figure 22.3 (left)
Landforms produced by alpine glaciers. (A) Before glaciation sets in, the region has smoothly rounded divides and narrow, V-shaped stream valleys. (B) After glaciation has been in progress for thousands of years, new erosional forms are developed. (C) With the disappearance of the ice a system of glacial troughs is exposed. (After A. K. Lobeck.)

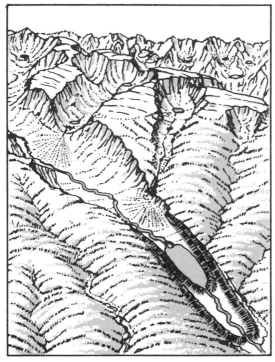

figure 22.5
Moraines of a former valley glacier appear as curved embankments marking successive positions of the ice margins. (After W. M. Davis.)

Glacial troughs and fiords

Many large glacial troughs now are filled with alluvium and have flat floors. Aggrading streams that issued from the receding ice front were heavily laden with rock fragments so that the deposit of alluvium extended far downvalley. Figure 22.6 shows a comparison between a trough with little or no fill and another with an alluvial-filled bottom.

When the floor of a trough open to the sea lies below sea level, the seawater enters as the ice front recedes. The result is a deep, narrow estuary known as a *fiord* (Figure 22.7). Fiords are observed to be opening up today along the Alaskan coast, where some glaciers are melting back rapidly and the ocean waters are being extended inland along the troughs. Fiords are found largely along mountainous coasts in lat.

50° to 70° N and S. On these coasts, glaciers were nourished by heavy snowfall of the orographic type, associated with the marine west coast climate (Chapter 14).

Environmental aspects of alpine glaciation

In past centuries the high, rugged terrain produced by alpine glaciation was sparsely populated and formed impassable barriers between settlements or between nations. This very difficulty of access protected the alpine terrain from the impact of Man and has left us a legacy of wilderness areas of striking scenic beauty. Mountainsides steepened by glaciation make ideal ski slopes. Ski resorts have mushroomed by the dozens in glaciated mountains, and this form of recreation is now a major industry. In summer, these same areas are invaded by tens of thousands of campers and hikers. Rock climbing draws hundreds more to scale the near-vertical rock walls of glacial troughs, cirques, and horns.

The floors of glacial troughs have served for centuries as major lines of access deep into the heart of glaciated alpine ranges. For example, in the Italian Alps, several major flat-floored glacial troughs extend from the northern plain of Italy deep into the Alps. The principal Alpine passes between Italy on the south and Switzerland and Austria on the north lie at the heads of these troughs. An example is the Brenner Pass, located at the head of the Adige trough.

Ice sheets of the present

Two enormous accumulations of glacial ice are the Greenland and Antarctic ice sheets. These are huge plates of ice, several thousand feet thick in the central areas, resting on landmasses of subcontinental size. The Greenland Ice Sheet has an area of 670,000 sq

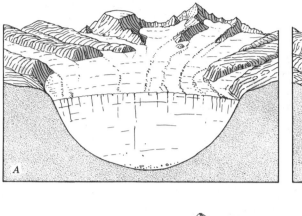
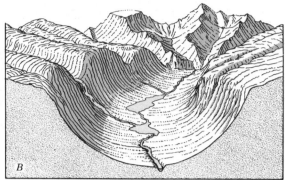
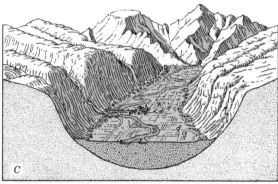

figure 22.6
Development of a glacial trough. (A) During maximum glaciation, the U-shaped trough is filled by ice to the level of the small tributaries. (B) After glaciation, the trough floor may be occupied by a stream and lakes. (C) If the main stream is heavily loaded, it may fill the trough floor with alluvium. (D) Should the glacial trough have been deepened below sea level, it will be occupied by an arm of the sea, or fiord. (After E. Raisz.)

figure 22.7
This Norwegian fiord has the steep rock walls of a deep glacial trough. (Mittet and Co.)

mi (1,740,000 sq km) and occupies about seven eighths of the entire island of Greenland (Figure 22.8). Only a narrow, mountainous coastal strip of land is exposed.

The Antarctic Ice Sheet covers 5 million sq mi (13 million sq km) (Figure 22.9). A significant point of difference between the two ice sheets is their position with reference to the poles. The antarctic ice rests almost squarely on the south pole, whereas the Greenland Ice Sheet is considerably offset from the north pole, with its center at about lat. 75° N. This position

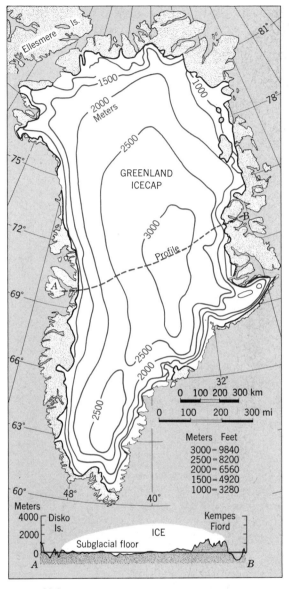

high point at an altitude of about 10,000 ft (3000 m), there is a gradual slope outward in all directions. The rock floor of the ice sheet lies near sea level under the central region, but is higher near the edges. Accumulating snows add layers of ice to the surface, while at great depth, the plastic ice slowly flows outward toward the edges. At the outer edge of the sheet the ice thins down to a few hundred feet. Continual loss through ablation keeps the position of the ice margin relatively steady where it is bordered by a coastal belt of land. Elsewhere the ice extends in long tongues, called outlet glaciers, to reach the sea at the heads of fiords. From the floating glacier edge huge masses of ice break off and drift out to open sea with tidal currents to become icebergs (Chapter 15).

Ice thickness in Antarctica is even greater than that of Greenland (Figure 22.9). For example, on Marie Byrd Land, a thickness of 13,000 ft (4000 m) was measured. Here the rock floor lies 6500 ft (2000 m) below sea level.

An important glacial feature of Antarctica is the presence of great plates of floating glacial ice, called *ice shelves* (Figure 22.9). The largest of these is the Ross Ice Shelf with an area of about 200,000 sq mi (520,000 sq km) and a surface height averaging about 225 ft (70 m) above the sea. Ice shelves are fed by the ice sheet, but they also accumulate new ice through the compaction of snow.

figure 22.8
The Greenland Ice Sheet. (After R. F. Flint, <u>Glacial and Pleistocene Geology</u>.)

illustrates a fundamental principle: that a large area of high land is essential to the accumulation of a great ice sheet. No land exists near the north pole; ice there exists only as sea ice.

The surface of the Greenland Ice Sheet has the form of a very broad, smooth dome. From a

Ice sheets of the Pleistocene Epoch

Much of North America and Europe and parts of northern Asia and southern South America were covered by enormous ice sheets in a period of time named the *Pleistocene Epoch*. This ice age ended 10,000 to 15,000 years ago with the rapid wasting away of the ice sheets. Landforms made by the last ice advance and

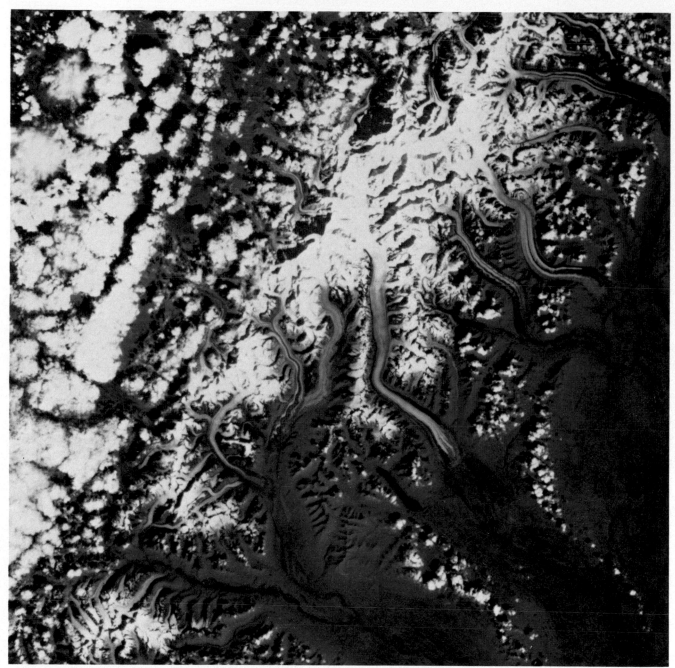

Alaskan Glaciers

Photographed by ERTS-1 orbiting satellite from a height of about 570 mi (920 km) glaciers of the Alaska Range in south central Alaska appear as blue curving bands, emerging from high collecting grounds on a snow-covered mountain axis trending from northeast to southwest across the center of the photo. The white patches at the left and right are clouds. Darker lines running down the length of a glacier are medial moraines, formed of rock debris on the ice surface. Where these moraines have been distorted into a sinuous pattern, the glacier has experienced a rapid downvalley surge at rates up to 4 ft (1.2 m) per hour. Those glaciers with smooth moraines, paralleling the banks, are experiencing very slow, uniform flow throughout their entire length. The group of high mountain peaks in the upper, central part of the area includes Mt. McKinley, highest point in North America. This is a false-color image, created by combining data of three narrow spectral bands, each assigned a primary color for printing. The area shown in about 65 mi (105 km) across; north is toward the top. (NASA — EROS Data Center, No. 81033210205A2)

(below) **Cirques and horns of the lofty Hindu Kush range, Afghanistan. Glaciers occupying the cirques hang far above the empty main glacial troughs. (Kenneth Boche)**

(left) **This wasting ice surface is in the lower, ablation zone of the Capps Glacier, Wrangell Mountains, Alaska. The man is standing in the channel of a meltwater stream. (Lynne Dozier)**

(above) **Looking down from a crest in the Hindu Kush range in Afghanistan we see a large, debris-covered alpine glacier (left) joined by small tributary glacier (center) with a deeply crevassed ice fall. (Shelly Sack)**

Alpine Glaciers

(right) **The Mer de Glace, in the Swiss Alps, seen from an aerial tramway. The glacier is greatly shrunken in both depth and width as compared with its full dimensions two centuries ago, during the Little Ice Age. A massive lateral moraine runs along the left side of the glacier. (Paul W. Tappan)**

(below) **Chamonix, Switzerland, occupies the bottom of a great glacial trough of the Swiss Alps. Alluvial deposits have formed a flat valley floor. (Paul W. Tappan)**

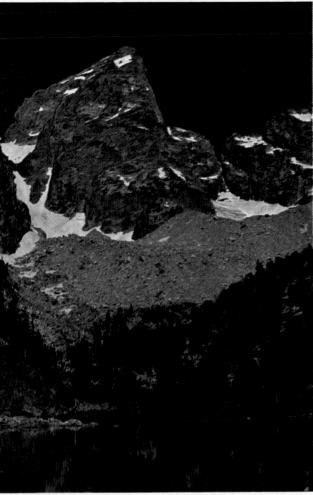

(right) **This abraded surface of granite has been striated and polished by an alpine glacier. (Mark A. Melton)**

(above) **Delta Lake occupies the floor of a cirque in the Grand Teton Range, Wyoming. In the background is a mass of morainal debris burying small relict glacial ice bodies. (Mark A. Melton)**

Alpine Glacial Landforms

(left) **A U-shaped glacial trough in the Beartooth Plateau, near Red Lodge, Montana. Talus cones have been built out from the steep trough walls. (Alan H. Strahler)**

(below) **Deep glacial grooves in limestone, Kelley's Island State Park, Lake Erie, (Orlo E. Childs)**

(above) **A Pleistocene continental ice sheet carved this finger-lake trough in granite on Mt. Desert Island, Maine. Beyond is the smoothly rounded ice-abraded profile of Mt. Sargent. (Arthur N. Strahler)**

(right) **This esker in southern Michigan is being excavated as a source of sand and gravel. The coarse boulders are screened out. (Orlo E. Childs)**

(above) **These striations on a glacially polished surface of marble, near West Rutland, Vermont, prove the former presence of an ice sheet over northern New England. (Orlo E. Childs)**

Landforms of Continental Glaciation

(below) **Seen close up, these layers of sand and gravel in a glacial delta kame are nicely sorted by grain size. Curvature of the upper layers shows flowage accompanying deposition. (Orlo E. Childs)**

(above) **Cattle are grazing on a glacial kame, adjacent to a steep-sided kettle, near Guilford, in southern New York State. (Arthur N. Strahler)**

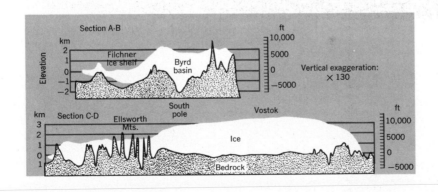

figure 22.9
The Antarctic Ice Sheet and
its ice shelves. (Based on
data of American
Geophysical Union.)

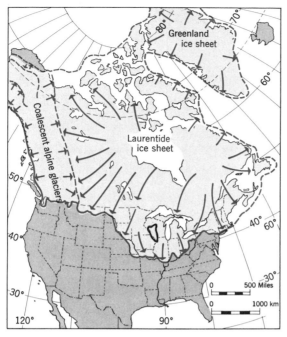

figure 22.10

Pleistocene ice sheets of North America at their maximum extent reached as far south as the present Ohio and Missouri Rivers. (After R. F. Flint.)

figure 22.11

The Scandinavian Ice Sheet dominated northern Europe during the Pleistocene glaciations. Solid line shows limits of ice in the last glacial stage; dotted line on land shows maximum extent at any time. (After R. F. Flint.)

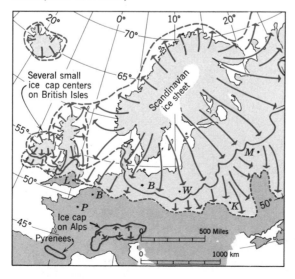

recession are very fresh today and show little modification by erosion.

Figures 22.10 and 22.11 show the extent to which North America and Europe were covered at the maximum known spread of the last advance of the ice. In the United States, most of the land lying north of the Missouri and Ohio rivers was covered, as well as northern Pennsylvania and all of New York and New England. In Europe, the ice sheet centered on the Baltic Sea, covering the Scandinavian countries and spreading as far south as central Germany. The British Isles were almost covered by a small ice sheet that had several centers on highland areas and spread outward to coalesce with the Scandinavian Ice Sheet. The Alps at the same time were heavily inundated by enlarged alpine glaciers, fused into a single body of ice. All high mountain areas of the world underwent greatly intensified alpine glaciation at the time of maximum ice-sheet advance. Today, only small alpine glaciers remain and, in less favorable locations, all the glaciers are gone.

Since the middle of the nineteenth century, when the great naturalist Louis Agassiz first defended the bold theory of continental glaciation, careful study of the deposits left by the ice has led to the knowledge that not one but four major advances and retreats occurred. These events were spaced over a total period of about 1 million years. Deposits of the last advance, known as the *Wisconsin Stage*, form fresh and conspicuous landforms. For the most part, then, our discussion of glacial landforms will concern these most recent deposits.

Glacial stages

The four principal glacial stages of North America are matched by corresponding stages in Europe. Between each glacial stage was an interglacial stage, a period of mild climate

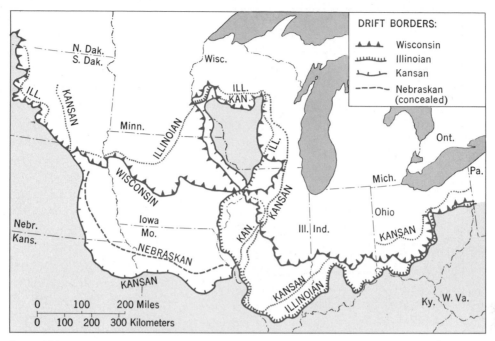

figure 22.12
**In each glacial stage, the ice sheet reached a different line of
maximum advance. (After R. F. Flint, <u>Glacial and Pleistocene Geology</u>.)**

during which the ice sheets disappeared. The American glacial stages in order of increasing age are:

Wiconsin glacial
    Sangamon interglacial
Illinoian glacial
    Yarmouth interglacial
Kansan glacial
    Aftonian interglacial
Nebraskan glacial

Figure 22.12 shows the limit of southerly spread of ice of each glacial stage in the north-central United States. Where older limits were overridden by younger, the boundaries are conjectural. Notice that an area in Wisconsin was surrounded by ice but never inundated.

The absolute age and duration in years of the Pleistocene Epoch and its stages are extremely difficult to fix, even though the relative order of events is well established. The last ice disappeared 10,000 to 15,000 years ago from the north-central United States. The Wisconsin Stage may have endured 60,000 years. The earliest onset of glacial advance (Nebraskan) may have occurred 300,000 to 600,000 years ago. The total duration of the Pleistocene Epoch was 1 to 2 million years.

Cause of continental glaciation

There is excellent geologic evidence of occasional periods of glaciation in the early part of the earth's geologic history. The underlying cause of climate change necessary to bring on glaciations is speculative. Obviously a prolonged period of colder climate with ample snowfall initiates the growth of ice sheets.

One possible contributing cause of glaciation is a decrease in the quantity of solar radiation intercepted by the earth. If this quantity of energy diminished somewhat at the beginning of the Pleistocene Epoch, the average temperature of the earth's atmosphere would have dropped to a lower level. Providing that this change did not also diminish the quantity of snowfall, the reduced ablation of snowfields would have increased the quantity of snow turned into glacial ice. As yet, no evidence has been found of the supposed reduction in output of solar energy.

A second factor, possibly working in harmony with the first, is the known increase in altitude of large parts of the continents during the Pliocene and early Pleistocene epochs. This height increase was a result of widespread mountain making, as well as broad uparching of continental interiors. We know that mountains intercept large quantities of precipitation by the orographic mechanism. The combined effects of reduced solar energy and increased altitude of continents could have brought on colder climates with increased snowfall over favorable highland areas of the continents, such as the Laurentian Upland of eastern Canada and the Scandinavian peninsula.

Another important theory attributes glaciation to a reduction of the carbon dioxide content of the atmosphere. The role of carbon dioxide in absorbing longwave radiation and warming the atmosphere has been explained in Chapter 3. It is estimated that if the carbon dioxide content of the atmosphere were reduced by half, the earth's average surface temperature would drop about 7 F° (4 C°). A reduction in carbon dioxide content coinciding by chance with an increase in continental altitude might bring on the growth of ice sheets.

Other theories invoke quite different mechanisms. It has been postulated that increased quantities of volcanic dust in the atmosphere might bring on glaciation because more solar energy would be reflected back into space, permitting less to enter the lower atmosphere. Along with the reduced air temperature, there would be an increase in number of tiny dust particles to serve as nuclei for the condensation of moisture, favoring increased precipitation.

Another theory proposes shifts in the positions of the continents with respect to the poles, bringing various parts of the landmasses into favorable geographical positions for the growth of ice sheets. Still another theory requires that changes in oceanic currents, specifically the diversion or blocking of warm currents such as the Gulf Stream, would have brought colder climates to the subarctic regions. Variations in the earth's orbit, causing changes in the amounts of solar energy received by the earth, have also been considered as the cause of glaciations.

Erosion by ice sheets

Like alpine glaciers, ice sheets are effective eroding agents. The slowly moving ice scraped and ground away much solid bedrock, leaving

figure 22.13
Glacial striations and fracture marks cover the smoothly rounded surface of this rock knob. The marks were made by the East Twin Glacier, Alaska. The ice moved in a direction away from the photographer. (Maynard M. Miller.)

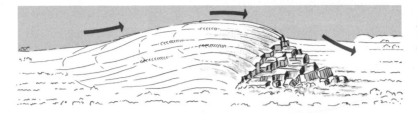

figure 22.14
A glacially abraded rock knob. (From A. N. Strahler, 1971, The Earth Sciences, 2nd ed., Harper & Row, New York.)

behind smoothly rounded rock masses bearing countless scratches tending in the general direction of ice movement (Figure 22.13). The evidences of ice abrasion are common throughout glaciated regions of North America. They may be seen on almost any exposed hard rock surface.

Conspicuous knobs of solid bedrock shaped by the moving ice are also common features (Figure 22.14). One side, that from which the ice was approaching, is characteristically smoothly rounded. The lee side, where the ice plucked out angular joint blocks, is irregular and blocky.

Vastly more important than the minor abrasion forms are enormous rock excavations made by the ice sheets. These excavations were localized where the bedrock was weak and the ice current was accentuated by the presence of a valley paralleling the direction of ice flow. Under such conditions the ice sheet behaved much as a valley glacier, scooping out a deep, U-shaped trough. The Finger Lakes of western New York State are fine examples (Figure 22.15). Here a set of former stream valleys lay parallel to southward spread of the ice, which scooped out a series of deep troughs. Blocked at the north ends by glacial debris, the basins now hold long, deep lakes.

Many hundreds of lake basins were created by glacial erosion and deposition over the glaciated portion of North America.

figure 22.15
Seen from an altitude of 10,000 ft (3000 m), Canandaigua Lake occupies a glacially-deepened trough within the northern fringe of the Appalachian Plateau in New York. In this infrared photograph the green vegetation of fields and forests is almost white in contrast to the lake water, which appears black. Lake Ontario lies in the distance. (Wahl's Photographic Service, Inc., Pittsford, N.Y.)

Deposits left by ice sheets

The term *glacial drift* includes all varieties of rock debris deposited in close association with glaciers. Drift is of two major types (1) *Stratified drift* consists of layers of sorted and stratified clays, silts, sands, or gravels deposited by melt-water streams or in bodies of water adjacent to the ice (2) *Till* is a heterogeneous mixture of rock fragments ranging in size from clay to boulders, deposited directly from the ice without water transport.

Over those parts of the United States formerly covered by Pleistocene ice sheets, glacial drift averages from 20 ft (6 m) thick over mountainous terrain such as New England to 50 ft (15 m) and more thick over the lowlands of the north-central United States. Over Iowa, drift is from 150 to 200 ft (45 to 60 m) thick; over Illinois, it averages more than 100 ft (30 m) thick. Locally, where deep stream valleys existed prior to glacial advance, as in Ohio, drift is several hundred feet deep.

To understand the form and composition of deposits left by ice sheets, you need to consider the conditions prevailing at the time of existence of the ice. Figure 22.16 *A* shows a region partly covered by an ice sheet with a stationary front edge. This condition occurs when the rate of ice ablation balances the amount of ice brought forward by spreading of the ice sheet. Although the Pleistocene ice fronts advanced and receded in many minor and major fluctuations, there were long periods when the front was essentially stable and thick deposits of drift accumulated.

The transportational work of an ice sheet is like that of a huge conveyor belt. Anything carried on the belt is dumped off at the end and, if not constantly removed, will pile up in increasing quantity. Rock fragments brought within the ice are deposited at the edge as the ice evaporates or melts. There is no possibility of return transportation.

Glacial till that accumulates at the immediate ice edge forms an irregular, rubbly heap, the terminal moraine. After the ice has disappeared (Figure 22.16*B*), the moraine appears as a belt of knobby hills interspersed with basinlike hollows, some of which hold small lakes. The name *knob and kettle* is often applied to morainal belts (Figure 22.17). Terminal moraines form great curving patterns; the convex curvature is southward and indicates that the ice advanced as a series of great *ice lobes*, each with a curved front (Figure 22.18). Where two lobes come together, the moraines curve back and fuse together into a single moraine pointed northward. This deposit is an *interlobate moraine* (Figure 22.16*B*). In its general recession accompanying disappearance, the ice front paused for some time along a number of lines, causing morainal belts similar to the terminal moraine belt to be formed. These belts are known as *recessional moraines* (Figures 22.16 and 22.18). They run roughly parallel with the terminal moraine but are often thin and discontinuous.

Figure 22.16 *A* shows a smooth, sloping plain lying in front of the ice margin. This is the *outwash plain*, formed of stratified drift left by braided streams issuing from the ice. The plain is built of layer upon layer of sands and gravels.

X figure 22.16 (right)

Marginal landforms of continental glaciers. (A) With the ice front stabilized and the ice in a wasting, stagnant condition, various depositional features are built by meltwater. (B) The ice has wasted completely away, exposing a variety of new landforms made under the ice.

A

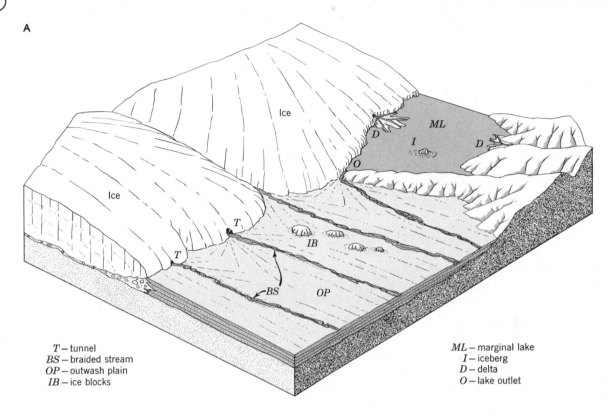

T — tunnel
BS — braided stream
OP — outwash plain
IB — ice blocks

ML — marginal lake
I — iceberg
D — delta
O — lake outlet

B

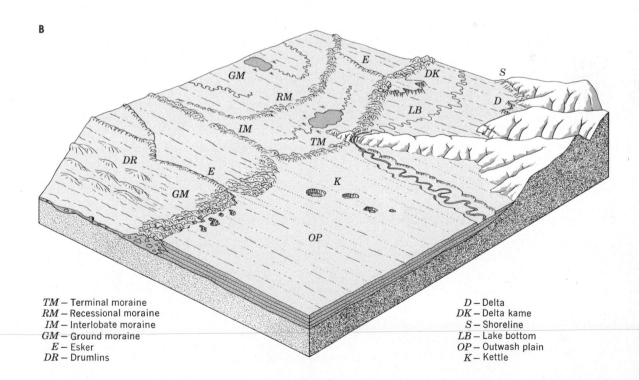

TM — Terminal moraine
RM — Recessional moraine
IM — Interlobate moraine
GM — Ground moraine
E — Esker
DR — Drumlins

D — Delta
DK — Delta kame
S — Shoreline
LB — Lake bottom
OP — Outwash plain
K — Kettle

figure 22.17
Rugged topography of small knobs and kettles characterizes this interlobate moraine northeast of Elkhart Lake, Wisconsin. (U.S. Geological Survey.)

Drumlins were formed under moving ice by a plastering action in which layer upon layer of bouldery clay was spread on the drumlin.

Between moraines, the surface overridden by the ice is overspread by a cover of glacial till known as *ground moraine*. This cover is often inconspicuous because it forms no prominent landscape feature. The ground moraine may be thick and may obscure or entirely bury the hills and valleys that existed before glaciation. Where thick and smoothly spread, the ground moraine forms a level *till plain*. Plains of this origin are widespread throughout the Middle West.

Between the ice front and rising ground, valleys that may have opened out northward

Large streams issue from tunnels in the ice, particularly when the ice for many miles back from the front has become stagnant, without forward movement. Tunnels then develop throughout the ice mass, serving to carry off the meltwater. After the ice has gone, the position of a former ice tunnel is marked by a long, sinuous ridge known as an *esker*. The esker is the deposit of sand and gravel formerly laid on the floor of the ice tunnel. After the ice has melted away, only the stream-bed deposit remains, forming a ridge (Figure 22.19). Eskers are often many miles long; a few are more than 100 mi (160 km) long.

Another common glacial form is the *drumlin*, a smoothly rounded, oval hill resembling the bowl of an inverted teaspoon. It consists of glacial till (Figure 22.20). Drumlins invariably lie in a zone behind the terminal moraine. They commonly occur in groups or swarms, which may number in the hundreds. The long axis of each drumlin parallels the direction of ice movement, and the drumlins thus point toward the terminal moraines and serve as indicators of direction of ice movement.

figure 22.18
Moraine belts of the north-central United States have a festooned pattern left by ice lobes. (After R. F. Flint and others, Glacial Map of North America.)

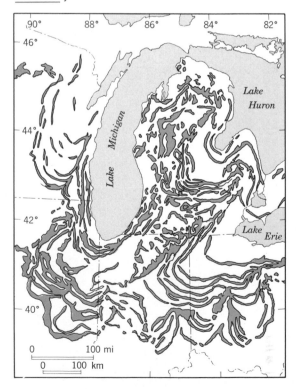

figure 22.19
Seen from the air, this esker in the Canadian shield looks like a narrow railroad embankment crossing a terrain of glacially eroded lake basins. (Geological Survey of Canada.)

were blocked by ice. Under such conditions, *marginal glacial lakes* formed along the ice front (Figure 22.16*A*). These lakes overflowed along the lowest available channel between the ice and the rising ground slope, or over some low pass along a divide. Streams of meltwater from the ice built *glacial deltas* into these marginal lakes. When the ice disappeared, the lake drained away, exposing the bottom. Here layers of fine clay and silt had accumulated. Glacial lake plains are extremely flat, with meandering streams and extensive areas of marshland.

Deltas, built with a flat top at what was formerly the lake level, are now curiously isolated, flat-topped landforms known as *delta kames* (Figure 22.16*B*). Built of well-washed and well-sorted sands and gravels, kames commonly show steeply dipping beds (Figure 22.21).

figure 22.21
Cross-bedded, sorted sands of a delta kame near North Haven, Connecticut. (R. J. Lougee.)

figure 22.20
This small drumlin, located south of Sodus, New York, shows a tapered form from upper right to lower left, indicating that the ice moved in that direction (north to south). (Ward's Natural Science Establishment, Inc., Rochester, N.Y.)

Environmental and resource aspects of glacial deposits

Because much of Europe and North America was glaciated by the Pleistocene ice sheets, landforms associated with the ice are of major environmental importance, and the deposits constitute a natural resource as well. Agricultural influences of glaciation are both favorable and unfavorable, depending on preglacial topography and whether the ice eroded or deposited heavily.

In hilly or mountainous regions, such as New England, the glacial till is thinly distributed and extremely stony. Podzolic soils that developed on glacial deposits of the northern United States and Canada are acid and low in fertility. Extensive bogs, unsuited to agriculture unless transformed by water drainage systems, are another unfavorable element. Early settlers found cultivation difficult because of countless boulders and cobbles in the soil. Till accumulations on steep mountain slopes are subject to mass movements in the form of earth flows. Clays in the till become weakened after absorbing water from melting snows and spring rains. Where slopes have been oversteepened by excavation for highways, movement of till is a common phenomenon.

Along moraine belts the steep slopes, irregularity of knob-and-kettle topography and abundance of boulders conspired to prevent crop cultivation but invited use as pasture. These same features, however, make morainal belts extremely desirable as suburban residential areas. Pleasing landscapes of hills, depressions, and small lakes made ideal locations for large estates.

Flat till plains, outwash plains, and lake plains, on the other hand, comprise some of the most productive agricultural land in the world. In this class belong the prairie lands of Indiana, Illinois, Iowa, Nebraska, and Minnesota. We must not lose sight of the fact that in these areas wind-deposited silt (loess) forms a blanket over clay-rich till and sandy outwash. Exposed glacial drift would be a poor parent base for soil.

Stratified drift deposits are of great economic value. The sands and gravels of outwash plains, delta kames, and eskers provide the aggregate necessary for concrete and the base courses beneath highway pavements. The purest sands may be used for molds, needed for metal castings.

Stratified drift, where thick, forms an excellent aquifer and is a major source of ground water supplies. Deep accumulations of stratified sands in preglacial valleys are capable of yielding ground water in adequate quantities for municipal and industrial uses. Water development of this type is widespread in Ohio, Pennsylvania, and New York.

Landforms Made by Waves

Chapter 23

THE CONTINENTAL SHORELINE, where the salt water of the oceans contacts fresh water and the solid mineral base of the continents, is a complex environmental zone of great importance to Man. Humans have occupied the shore zone for a number of reasons. First, there are food resources of shellfish, finfish, and waterfowl to be had in the shallow waters and estuaries. Second, the shoreline is a base from which ships embark to seek marine food resources further from land and to transport people and goods between the continents. In time of war the shoreline is a critical barrier to be defended from invading forces arriving by sea. Third, the coastal zone is a recreational facility with its sea breezes and bathing beaches and its opportunities for surfing, skin diving, sport fishing, and boating.

Along with its opportunities, the shore zone imposes restraints and hazards on Man and man-made structures. Some coasts are rocky and cliffed; they provide little or no shelter in the form of harbors. Along other coasts the enormous energy of storm wave can cut back the shore, undermining buildings and roads. High water levels in time of storm can cause inundation of low-lying areas and bring the force of breaking waves to bear on ground many feet above the normal levels reached by seawater. Storm surges and seismic sea waves, which cause such inundations, have already been described in earlier chapters. For centuries, Man has been at war with the sea; building fortresslike walls to keep out the sea, and even forcing the sea to give up coastal land so as to produce more crops for food and forage.

Great segments of our coastlines face environmental degradation and destruction as urbanization of the coastal zone demands more land and expanded port facilities. The pressures of a growing population to seek its holiday pleasures along the shore threaten to destroy the very benefits that the shore offers. Management of the environment of the continental shorelines requires a knowledge of the natural forms and processes of this sensitive zone; the purpose of this chapter is to provide that basic knowledge.

Throughout this chapter we will use the term *shoreline* to mean the shifting line of contact between water and land. The broader term *coastline*, or simply *coast*, refers to a zone in which coastal processes operate or have a strong influence. The coastline includes the shallow water zone in which waves perform their work, as well as beaches and cliffs shaped by waves, and coastal dunes.

Wave erosion

Waves travel across the deep ocean with little loss of energy. When waves reach shallow water, the drag of the bottom slows and steepens the wave until it leaps forward and collapses as a *breaker* (Figure 23.1). Many tons of water surge forward, riding up the beach slope. Where a cliff lies within reach of the moving water, it is impacted with enormous

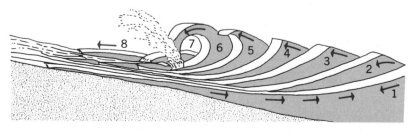

figure 23.1
A breaking wave.

figure 23.2
Tremendous forward thrust is evident in these storm waves breaking against a sea wall at Hastings, England.

force (Figure 23.2). Rock fragments of all sizes, from sand to cobbles, are carried by the surging water and thrust against bedrock of the cliff. The impact breaks away new rock fragments and the cliff is undercut at the base.

figure 23.3
Storm waves breaking against a coast underlain by weak sand quickly undermined this shore home at Seabright, New Jersey. The barrier of wooden pilings (right) proved ineffective in preventing cutting back of the cliff. (Douglas Johnson.)

Where weak, incoherent materials make up the coastline — glacial till or outwash, or alluvium — the force of the moving water alone easily cuts into the coastline. Here, erosion is rapid, and the shoreline may recede many feet in a single storm, undermining nearby buildings and roads (Figure 23.3).

Some details of a wave-cut cliff, or *sea cliff*, are shown in Figures 23.4 and 23.5. A deep basal indentation, the wave-cut notch, marks the line of most intense wave erosion. The waves find points of weakness in the bedrock, penetrating deeply to form crevices and sea caves. More resistant rock masses project seaward and are cut through to make picturesque arches. After an arch collapses the remaining rock column forms a stack, but this is ultimately leveled.

As the sea cliff retreats landward, continued wave abrasion forms an *abrasion platform*. This sloping rock floor continues to be eroded and widened by abrasion beneath the breakers. If a beach is present, it is little more than a thin layer of gravel and cobblestones.

Sea cliffs are features of spectacular beauty as well as habitats for many forms of life, including sea mammals and shore birds. Only in recent years has the need to preserve these cliffed coasts in their natural state been fully appreciated. Although the strength of these rocky features resists manmade alterations, the cliff line is vulnerable to heavy use for summer homes, motels, and restaurants. Intensive use not only destroys the pristine scenery, but it also adds pollution.

Beaches

Where sand is in abundant supply, it accumulates as a thick, wedge-shaped deposit, or *beach*. Beaches absorb the energy of breaking waves. During short periods of storm, the beach is cut back, but the sand is restored during long periods when waves are weak.

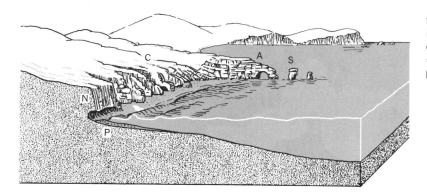

figure 23.4
Landforms of sea cliffs. A = arch; S = stack; C = cave; N = notch; P = abrasion platform. (After E. Raisz.)

In this way a beach may retain a stable configuration, or equilibrium state, over many years' time.

Beaches are shaped by alternate landward and seaward currents of water generated by breaking waves. After a breaker has collapsed, a foamy, turbulent sheet of water rides up the beach slope. This _swash_ is a powerful surge causing a landward movement of sand and gravel on the beach. When the force of the swash has been spent against the slope of the beach, a return flow, or _backwash_, pours down the beach (Figure 23.6). Sands and gravels are swept seaward by the backwash.

figure 23.5
The chalk cliffs of Normandy, along the French channel coast, show stacks, arches, and sea caves. A pocket beach (left) is intensively used by both fishermen and vacationers.

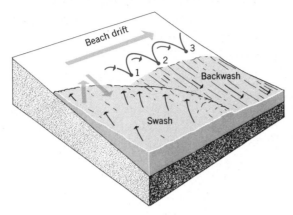

figure 23.6
Oblique approach of waves allows the swash and backwash to move sand grains along the shoreline in a series of arched paths.

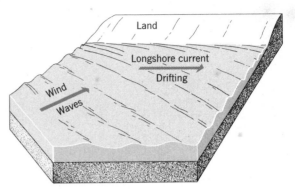

figure 23.7
Longshore current drifting.

Littoral drift

The unceasing shifting of materials with swash and backwash of breaking waves also results in a sidewise movement known as *beach drifting* (Figure 23.6). Wave fronts usually approach the shore obliquely rather than directly head on. The swash rides obliquely up the beach, and the sand is moved obliquely up the slope. After the water has spent its energy, the backwash flows down the slope of the beach in the most direct downhill direction. The particles are dragged directly seaward and come to rest at a position to one side of the starting place. On a particular day, wave fronts approach consistently from the same direction, so that this movement is repeated many times. Individual rock particles travel long distances along the shore. Multiplied many thousands of times to include the numberless particles of the beach, beach drifting becomes a leading form of sediment transport.

When waves approach a shoreline under the influence of strong winds, the water level is slightly raised near shore by a slow shoreward drift of water. The excess water pushed shoreward must escape. A longshore current is set up parallel to shore in a direction away from the wind (Figure 23.7). When wave and

wind conditions are favorable, this current is capable of moving sand along the sea bottom in a direction parallel to the shore. The process is called *longshore drifting*.

Both beach drifting and longshore drifting move particles in the same direction for a given set of onshore winds, and they supplement each other's influence. The total process is called *littoral drift*.

Littoral drift operates to shape shorelines in two quite different situations. Where the shoreline is straight or broadly curved for many miles at a stretch, littoral drift moves the sand along the beach in one direction for a given set of

figure 23.8
Along a straight coast littoral drift carries the sand in one direction to reach the mouth of a bay, where it is formed into a sandspit. (From A. N. Strahler, 1971, The Earth Sciences, 2nd ed., Harper & Row, New York.)

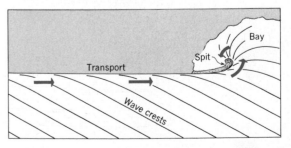

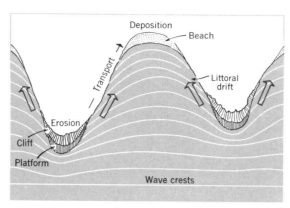

figure 23.9

On an embayed coast, sediment is carried from eroding headlands to the bayheads, where pocket beaches can accumulate. (From A. N. Strahler, 1971, The Earth Sciences, 2nd ed., Harper & Row, New York.)

prevailing winds. This situation is shown in Figure 23.8. Where a bay exists, the sand is carried out into open water as a long finger, or *sandspit*. As the sandspit grows, it forms a barrier, called a *bar*, across the mouth of the bay (see Figure 23.17).

A second situation is shown in Figure 23.9. Here the coastline consists of prominent headlands, projecting seaward, and deep bays. Erosional energy of the waves is concentrated on the exposed headlands. Here sea cliffs develop. Sediment from the eroding cliffs is carried by littoral drift along the sides of the bay, converging on the head of the bay. The result is a crescent-shaped beach, often called a *pocket beach*. A pocket beach provides a place to launch fishing boats and to beach them safely.

Littoral drift and shore protection

When sand is arriving at a particular section of the beach more rapidly than it can be carried away, the beach is widened and built shoreward. This change is called *progradation*.

When sand is leaving a section of beach more rapidly than it is being brought in, the beach is narrowed and the shoreline moves landward, a change called *retrogradation*.

Along stretches of shoreline affected by retrogradation, the beach may be seriously depleted or even entirely destroyed. When this occurs, cutting back of the coast can be rapid in weak materials, destroying valuable shore property. Protective engineering structures, such as sea walls, designed for direct resistance to frontal wave attack, are prone to failure and are extremely expensive, as well. In some circumstances, a successful alternative strategy is to install structures that will cause progradation, building a broad protective beach. The principle here is that the excess energy of storm waves will be dissipated in reworking the beach deposits. Cutting back of the beach in a single storm will be restored by beach-building between storms.

Progradation requires that sediment moving as littoral drift be trapped by the placement of baffles across the path of transport. To accomplish this result, groins are installed at close intervals along the beach. A *groin* is simply a wall or embankment built at right angles to the shoreline; it may be constructed of huge rock masses, of concrete, or of wooden pilings. Figure 23.10 shows the shoreline

figure 23.10

Construction of a groin causes marked changes in configuration of the sand beach. (From A. N. Strahler, 1972, Planet Earth, Harper & Row, New York.)

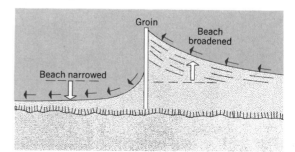

(above) Slabs of pink granite dip seaward on the Atlantic coast of Mt. Desert Island, Maine. Sea caves in the middle distance; a cobblestone pocket beach in the far distance. (Arthur N. Strahler)

Coastal Landforms

(right) Each property owner has built his own protective sea wall to suit his style along this stretch of Lake Michigan shoreline of northern Indiana. (Ned L. Reglein)

(below) A white sandspit, growing in a direction toward the observer, leaves only a narrow inlet (foreground) for tidal currents. Sandy shoals lie in the bay at the right. Martha's Vineyard, Massachusetts. (Donald W. Lovejoy)

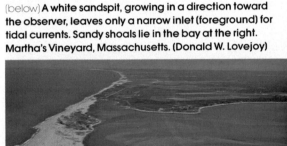

(above) Rock arches have been carved by waves from this sea cliff of horizontal sandstone strata on the Lake Superior shore. (Orlo E. Childs)

(above) The flat land surface at the left, above the rocky cliffs, is a marine terrace, a former abrasion platform now elevated above the sea. South of Lucia, California. (Orlo E. Childs)

(left) The ria coastline of Tasmania stretches beyond the city of Hobart. Submergence, partly a result of postglacial rise of sea level, has drowned many stream valleys. (Paul W. Tappan) (right) **The barrier island coast of North Carolina, seen from Apollo 9 spacecraft. The white barrier beach of sand — the Outer Banks — projects sharply seaward in two points: Cape Hatteras (H) and Cape Lookout (L). Pamlico Sound (P) lies between the barrier and the mainland shore, which is deeply embayed as a result of postglacial submergence. Extensive areas of salt marsh (S) make up the low coastal zone. Through two inlets (I) between Hatteras and Lookout sediment is being carried seaward in plumes of lighter color. The Gulf Stream boundary (G) is sharply defined. The area shown is about 100 mi (160 km) across. (NASA — No. AS9-20-3128)**

Barrier Island Coast

(below) **A close look at the barrier beach, Cape Hatteras. A dune ridge, well protected by grass cover, lies to the left of the beach. The view is north; Atlantic Ocean at the right. (Mark A. Melton)**

(left) **Near Cape Hatteras this stretch of barrier beach has been swept over by storm swash, topping the dune barrier and carrying sand in sheets toward the bay at the left (Mark A. Melton)** (right) **The Nile Delta** appears in true color as a dark blue-green triangle bounded by pale desert areas. Two major distributaries (N) of the Nile River end in prominent cusps: the Damietta Mouth (D) and the Rosetta Mouth (R). Alexandria (A) is just within view in the foreground. Sand carried by littoral drift from the river mouths has accumulated in a broad barrier beach (B), separated from the delta plain by an open lagoon (L). The Suez Canal (C) connects the Mediterranean Sea (left) with the Gulf of Suez (right). Beyond lies the Sinai Peninsula. (NASA — No. S-65-34776)

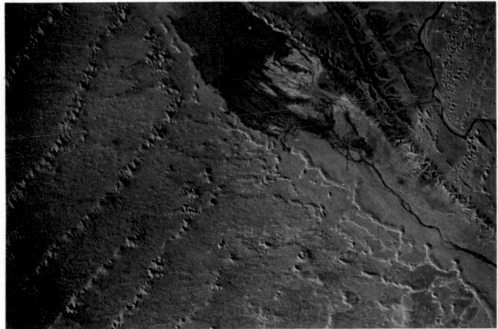

(left) **Star dunes of** reddish-yellow sand cover much of the area in this Gemini VII photo of the Sahara Desert in western Algeria. The upper bluish zone is a shallow lake. Water and sediment are brought to the lake basin by a major stream, Wadi Saura, entering from the lower right. Bedrock is exposed in narrow hogback ridges representing a deeply eroded fold structure in sedimentary rocks. The area shown is about 30 mi (50 km) across. (NASA — No. S65-6380)

Dunes and Loess

(left) **A great sand sea of transverse dunes near Yuma, Arizona. Slip faces tell us that the prevailing wind blows from right to left. (Arthur N. Strahler)**

(below) **This vertical road cut in thick loess, east of Vicksburg, Mississippi, has sustained little erosion over many years' time. (Orlo E. Childs)**

(right) **Closeup of the loess in the same cliff as in the photograph at right. The material is a soft, friable silt, deposited by wind during the late Pleistocene. (Orlo E. Childs)**

(above) **Deflation of this desert floor winnowed away the fines, leaving a pavement of pebbles, closely fitted together to protect the soft material beneath. Big Bend National Park, Texas. (Mark A. Melton)**

(above) **Coastal dunes advance landward along the Pacific shore near Morro Bay, California. Morro Rock, marine stack of plutonic rock, rises offshore in the distance. (Orlo E. Childs)**

(right) **This coastal blowout dune is advancing over a forest, the slip face gradually burying the tree trunks. Lake Michigan coast, near Michigan City, Indiana. (Barry Voight)**

figure 23.11
A system of groins for trapping of beach sand, Willoughby Spit, Virginia. Littoral drift is from lower left to upper right. (Department of the Army, Corps of Engineers.)

changes induced by groins. Sand accumulates on the up-drift side of the groin, developing a curved shoreline. On the down-drift side of the groin, the beach will be depleted because of the cutting off of the normal supply of drift sand. The result may be harmful retrogradation and cutting back of the coast. For this reason groins must be closely spaced so that the trapping effect of one groin will extend to the next (Figure 23.11). Ideally, when the groins have trapped the maximum quantity of sediment, beach drift will be restored to its original rate for the shoreline as a whole.

In some instances the source of beach sand is from the mouth of a river. Construction of dams far upstream on the river may drastically reduce the sediment load and therefore also cut off the source of sand for littoral drift. Retrogradation may then occur on a long stretch of shoreline. The Mediterranean shoreline of the Nile Delta has suffered retrogradation because of reduction in sediment supply following dam construction far up the Nile.

Tidal currents

Most marine coastlines are influenced by the *ocean tide,* a rhythmic rise and fall of sea level under the influence of changing attractive forces of moon and sun on the rotating earth. Where tides are great, the effects of changing water level and the currents set in motion are of major importance in shaping coastal landforms.

The tidal rise and fall of water level is graphically represented by the *tide curve*. We can make half-hourly observations of the position of water level against a measuring stick attached to a pier or sea wall. We then plot the changes of water level and draw the tide curve. Figure 23.12 is a tide curve for Boston Harbor covering a day's time. The water reached its maximum height, or *high*

figure 23.12
Height of water at Boston Harbor measured every half hour.

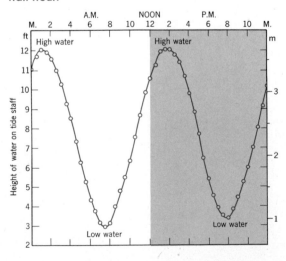

water, at the 12-ft (3.7-m) mark on the tide staff, then fell to its minimum height, or *low water,* occurring about 6¼ hours later. A second high water occurred about 12½ hours after the previous high water, completing a single tidal cycle. In this example the range of tide, or difference between heights of successive high and low waters, is 9 ft (2.7 m).

The rising tide sets in motion in bays and estuaries currents of water known as *tidal currents.* The relationships between tidal currents and the tide curve are shown in Figure 23.13. When the tide begins to fall, an *ebb current* sets in. This flow ceases about the time when the tide is at its lowest point. As the tide begins to rise, a landward current, the *flood current,* begins to flow.

Tidal current deposits

Ebb and flood currents generated by tides perform several important functions along a shoreline. First, the currents that flow in and out of bays through narrow inlets are very swift and can scour the inlet strongly to keep it open despite the tendency of shore drifting processes to close the inlet with sand.

figure 23.13

The ebb current flows seaward as the tide level falls; the flood current flows landward as the tide level rises.

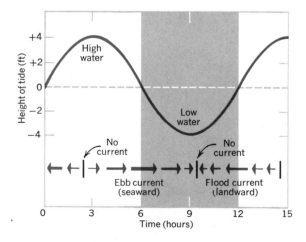

Second, tidal currents carry much fine silt and clay in suspension, derived from streams that enter the bays or from bottom muds agitated by storm wave action. This fine sediment settles to the floors of the bays and estuaries where it accumulates in layers and gradually fills the bays. Much organic matter is present in this sediment.

In time, tidal sediments fill the bays and produce *mud flats,* which are barren expanses of silt and clay exposed at low tide but covered at high tide. Next, a growth of salt-tolerant plants takes hold on the mud flat. The plant stems entrap more sediment and the flat is built up to approximately the level of high tide, becoming a *salt marsh* (Figure 23.14). A thick layer of peat is eventually formed at the surface. Tidal currents maintain their flow through the salt marsh by means of a highly complex network of sinuous tidal streams (Figure 23.15).

Salt marsh is important to geographers because it is land that can be drained and made agriculturally productive. The salt marsh is first cut off from the sea by construction of an embankment of earth (a dike), in which gates are installed to allow the freshwater drainage of the land to exit during ebb flow. Gradually, the salt water is excluded and soil water of the diked land becomes fresh. Such diked lands are intensively developed in Holland (poiders) and southeast England (fenlands).

Over many decades the surface of reclaimed salt marsh subsides because of compaction of the underlying peat layers and may come to lie well below mean sea level. The threat of flooding by salt water, when storm waves breach the dikes, hangs constantly over the inhabitants of such low areas. The reclamation of salt marsh by dike construction and drainage was practiced by New World settlers in New England and Nova Scotia. In the industrial era, large expanses of salt marsh have been destroyed by land fill, a practice only recently inhibited by new legislation.

figure 23.14
Salt marsh at South Wellfleet, Massachusetts. Late winter stubble and dead leaves of salt-marsh grass cover the peat layer. A small tidal channel lies drained empty at low tide. (A. N. Strahler.)

figure 23.15
This broad tidal marsh along the east coast of Florida is laced with serpentine tidal channels. (Laurence Lowry.)

Common kinds of coastlines

There are many different kinds of coastlines, each kind unique because of the distinctive landmass against which the ocean water has come to rest. One group of coastlines derives its qualities from *submergence,* the partial drowning of a coast by a rise of sea level or a sinking of the crust. Another group derives its qualities from *emergence,* the exposure of submarine landforms by a falling of sea level or a rising of the crust. Another group of coastlines results when new land is built out into the ocean by volcanoes and lava flows, by the growth of river deltas, or by the growth of coral reefs.

A few important types of coastlines are illustrated in Figure 23.16. The *ria coast (A)* is a deeply embayed coast resulting from submergence of a landmass dissected by streams. This coast has many offshore islands. A *fiord coast (B)* is deeply indented by steep-walled fiords, which are submerged glacial troughs (Chapter 22). The *barrier-island coast (C)* is associated with a recently emerged coastal plain. The offshore slope is very gentle, and a barrier island of sand is usually thrown up by wave action at some distance offshore. Large rivers build elaborate deltas, producing *delta coasts (D).* The *volcano coast (E)* is formed by eruption of volcanoes and lava flows, partly constructed below water level. Reef-building corals create new land and make a *coral-reef coast (F).* Down-faulting of the coastal margin of a continent can allow the shoreline to come to rest against a fault scarp, producing a *fault coast (G).*

Development of a ria coast

The ria coast is formed when a rise of sea level or a crustal sinking (or both) brings the shoreline to rest against the sides of valleys previously carved by streams. This event is illustrated in Figure 23.17A. Soon wave attack

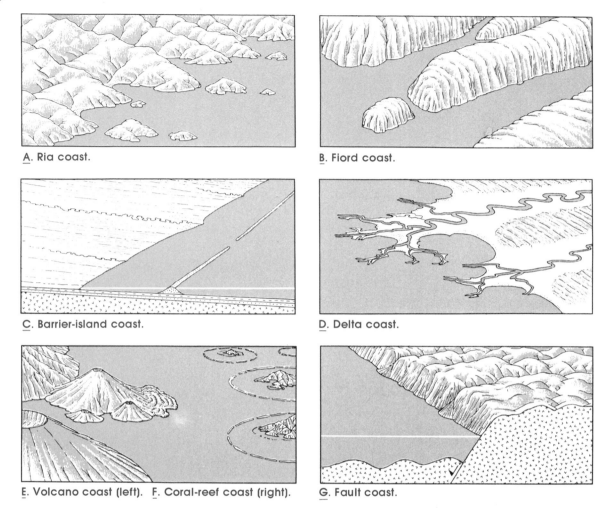

A. Ria coast.

B. Fiord coast.

C. Barrier-island coast.

D. Delta coast.

E. Volcano coast (left). F. Coral-reef coast (right).

G. Fault coast.

figure 23.16
Seven common kinds of coastlines are illustrated here. These examples have been selected to illustrate a wide range in coastal features.

forms cliffs on the exposed seaward sides of the islands and headlands (*B*). Sediment produced by wave action then begins to accumulate in the form of beaches along the cliffed headlands and at the heads of bays. This sediment is carried by littoral drift and is built into sandspits across the bay mouths and as connecting links between islands and mainland (*C*). Finally, all outlying islands are planed off by wave action and a nearly straight shoreline develops in which the sea cliffs are fully connected by baymouth bars (*D*). Now the bays are sealed off from the open ocean, although narrow tidal inlets may persist, kept open by tidal currents. Frame *E* shows a much later stage in which the coastline has receded beyond the inner limits of the original bays.

The influence of ria coastlines on human activity has been strong down through the ages. The deep embayments of the ria shoreline make splendid natural harbors. Much of the ria coastline of Scandinavia, France, and the British Isles is provided with such harbor facilities. Consequently, these peoples have a strong tradition of fishing, shipbuilding, ocean commerce, and marine activity generally. Mountainous relief of ria and fiord coasts made agriculture difficult or impossible, forcing the people to turn to the sea for a livelihood. New England and the Maritime Provinces of Canada have a ria coastline with abundant good harbors. The influence of this environment was to foster the same development of fishing, whaling, ocean commerce, and shipbuilding seen in the British Isles and Scandinavian countries.

Barrier-island coasts

In contrast to ria and fiord coasts, with their bold relief and deeply embayed outlines, we find low-lying coasts from which the land slopes gently beneath the sea. The coastal plain of the Atlantic and Gulf coasts of the

figure 23.17
Stages in the evolution of a ria coastline.

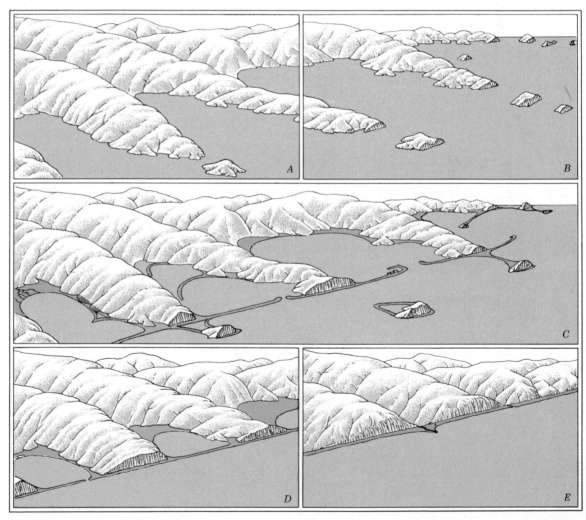

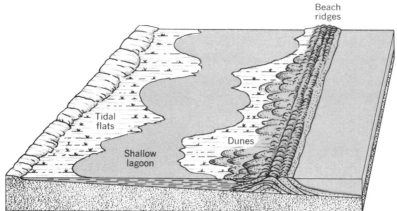

figure 23.18
A barrier island is separated from the mainland by a wide lagoon. Sediments fill the lagoon, while dune ridges advance over the tidal flats.

United States presents a particularly fine example of such a gently sloping surface. As we explained in Chapter 21, this coastal plain is a belt of relatively young sedimentary strata, formerly accumulated beneath the sea as deposits on the continental shelf. Emergence as a result of repeated crustal uplifts has characterized this coastal plain during the latter part of the Cenozoic era and into recent time.

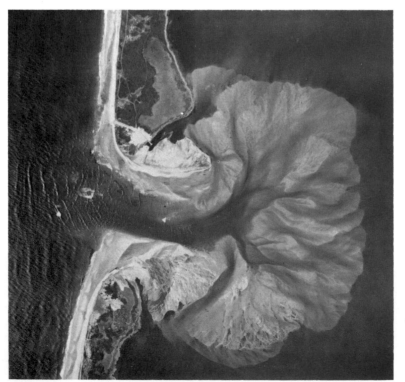

figure 23.19
East Moriches Inlet was cut through Fire Island, a barrier island off the Long Island shoreline, during a severe storm in March 1931. This aerial photograph, taken a few days after the breach occurred, shows the underwater tidal delta being built out into the lagoon (right) by currents. The entire area shown is about 1 mi (1.6 km) long. North is to the right; the open Atlantic Ocean on the left. (U.S. Army Air Force.)

Along much of the Atlantic and Gulf coast, a _barrier island_ exists, which is a low ridge of sand built by waves and further increased in height by the growth of sand dunes (Figure 23.18). Behind the barrier island lies a _lagoon_, which is a broad expanse of shallow water, often several miles wide, and in places largely filled with tidal deposits. The salt marsh shown in Figure 23.15 is a filled lagoon.

A characteristic feature of most barrier islands is the presence of gaps, known as _tidal inlets_. Through these gaps strong currents flow alternately seaward and landward as the tide rises and falls. In heavy storms, the barrier may be breached by new inlets (Figure 23.19). Tidal currents will subsequently tend to keep a new inlet open, but it may be closed by shore drifting of sand.

Shallow water results in generally poor natural harbors along barrier-island shorelines. The lagoon itself may serve as a harbor if channels and dock areas are dredged to sufficient depths. Ships enter and leave through one of the passes in the barrier island, but artificial sea walls and jetties are required to confine the current and keep sufficient channel depth. Many of the major port cities are located where a large river empties into the lagoon. The lower courses of large rivers provide tidal channels that may be dredged to accommodate large vessels and thus make seaports of cities many miles inland.

One of the finest examples of a barrier island and lagoon is along the Gulf Coast of Texas (Figure 23.20). Here the island is unbroken for as much as 100 mi (160 km) at a stretch, and passes are few. The lagoon is 5 to 10 mi (8 to 16 km) wide. Galveston is built on the barrier island adjacent to an inlet connecting Galveston Bay with the sea. Most other Texas ports, however, are located on the mainland shore. Corpus Christi, Rockport, Texas City, Lavaca, and other ports are located along the shores of river embayments.

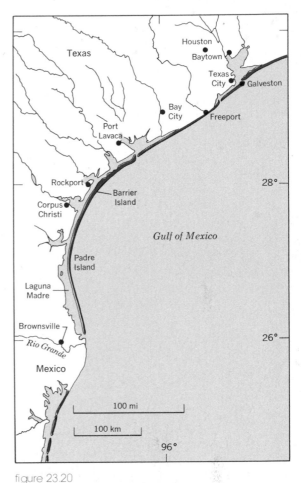

figure 23.20
The Gulf Coast of Texas is dominated by its offshore barrier island.

Delta coasts

The deposit of clay, silt, and sand made by a stream where it flows into a body of standing water is known as a _delta_ (Figure 23.21). Deposition is caused by rapid reduction in velocity of the current as it pushes out into the standing water. Typically, the river channel divides and subdivides into lesser channels called _distributaries_. The coarse particles settle out first; the fine clays continue out farthest and eventually come to rest in fairly deep water. Contact of fresh with salt water causes the finest clays to clot into larger aggregates, which settle to the sea floor.

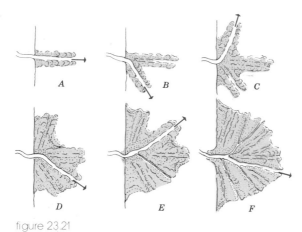

figure 23.21

Stages in the formation of a simple delta. (After G. K. Gilbert.)

Deltas show a variety of shapes, both because of the configuration of the coastline and because of wave action. The Nile delta, whose resemblance to the Greek letter "delta" suggested the name for this type of landform, has many distributaries that branch out in a radial arrangement (Figure 23.22A). Because of its broadly curving shoreline, causing it to resemble in outline an alluvial fan, this type may be described as an arcuate delta. The Mississippi River delta presents a very different sort of picture (Figure 23.22B). It is of the bird-foot type, with long, projecting fingers growing far out into the water at the ends of each distributary.

Where wave attack is vigorous, the sediment brought out by the stream is spread along the shore in both directions from the river mouth, giving a pointed delta with curved sides. Because of its resemblance to a sharp tooth, this type is called a cuspate delta (Figure 23.22C). Where a river empties into a long, narrow estuary, the delta is confined to the shape of the estuary (Figure 23.22D). This type can be called an estuarine delta.

Deltas of large rivers have been of environmental importance from earliest historical times because their extensive flat fertile lands support dense agricultural populations. Important coastal cities, linking ocean and river traffic, are often situated on or near deltas. Examples are Alexandria on the Nile, Calcutta on the Ganges-Brahmaputra, Amsterdam and Rotterdam on the Rhine, Shanghai on the Yangtze, Marseilles on the Rhone, and New Orleans on the Mississippi.

Delta growth is often rapid, ranging from about 10 ft (3 m) per year for the Nile to 200 ft (60 m) per year for the Po and Mississippi rivers. Some cities and towns that were at river mouths several hundred years ago are today several miles inland. An important engineering problem is to keep an open channel for ocean-going vessels that have to enter the delta distributaries to reach port. The mouths of the Mississippi River delta distributaries, known as passes, have been extended by the construction of jetties. The narrowed stream is forced to move faster, scouring a deep channel.

Coral-reef coasts

Coral-reef coasts are unique in that the addition of new land is made by organisms: corals and algae. Growing together, these organisms secrete rocklike deposits of mineral carbonate, called _coral reefs_. As coral colonies die, new ones are built on them, accumulating as limestone. Coral fragments are torn free by wave attack, and the pulverized fragments accumulate as sand beaches.

Coral-reef coasts occur in warm, tropical and equatorial waters between the limits lat. 30° N and 25° S. Water temperatures above 68° F (20° C) are necessary for dense reef coral growth. Reef corals live near the water surface. Water must be free of suspended sediment and well aerated for vigorous coral growth. For this reason corals thrive in positions exposed to wave attack from the open sea. Because

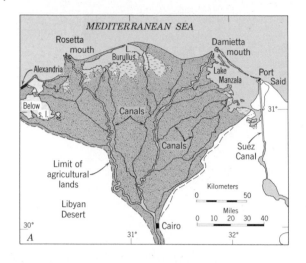

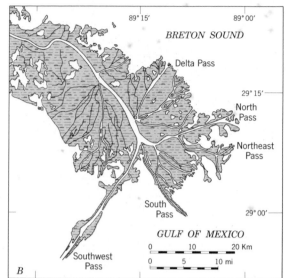

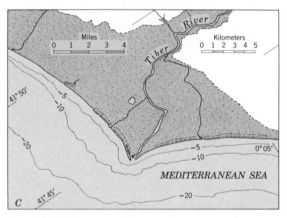

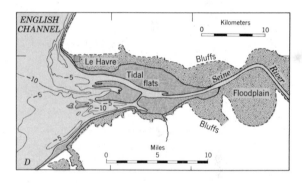

figure 23.22

Deltas. (A) The Nile delta has an arcuate shoreline and is triangular in plan. (B) The Mississippi delta is of the branching, bird-foot type with long passes. (C) The Tiber delta on the Italian coast is pointed, or cuspate, because of strong wave and current action. (D) The Seine delta is filling in a narrow estuary.

muddy water prevents coral growth, reefs are missing opposite the mouths of muddy streams. Coral reefs are remarkably flat on top (Figure 22.23). They are exposed at low tide and covered at high tide.

Three general types of coral reefs may be recognized: (1) fringing reefs, (2) barrier reefs, and (3) atolls. *Fringing reefs* are built as platforms attached to shore (Figure 23.23). They are widest in front of headlands where wave attack is strongest, and the corals receive clean water with abundant food supply.

Barrier reefs lie out from shore and are separated from the mainland by a lagoon (Figure 23.24). Narrow gaps occur at intervals in barrier reefs. Through these openings excess water from breaking waves is returned from the lagoon to the open sea.

Atolls are more or less circular coral reefs enclosing a lagoon, but without any land inside (Figure 23.25). On large atolls, parts of the reef have been built up by wave action and wind to form low island chains, connected by

figure 23.23
A fringing reef on the south coast of Java forms a broad bench between surf zone (left) and a white coral-sand beach. Inland is rainforest. (Luchtvaart-Afdeeling, Bandung.)

the reef. Most atolls have been built on a foundation of volcanic rock. These foundations are thought to have been basaltic volcanoes, built as seamounts on the deep ocean floor. After being planed off by wave erosion, the extinct volcanoes slowly subsided. At the same time the fringing coral reef continued to build upward (Figure 23.26).

The environmental aspects of atoll islands are unique in some respects. First, there is no rock other than coral limestone, composed of calcium carbonate. This means that trees

figure 23.24
A barrier reef is separated from the mainland by a shallow lagoon. (After W. M. Davis.)

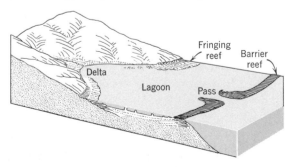

requiring other minerals, such as silica, cannot be cultivated without the aid of fertilizers or some outside source of rock from a larger island composed of volcanic or other igneous rock.

The palm tree is native to atoll islands because it thrives on brackish water, and the seed, or palm nut, is distributed widely by floating from one island to another. Native inhabitants cultivated the cocoanut palm to provide food, clothing, fibers, and building materials. Fresh water is scarce on small atoll islands. Rainfall must be caught in open vessels or catchment basins and carefully conserved. Fish and other marine animals are an important part of the human diet on atoll islands. Calm waters of the lagoon make a good place for fishing and for beaching canoes.

Coral islands of the western Pacific stand in continual danger of devastation by tropical cyclones (typhoons). Breaking waves wash over the low-lying ground, sweeping away palm trees and houses and drowning the inhabitants. There is no high ground for refuge. Great seismic sea waves of unpredictable occurrence also inundate atoll islands.

figure 23.25
Photographed from an altitude of about 150 mi (240 km) by astronauts aboard <u>Gemini V</u> spacecraft, Rongelap Atoll appears as a closed loop among cloud patches, Marshall Islands, Pacific Ocean. (NASA photograph.)

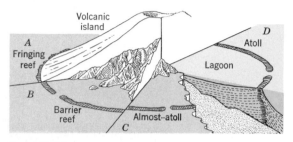

figure 23.26
The subsidence theory of barrier-reef and atoll development is shown in four stages, beginning with a fringing reef attached to a volcanic island and ending with a circular reef. (After W. M. Davis.)

Raised shorelines and marine terraces

The active life of a shoreline is sometimes cut short by a sudden rise of the coast. When this event occurs, a *raised shoreline* is formed. The marine cliff and abrasion platform are abruptly raised above the level of wave action. The former abrasion platform has now become a *marine terrace* (Figure 23.27). Fluvial denudation begins to destroy the terrace, and it may also undergo partial burial under alluvial fan deposits.

Marine terraces are important to Man along mountainous coasts because they offer strips of flat ground extending for tens of miles parallel with the shoreline. Highways and railroads follow these terraces, and they are excellent sites for coastal towns and cities. Agriculture makes use of the flat terrace surfaces where the soil is good.

Raised shorelines are common along the continental and island coasts of the Pacific Ocean, because here tectonic processes are active along the mountain and island arcs. Repeated uplifts result in a series of raised shorelines in a steplike arrangement. Fine examples of these multiple marine terraces are seen on the western slope of San Clemente Island, off the California coast (Figure 23.28).

figure 23.27
A raised shoreline becomes a cliff parallel with the newer, lower shoreline. The former abrasion platform is now a marine terrace.

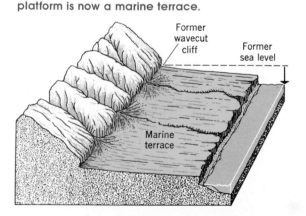

figure 23.28
Marine terraces on the western slope of San Clemente Island, off the southern California coast. More than twenty terraces have been identified in this series; the highest has an elevation of about 1300 ft (400 m). (John S. Shelton.)

Coastal landforms in review

Variety is the key word in describing the landforms produced by waves, currents, and organisms along the world's shorelines. We have dealt with only a few representative examples of the kinds of coastlines to be found on our planet. Each kind of coastal zone offers a different habitat for life forms, and each offers a different situation to confront Man in his expanding development of coastal resources.

In past centuries the coastal zone was used principally as a source of food or as a place from which to embark on the ocean in search of food or to trade with other lands and peoples. Today the situation is changing. The coastal zone is now in great demand as a recreation zone, and its economic value lies mostly in the worth of the waterfront land as real estate. Marinas spring up at every available sheltered spot. Here, recreational boating and sport fishing dwarf the shrinking commercial fishing activity. Luxury condominiums rise on filled land where the tidal flat and salt marsh once supported a complex marine ecosystem. Industry, too, presses for its share of the coastal zone for use as sites of nuclear power plants and oil refineries.

Wise decisions on coastal-zone management often depend partly on accurate knowledge of the operation of coastal processes, such as wave erosion and littoral drift of sediment. Man-made shoreline changes often set off unwanted retrogradation or progradation elsewhere along the coast, or result in filling of channels and harbors. These changes also profoundly affect the shallow-water ecosystems.

Transportation and deposition of sand by wind is an important process in shaping coastal landforms. We have made references in this chapter to coastal sand dunes derived from beach sand. In the next chapter we shall investigate the transport of sand by wind and the shaping of dune forms. In so doing we complete the linkage between wind action and wave action in controlling coastal environments.

Landforms Made by Wind

Chapter 24

WIND BLOWING OVER the solid surface of the lands is another of the active agents of landform development. Ordinarily, wind is not strong enough to dislodge mineral matter from the surfaces of tightly knit rock or of moist, clay-rich soils, or of soils bound by a dense plant cover. Instead, the action of wind in eroding and transporting sediment is limited to land surfaces where small mineral and organic particles are in the loose state. Such areas are typically deserts and semiarid lands (steppes). An exception is the coastal environment, where beaches provide abundant supplies of loose sand, even where the climate is humid and the land surface inland from the coast is well protected by a plant cover.

Landforms shaped and sustained by wind erosion and deposition represent distinctive life environments, often highly specialized with respect to the communities of animals and plants they support. In climates with barely sufficient soil water, there is a contest between wind action and the growth of plants that tend to stabilize landforms and protect them from wind action. We shall find precarious balances in the ecosystems of certain marginal climatic zones. These balances are not only altered by natural changes in climate, but are easily upset by Man's activities, and often with serious consequences. To understand these environmental changes, we need to acquire a working knowledge of the physical processes of wind action on the land surfaces.

Erosion by wind

Wind performs two kinds of erosional work. Loose particles lying on the ground surface may be lifted into the air or rolled along the ground. This process is *deflation*. Where the wind drives sand and dust particles against an exposed rock or soil surface, causing it to be worn away by the impact of the particles, the process is *wind abrasion*. Abrasion requires cutting tools carried by the wind; deflation is accomplished by air currents alone.

Deflation acts wherever the ground surface is thoroughly dried out and is littered with small, loose particles of soil or regolith. Dry river courses, beaches, and areas of recently formed glacial deposits are highly susceptible to deflation. In dry climates, almost the entire ground surface is subject to deflation because the soil or rock is largely bare. Wind is selective in its deflational action. The finest particles, those of clay and silt sizes, are lifted most easily and raised high into the air. Sand grains are moved only by moderately strong winds and travel close to the ground. Gravel fragments and rounded pebbles can be rolled over flat ground by strong winds, but they do not travel far. They become easily lodged in hollows or between other large grains. Consequently, where a mixture of sizes of particles is present on the ground, the finer sizes are removed; the coarser particles remain behind.

A landform produced by deflation is a shallow depression called a *blowout*. This depression may be from a few yards to a mile or more in diameter, but it is usually only a few feet deep. Blowouts form in plains regions in dry climates. Any small depression in the surface of the plain, particularly where the grass cover is broken through, may develop into a blowout. Rains fill the depression, creating a shallow pond or lake. As the water evaporates, the mud bottom dries out and cracks, forming small scales or pellets of dried mud that are lifted out by the wind. In grazing lands, animals trample the margins of the depression into a mass of mud, breaking down the protective grass-root structure and facilitating removal when dry. In this way the depression is enlarged (Figure 24.1). Blowouts are also found on rock surfaces where the rock is being disintegrated by weathering.

In the great deserts of the southwestern United States, the floors of intermontane basins are

figure 24.1
A blowout hollow on the plains of Nebraska. The ground is well trampled by the hooves of cattle. The remnant column of the original soil provides a natural yardstick for the depth of material removed by deflation. (U.S. Geological Survey.)

vulnerable to deflation. The flat floors of shallow playas have in some places been reduced by deflation as much as several feet over areas of many square miles.

Where deflation has been active on a ground surface littered with loose fragments of a wide range of sizes, the pebbles that remain behind accumulate until they cover the entire surface. By rolling or jostling about as the fine particles are blown away, the pebbles become closely fitted together, forming a _desert pavement_ (Figure 24.2). In North Africa such a pebble-covered surface is called a _reg_. The armored surface is well protected against further deflation, but it is easily disturbed by the wheels of trucks or motorcycles.

The sandblast action of wind against exposed rock surfaces is limited to the basal few feet of a rock mass rising above a flat plain, because sand grains do not rise high into the air. Wind abrasion produces pits, grooves, and hollows in the rock. Telephone poles on windswept sandy plains are quickly cut through at the base unless a protective metal sheathing or heap of large stones is placed around the base.

Dust storms

Strong, turbulent winds, blowing over desert surfaces, lift great quantities of fine dust into the air, forming a dense, high cloud called a _dust storm_. In semiarid grasslands a dust storm is generated where ground surfaces have been stripped of protective vegetative cover by cultivation or grazing. A dust storm approaches as a dark cloud extending from the ground surface to heights of several thousand feet (Figure 24.3). Within the dust cloud there is deep gloom or even total darkness. Visibility is cut to a few yards, and a fine choking dust penetrates everywhere.

figure 24.2
This desert pavement of quartzite fragments was formed by action of both wind and water on the surface of an alluvial fan in the desert of southeastern California. Silt underlies the layer of stones. (C. S. Denny, U.S. Geological Survey.)

figure 24.3
Front of an approaching dust storm, Coconino Plateau, Arizona. (D. L. Babenroth.)

It has been estimated that as much as 4000 tons of dust may be suspended in a cubic mile of air (875 metric tons per cubic kilometer). On this basis, a large dust storm can carry more than 100 million tons of dust — enough to make a hill 100 ft (30 km) high and 2 mi (3 km) across the base. Dust travels long distances in the air. Dust from a single desert dust storm is often traceable as far as 2500 mi (4000 km).

Sand dunes

A *dune* is any hill of sand shaped by the wind. Dunes may be active, when bare of vegetation and constantly changing form under wind currents. They may be inactive, covered by vegetation that has taken root and prevents further shifting of the sand.

Dune sand is most commonly composed of the mineral quartz, which is extremely hard and largely immune to chemical decay. The grains are beautifully rounded (see Figure 16.12) and behave as tiny elastic spheres. Under the force of strong winds the grains move in long, low leaps, rebounding after impact with other grains. Rebounding grains rarely rise more than a few inches above the dune surface. Grains struck by leaping grains are pushed forward and, in this way, the surface sand layer creeps downwind (Figure 24.4).

One common type of sand dune is an isolated heap of free sand called a *crescent dune*. As the name suggests, this dune has a crescentic outline; the points of the crescent are directed downwind (Figure 24.5). On the windward side of the crest the sand slope is gentle and smoothly rounded. On the lee side of the dune, within the crescent, is a steep curving dune slope, the *slip face*. This face maintains an angle of about 35° from the horizontal (Figure 24.6). Sand grains slide down the steep face after being blown free of the sharp crest. When a strong wind is blowing, the flying sand makes a perceptible cloud at the crest.

Crescent dunes rest on a flat, pebble-covered ground surface. The sand heap may originate as a drift in the lee of some obstacle, such as a small hill, rock, or clump of brush. Once a

figure 24.4
Sand particles travel in a series of long jumps.

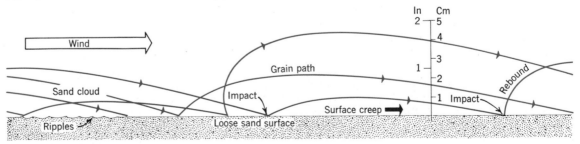

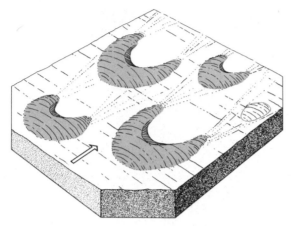

figure 24.5
Crescent dunes. The arrow indicates wind direction.

sufficient mass of sand has formed, it begins to move downwind, taking the form of a crescent dune. For this reason the dunes are usually arranged in chains extending downwind from the sand source.

Where sand is so abundant that it completely covers the ground, dunes take the form of wavelike ridges separated by troughlike furrows. The dunes are called *transverse dunes* because, like ocean waves, their crests trend at right angles to direction of wind (Figure 24.7). The entire area may be called a *sand sea,* because it resembles a storm-tossed sea suddenly frozen to immobility. The sand ridges

have sharp crests and are asymmetrical, the gentle slope being on the windward, the steep slip face on the lee side. Deep depressions lie between the dune ridges. Sand seas require huge quantities of sand, often derived from weathering of a sandstone formation underlying the ground surface, or from adjacent alluvial plains. Still other transverse dune belts form adjacent to beaches that supply abundant sand and have strong onshore winds (see Figure 24.10).

Another group of dunes belongs to a family in which the curve of the dune crest is bowed convexly downwind, the opposite of the curvature of crests in the crescent dune. These dunes are parabolic in form. A common representative of this class is the *coastal blowout dune,* formed adjacent to beaches. Here large supplies of sand are available and are blown landward by prevailing winds (Figure 24.8A). A saucer-shaped depression is formed by deflation; the sand is heaped in a curving ridge resembling a horseshoe in plan. On the landward side is a steep slip face that advances over the lower ground and buries forests, killing the trees (Figure 24.9). Coastal blowout dunes are well displayed along the southern and eastern shore of Lake Michigan. Dunes of the southern shore have been protected for public use as the Indiana Dunes State Park.

figure 24.6
Crescent dunes at Biggs, Oregon. The slip faces are in deep shadow. Wind direction is from left to right. (U.S. Geological Survey.)

figure 24.7
A great sand sea of transverse dunes between Yuma, Arizona and Calexico, California. A field of crescent dunes lies at the lower right. (Spence Air Photos.)

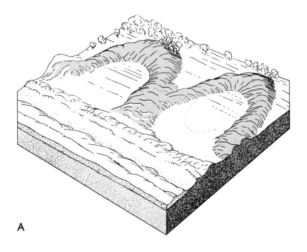

A

figure 24.8
Parabolic dunes. In all three diagrams, the prevailing wind blows from lower left to upper right.

A. Coastal blowout dunes move landward to override a forest.
B. Parabolic blowout dunes are common on semiarid plains.
C. These parabolic blowout dunes have become elongated into hairpin forms.

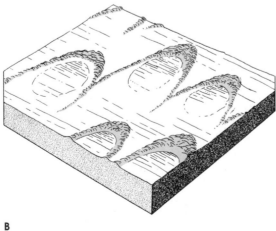

B

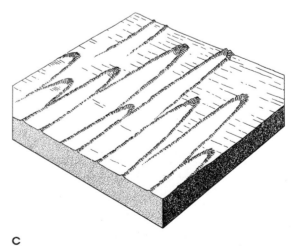

C

figure 24.9
Slip face of a coastal blowout dune advancing over a forest. Cape Henry, Virginia. (Douglas Johnson.)

On semiarid plains, where vegetation is sparse and winds strong, groups of *parabolic blowout dunes* develop to the lee of shallow deflation hollows (Figure 24.8B). Sand is caught by low bushes and accumulates in a broad, low ridge. These dunes have no steep slip faces and may remain relatively immobile. In some cases the dune ridge migrates downwind, drawing the

parabola into a long, narrow form with parallel sides resembling a hairpin in outline (Figure 24.8C). Hairpin dunes stabilized by vegetation are seen in Figure 24.10.

In the vast deserts of North Africa, Arabia, and Iran there are large, complex dune forms not represented in North America. One distinctive type is the *star dune,* a great hill of sand whose base resembles a many-pointed star in plan (Figure 24.11). Ridges of sand rise toward the dune center, culminating in sharp peaks as high as 300 ft (100 m) or more above the base. Star dunes remain fixed in position for centuries and have served as reliable landmarks for desert travelers.

Coastal dunes and Man

Landward of sand beaches we usually find a narrow belt of dunes in the form of irregularly shaped hills and depressions; these are the *foredunes.* They normally bear a cover of beachgrass and a few other species of plants capable of survival in the severe environment.

On coastal foredunes, the cover of beachgrass and other small plants, sparse as it seems to be, acts as a baffle to trap sand moving landward

figure 24.10
The arrows on this photograph point to parabolic blowout dunes of hairpin form. These dunes once advanced from the beach and have since become stabilized by vegetation. Active transverse dunes are overriding the blowout dunes in a fresh wave, San Luis Obispo Bay, California. (Spence Air Photos.)

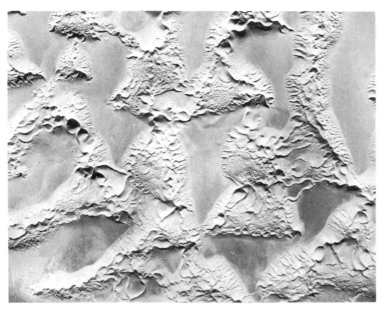

figure 24.11
Star dunes seen from an altitude of 6 mi (10 km) above the Libyan desert. The star-shaped central peaks rise 300 to 600 ft (90 to 180 m) higher than the intervening flat ground. Width of the photograph represents about 7 mi (11 km). (Aero Service Corporation. Western Geophysical Company of America.)

figure 24.12
Foredunes protected by beachgrass, Provincelands of Cape Cod. (A. N. Strahler.)

from the adjacent beach (Figure 24.12). As a result, the foredune ridge is built up as a barrier rising many feet above high tide level. For example, dune summits of the Landes coast of France reach heights of 300 ft (90 m) and span a belt 6 mi (10 km) wide.

The swash of storm waves cuts away the upper part of the beach, and the dune barrier is attacked. Between storms the beach is rebuilt and, in due time, the dune ridge is also restored if plants are maintained. In this way the foredunes form a protective barrier for tidal lands lying on the landward side.

If the plant cover of the dune ridge is depleted by vehicular and foot traffic or by bulldozing of sites for approach roads and buildings, a blowout will rapidly develop. The new cavity may extend as a trench across the dune ridge. With the onset of a storm with high water levels, swash is funneled through the gap and spreads out on the tidal marsh or tidal lagoon behind the ridge. Sand swept through the gap is spread over the tidal deposits.

For many coastal communities of the eastern United States seaboard, the breaching of a dune ridge with its accompanying overwash brings a certain measure of environmental damage to the tidal marsh or estuary, and there may be extensive property damage as well.

Another form of environmental damage related to dunes is the rapid downwind movement of sand when the dune status is changed from one of fixed, plant-controlled forms to that of active dunes of free sand. When plant cover is depleted, wind rapidly reshapes the dunes to produce crests and slip faces. Blowout dunes are developed, and the free sand slopes advance on forests, roads, buildings, and agricultural lands. In the Landes region of coastal dunes on the southwestern coast of France, landward dune advance has overwhelmed houses and churches and even caused entire towns to be abandoned.

Loess

In several large midlatitude areas of the world the ground is underlain by deposits of wind-transported silt, which settled out from dust storms over many thousands of years. This material is known as _loess_. It generally has a uniform yellowish to buff color and lacks any visible layering. Loess has a tendency to break away along vertical cliffs wherever it is exposed by the cutting of a stream or by Man (Figure 24.13).

The thickest deposits of loess are in northern China, where a layer over 100 ft (30 m) thick is common and a maximum of 300 ft (90 m) has been measured. It covers many hundreds of square miles and appears to have been derived as dust from the interior of Asia. Loess deposits are also of major importance in the United States, central Europe, central Asia, and Argentina.

In the United States, thick loess deposits lie in the Missouri-Mississippi valley (Figure 24.14). Much of the prairie plains region of Indiana, Illinois, Iowa, Missouri, Nebraska, and Kansas is underlain by a loess layer ranging in thickness from 3 to 100 ft (1 to 30 m). There are also extensive deposits along the lands

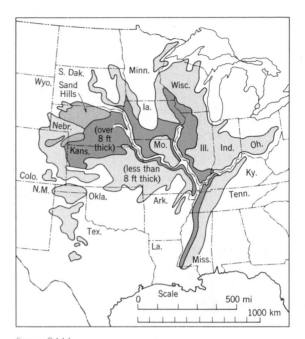

figure 24.14

Map of loess distribution in the central United States.

bordering the lower Mississippi River floodplain on its east side, throughout Tennessee and Mississippi. Still other loess deposits are in northeast Washington and western Idaho.

The American and European loess deposits are directly related to the continental glaciers of the Pleistocene Epoch. At the time when the ice covered much of North America and Europe, a generally dry winter climate prevailed in the land bordering the ice sheets. Strong winds blew southward and eastward over the bare ground, picking up silt from the floodplains of braided streams that discharged the meltwater from the ice. This dust settled on the ground between streams, gradually building up to produce a smooth, level ground surface. The loess is particularly thick along the eastern sides of the valleys because of prevailing westerly winds, and it is well exposed along the bluffs of most streams flowing through the region today.

figure 24.13

Road sunken deeply into loess, Shensi, China. (Frederick G. Clapp, The Geographical Review.)

Loess is of major importance in world agri-cultural resources. Loess forms the parent matter of rich, black soils especially suited to cultivation of grains. These are the prairie, chernozem, chestnut, and brown soils described in Chapter 10. The highly productive plains of southern Russia, the Argentine pampa, and the rich grain region of north China are underlain by loess. In the United States, corn is extensively cultivated on the loess plains in states such as Iowa and Illinois, where rainfall is sufficient; wheat is grown father west on loess plains of Kansas and Nebraska and in the Palouse region of eastern Washington.

Loess forms vertical walls along valley sides and is able to resist sliding or flowage. Because it is easily excavated, loess has been widely used for cave dwellings both in China and in central Europe.

Man as an agent in inducing deflation

Cultivation of vast areas of short-grass prairie under a climate of substantial seasonal water shortage is a practice inviting deflation of soil surfaces. Much of the Great Plains region of the United States is such a marginal region. In past centuries these plains have experienced many dust storms generated by turbulent winds. Strong cold fronts frequently sweep over this area, lifting dust high into the troposphere at times when soil moisture is low.

Deflation and soil drifting reached disastrous proportions during a series of drought years in the middle 1930s, following a great expansion of wheat cultivation. These former grasslands are underlain by humus-rich brown and chestnut soils. During the drought a sequence of exceptionally intense dust storms occurred. Within their formidable black clouds visibility declined to nighttime darkness, even at noonday. The area affected became known as the Dust Bowl. Many inches of topsoil were removed from fields and transported out of the region as suspended dust, while the coarser silt and sand particles accumulated in drifts along fence lines and around buildings. The combination of environmental degradation and repeated crop failures caused widespread abandonment of farms and a general exodus of farm families.

Among scientists who have studied the Dust Bowl phenomenon, there is a difference of opinion as to how great a role soil cultivation and livestock grazing played in augmenting deflation. The drought was a natural event over which Man had no control, but it seems reasonable that the natural grassland would have sustained far less soil loss and drifting if it had not been destroyed by the plow.

Man's activities in the very dry, hot deserts have contributed measurably to raising of dust clouds. In the desert of northwest India and Pakistan (the Thar Desert bordering the Indus River), the continued trampling of fine-textured soils by hooves of grazing animals and by human feet produces a blanket of dusty hot air that hangs over the region for long periods and extends to a height of 30,000 ft (9 km).

In other deserts, such as those of North Africa and the southwestern United States, ground surfaces in the natural state contribute comparatively little dust because of the presence of desert pavements and sheets of coarse sand from which fines have already been winnowed. This protective layer is easily destroyed by wheeled vehicles, exposing finer-textured materials and allowing deflation to raise dust clouds. The disturbance of large expanses of North African desert by tank battles during World War II caused great dust clouds; dust from this source was identified as far away as the Caribbean region.

Man and the
Physical Environment

Epilogue

OUR SURVEY OF physical geography has brought together four major ingredients that define the physical environment: climate, soils, vegetation, and landforms. We used climate as the basis for recognizing global environmental regions. We made this choice because climate provides the basic input of heat and water into the life layer where soil is formed and plants develop their distinctive formation classes.

Landforms add another dimension of variation within the broad climatic regions. Geologic processes play an important, independent role in landform evolution through tectonic and volcanic activities that have shaped the continents and the major landform units within the continents. Climate is also a powerful factor in shaping landforms. In this way, each of the climatic environmental regions has some distinctive quality of landform instilled by climate through the same inputs of heat and water that vary the quality of soil and plant cover.

Interaction of Man with the natural environment has been a persistent theme of our study of physical geography. Here we are in the heartland of geography as a science — a study in human ecology and the spatial distribution of the varied ingredients. Many forms of interaction appear as harmful to the natural environment and pose threats to the very sources of water and food on which humans and lesser life forms depend. In closing this survey of physical geography we turn to some broad problems confronting the human race. Physical geography can provide an important part of the science background needed to analyze these problems in a sound perspective. Among the basic questions we ask are these: Do we face an environmental crisis? Do we face a shortage in supplies of materials and energy? What is our present status with respect to these problems? What will the future bring for mankind?

Is there an environmental crisis?

Many voices have been raised in alarm over the degraded state of the environment and threats to the well-being of ecosystems. Is mankind really threatened by the consequences of his own action? Degrees of alarm vary substantially. Which voice do we heed?

Throughout innumerable recent publications we find statements referring to a crisis or to multiple crises arising to plague our planet. Often the authors are not explicit in giving any clear picture of what constitutes a crisis, what part of the biosphere will be affected, and whether the crisis exists now or will arise in the future. One basic form of crisis deals with shortages of food or mineral resources, or both, in the face of a rising global population of humans. Another form of crisis deals with the unfavorable impact of pollution — broadly defined to include all forms of environmental deterioration — on the biosphere. In other words, crises can take the form of biospheric destruction by processes such as poisoning by toxic substances, genetic damage from ionizing radiation, reduction of nutrient substances from the primary producers, and physical destruction of ecosystems and their habitats by release of mechanical or heat energy. A third crisis, in a class by itself, is catastrophic global disruption by use of nuclear weapons.

If we define an environmental crisis as a situation demanding immediate corrective action, if such be possible, to avert deterioration, damage, or destruction of the affected system, we can point to crises on various scales. Some crises will be current and local. A simple example would be a city beset by worsening smog and a static weather situation. Signs of eutrophication of a lake signal an approaching crisis in the environment of an aquatic ecosystem.

Perhaps what most environmentalists mean when they refer to an environmental crisis of global proportions is the total situation reached at that point in time when the global demand for vital forms of energy or matter catches up with the available supply. Energy, in this sense, can take the form of food for organisms, or heat and power for industrial uses. Matter can also take the form of food for organisms, or it may consist of essential materials for industrial uses.

A global crisis involving energy or matter, or both, is inherent in one basic fact that few will dispute: the earth's resources of energy and matter are finite, whereas the demand for these resources is rising, with no limit in sight except the total finite resource itself.

[To sum it up, the environmental crisis, of which so many voices warn us, is an event of many dimensions. Crisis presents itself on an areal scale ranging from local to global, and on a time scale ranging from the present to an indeterminate future point in time. The threat of crisis ranges in magnitude from deterioration of small ecosystems to total destruction of the biosphere]

While population growth of the human species is not a topic of physical geography, that growth is a vital ingredient in a global crisis, since greater population means both greater demand for usable energy and materials and a greater potential to pollute and degrade the environment. Therefore we turn next to a brief review of some current views on population growth in relation to environment.

Population, food resources, and environment

The current debate over population growth and its consequences can scarcely have escaped your attention. "Zero Population Growth" has joined the long list of slogans of a growing ecology-conscious group of citizenry. Perhaps in no other debate on great issues of mankind has so much emotionalism been injected by so many participants. We have rarely witnessed such bitterness as is expressed by those whose positions lie at the two extremes of the debate. On the one hand there are some who claim we can sustain an almost indefinite rise in global population, since our ingenuity and technology will rise to meet any challenge. After all, they say, are there not enormous untapped food resources in the sea? And are we not experiencing a green revolution in crop production? At the other extreme are those who warn of wholesale destruction of human life by starvation in several of the developing nations within a few decades, despite the green revolution. They have gone so far as to warn that once mass starvation sets in, it will be futile for food-rich industrial nations even to attempt to provide relief supplies.

The phenomenal rise in total world population, together with a sharp increase in the rate of that rise, is a fact on which there can be no debate. When population is plotted against time on arithmetic scales the rising curve steepens so sharply that it becomes almost perpendicular in the time interval of the past half-century. In contrast, that part of the curve representing the first millenium A.D. appears almost flat, so slow is the increase. Consider that during the entire course of human history, up to the year 1850, human population had reached only one billion persons. The present world population is about 3.6 billion. Currently the rate of increase is such that a doubling of the population can take place in about thirty-five years. One recent summary of population trends states that, allowing for some decline in birth rates, the world population for the year 2000 will be 6.5 to 7.0 billion.[2] This same source goes on to say:

> Failing that reduction we may have 7.5 billion. A continued decline in death rates and progress toward a net reproduction rate of one* by the year 2040 (an optimistic assumption) would still

*Synonymous with *zero population growth*.

yield a population leveling off at 15 billion late in the next century. Thus, considering demographic factors alone and leaving aside questions of resource adequacy or environmental tolerances, our best prospect would be for a stationary world population of between 15 and 20 billions. . . . The best estimate of demographers is that rapid growth is sure to continue for some time, probably more than a hundred years, even on optimistic assumptions about controlling fertility rates.[1]

Predictions relating to leveling off of population growth in developing nations presently experiencing rapid growth are subject to considerable disagreement.

On the subject of the ability of the earth to furnish enough food to sustain a much larger population, Professor Preston Cloud, a noted biogeologist, has stated:

The Committee on Resources and Man of the National Academy of Sciences, which I chaired, has examined the ultimate capability of an efficiently managed world to produce food, given the cultivation of all potentially arable lands and an optimal expression of scientific and technological innovation. Its results imply that 1968 world food supplies might eventually be increased by as much as nine times, provided that sources of protein were essentially restricted to plants, and to seafood mostly from a position lower in the marine food chain than customarily harvested and provided metal resources, agricultural water, and risky pesticides and mineral fertilizers are equal to the task and do not create intolerable side effects, and that agricultural land is not unduly preempted for other purposes. Such an ultimate level of productivity might, therefore, sustain a world population of 30 billion. Most of these, of course, would live at a level of chronic malnutrition unless distribution were so regulated as to be truly equitable, in which case all would live at a bare subsistence level, although with protein deficiency. . . .

That seems to place a maximal limit on world populations, at a figure that could be reached by little more than a century from now at present rates of increase. Of course, many students of the problem do not believe that it is possible

to sustain 30 billion people — especially considering the variety and extent of demands on water, air, and earth materials, and the psychic stresses that would arise with such numbers. I have no dispute with them. Here I seek simply to establish some theoretical outside limit, based on optimistic assumptions about what might be possible. More important attributes than a starvation diet would be sacrificed by a world that full![2]

An increasing world population places a heavy strain on the planet in two ways. First, more people cause more pollution and more environmental degradation. Second, more people require more of the earth's resources besides food (e.g., water supplies and plant fiber); they place heavier demands on the nonrenewable mineral resources. So, we turn next to some considerations of the environmental impact of a greatly increased population.

Resources and their management

In Chapter 16 we noted that mineral resources — including the fossil fuels — represent the results of accumulated geological activity of hundreds of millions of years. Consumption of these resources has skyrocketed in the past century, just as the world population itself has shot precipitously upward. Let us examine the consequences of this prodigious consumption of resources that can never be renewed in the lifetime of Man on earth.

A number of authorities in a good position to know the facts have attempted to appraise the public of what the future holds in store. Walter R. Hibbard, Jr., formerly Director of the U.S. Bureau of Mines, puts it this way.

A requisite for affluence, now or in the future, is an adequate supply of minerals — fuels to energize our power and transportation; non-metals, such as sulfur and phosphates to fertilize farms; and metals, steel, copper, lead, aluminum, and so forth, to build our machinery, cars, buildings, and bridges. These are the materials basic to our economy, the multipliers in our gross national product. But the needed materials which can be recovered by known methods at reasonable cost from the earth's crust are limited, whereas their rates of exploitation are not. This situation cannot continue.[3]

Differences exist, as one would expect, in the best estimates of reserves of mineral resources remaining in the earth's crust. More optimistic than most authorities in the field is Vincent E. McKelvey, Director of the U.S. Geological Survey. In a paper presented at Harvard University in 1971, McKelvey pointed out that dire predictions of mineral shortages have been made at intervals over the past 60 years. Nevertheless, as economic growth has moved rapidly ahead and mineral consumption has risen, the intensified research for new mineral reserves has been thus far able to keep pace with demands. McKelvey stated:

> Personally, I am confident that for millennia to come we can continue to develop the mineral supplies needed to maintain a high level of living for those who now enjoy it and to raise it for the impoverished people of our own country and the world. My reasons for thinking so are that there is a visible underdeveloped potential of substantial proportions in each of the processes by which we create resources and that our experience justifies the belief that these processes have dimensions beyond our knowledge and even beyond our imagination at any given time.[4]

However, McKelvey takes a cautious position. He recognizes that many do not share his views. He feels that a searching review of resource adequacy may be required. He then goes on to say:

> If our supply of critical materials is enough to meet our needs for only a few decades, a mere tapering off in the rate of increase of their use, or even a modest cutback, would stretch out these supplies for only a trivial period. If resource adequacy cannot be assured into the far distant future, a major reorientation of our philosophy, goals, and way of life will be necessary. And if we do need to revert to a low resource-consuming economy, we will have to begin

the process as quickly as possible in order to avoid chaos and catastrophe.[4]

Although finite, our nonrenewable resources, including fossil fuels and uranium, the principal energy sources, allow us some options with respect to the program of depletion. Preston Cloud analyzes this concept in the following words.

> Thus it is possible to think of man in relation to his renewable resources of food, water, and breathable air as limited in numbers by the sustainable annual crop. His non-renewable resources — metals, petrochemicals, mineral fuels, etc. — can be thought of, in somewhat over-simplified but valid terms, as some quantity which may be withdrawn at different rates, but for which the quantity beneath the curve of cumulative total production, from first use to exhaustion of primary sources, is largely independent of the shape of the curve. In other words, the depletion curve of a given non-renewable raw material, or of that class of resources, may rise and decline steeply over a relatively short time, or it may be a flatter curve that lasts for a longer time, depending on use rates and conservation measures such as recycling. This is a choice that civilized, industrialized man expresses whenever his collective aspirations, efforts, and ideals express themselves in a given general density of population, per capita rate of consumption, and conservation practice.[2]

Stretching of our limited earth resources to the utmost is a conservation procedure strongly endorsed by all who study resource problems. Pollution and environmental degradation are usual by-products of development and use of raw materials. Thus the various means available to conserve mineral use simultaneously serve two vital functions. One is to make the resource last as long as possible; the other is to minimize environmental impact.

Energy resources, together with matter resources, comprise the total natural resource picture. While various aspects of energy resources have been included in the preceding paragraphs, it is useful to make a further analysis of energy consumption problems.

Consumption of energy

Do we face a long-term energy crisis? Along with other natural resources, the consumption of energy has taken the same sharp upturn as the global population. In 1972, testifying before the House Committee on Interior and Insular Affairs, Vice Admiral Hyman G. Rickover, a leading authority on technical applications of nuclear energy, had this to say.

> I can think of no public issue confronting us today that will touch the lives of our children and grandchildren — if not ourselves — more closely than the ability of the U.S. to command the energy resources that are needed to sustain our economy. . . . A realistic assessment of our energy resources position would stress the fact that it is not important exactly *when* our fossil fuels give out. We may find better ways to extract oil and gas from coal, tar sands, shale; we may discover more reserves than we know of at present on the Continental Shelf surrounding our country or elsewhere in the sea. What is important is to comprehend that some day the fossil fuels will be gone; that renewable energy resources are unlikely to provide more than a small percentage of our needs; that even atomic energy, since it requires uranium, is finite; and that we cannot be certain that some man-made alternative as yet undeveloped will arrive in time — or ever — to supply our energy needs.[5]

As in the case of materials resources, Vincent E. McKelvey, Director of the U.S. Geological Survey, presents a more optimistic outlook on the future of energy resources.

> Most important to secure our future is an abundant and cheap supply of energy, for if that is available we can obtain materials from low-quality sources, perhaps even country rocks, as Harrison Brown has suggested. Again, I am personally optimistic on this matter, with respect both to the fossil fuels and particularly to the nuclear fuels. Not only does the breeder reactor appear to be near enough to practical reality to justify the belief that it will permit the use of extremely low-grade sources of uranium and thorium that will carry us far into the future, but during the last couple of years there have been exciting new developments in the prospects for commercial energy from fusion.[4]

Whatever may be the shades of opinion on this question of future supplies of energy, there is no question that the prodigious increase in energy consumption poses many problems, including serious threats to the environment. The present rate of increase of energy consumption is such that the doubling time is about 15 years. We can anticipate a corresponding increase in the frequency and severity of environmental problems in years to come.

Discussing the environmental impact of energy use, Philip H. Abelson, had this to say in an editorial in the journal *Science*.

> Consumption of energy is the principal source of air pollution, and energy production, transportation, and consumption are responsible for an important fraction of all our environmental problems. Use of energy continues to rise at the rate of 4.5 percent per year. Even if fuel supplies were infinite, such an increase could not be tolerated indefinitely. But fuel supplies are not inexhaustible, and this combined with the need to preserve the environment will force changes in patterns of energy production and use.
>
> Our economy has been geared to profligate expenditure of energy and resources. Much of our pollution problem would disappear if we drove 1-ton instead of 2-ton automobiles. Demand for space heat and cooling could be reduced if buildings were properly insulated. Examples of needless use of electricity are everywhere. Promotional rates and advertising tend to encourage excessive consumption.
>
> Slowing down the rate of increase in use of energy will not be easy. Public habits of energy consumption will not be quickly altered, and a sudden change in the rate of growth of energy consumption would cause major additional unemployment.[6]

What are the prospects of achieving success in reducing the environmental impact of energy consumption? Abelson concludes his editorial in this way.

> Measures to cut excessive use of energy are likely to come only after a long time, if ever. We should face the possibility that increased

consumption of energy will continue and prepare to meet that possibility. Atmospheric pollution is not an inevitable consequence of production of energy. In the use of fossil fuels, production of sulfur dioxide is not an essential by-product. Destruction of the environment is not a necessary consequence of strip mining. Pollution from almost every method of producing and utilizing energy could be sharply attenuated either through better practices or through development of new methods. In view of the importance of energy to society, present expenditures on research and development related to energy are small and these are not well apportioned.[6]

The prospect for a steady-state existence

[Increasingly, in the past few years, members of the scientific community have been voicing the concept that the survival of mankind on our planet will be contigent on achievement of an overall steady state in the total global system of flux of energy and matter, including Man and his institutions within the framework of the biosphere.]

The concept of a steady state in the total global system is a fitting one with which to close this book. Let us read what others have had to offer on this subject.

> We need to concentrate our orientation of knowledge on the human-society-plus-environment level of integration because of its relevance to the central world problem of achieving a reasonably steady state between human societies and the finite resources of our planet. We are as much concerned with human society itself as with the environments in which men live; both are parts of an interacting whole that evolves as a unit through time. -*S. Dillon Ripley and Helmut K. Beuchner.*[7]

The use of the environment is a necessary and acceptable concept. The difference is that future use must be in the recycle context of perpetual renewal and reuse, not in the old pattern of use and discard. A sort of stable state between civilization and the environment is called for — not a balance of nature (for nature is always changing in its own right) but a harmony of society and the environment within natural laws of physics, chemistry, and biology. -*Committee on Scientific Astronautics.*[8]

Very likely a steady-state economy may come about eventually as the result of increased scarcity and higher costs of environmental protection, which will also encourage the recovery and recycling of materials. If a more or less steady level of consumption can be maintained and can be accompanied by a stable population, technological advance may permit some small but steady gain in level of living. The way in which we approach a steady-state economy is most important. If it can be done through the operation of market forces, the steady state would bring a high quality of life for a maximum population. On the other hand, if the steady state is approached by way of social upheavals, sudden shortages of energy and mineral resources may produce a catastrophe that will find us with a steady-state economy but at a much lower level. If the catastrophe results in world-wide nuclear war then the level may well be at or close to zero. -*V. E. McKelvey and S. Fred Singer.*[9]

Over three billion years of history, read from the geological record, make it clear that the question is not whether man will come into balance with nature or not. The only question is whether this will happen as a result of self-imposed restraints of natural catastrophes over which he has no control. Nor is there any question that man can, if he chooses, determine his own destiny — at least to a very large degree and for a very long time. The question is, "will he choose to do so?". . . [In all his long history man has never faced a worthier or a more critical challenge than that of achieving a lasting balance with his environment — and I mean the *total* human ecosystem, including the city, the sea, and the wilderness. That is the challenge primarily of the generation now entering maturity. Population control, sensible resource management, and continuous surveillance of all components of a thoroughly researched global ecosystem, including its various sociopolitical components, are the essential steps toward such a balance.] A principal goal of higher education should be to discover and to communicate the basic knowledge and comprehensive understanding that will lead to those steps being taken in a well-informed, humane, and orderly manner, and as soon as practicable. -*Preston Cloud.*[2]

References cited

1. Sterling Brubaker, 1972, *To Live on Earth; Man and His Environment in Perspective,* published for Resources for the Future, Inc., by The Johns Hopkins Press, Baltimore and London, p. 38.

2. Preston Cloud, 1971, "This Finite Earth," *A. & S. Review,* Indiana University, Spring 1971, pp. 17-32.

3. Walter R. Hibbard, Jr., 1968, "Mineral Resources: Challenge or Threat?," *Science,* Vol. 160, p. 143. Copyright 1968 by the American Association for the Advancement of Science.

4. Vincent E. McKelvey, 1972, "Mineral Resource Estimates and Public Policy," *American Scientist,* Vol. 60, p. 36.

5. Hyman G. Rickover, 1972, Statement to House Committee on Interior and Insular Affairs regarding fuel and energy resources, April 18, 1972.

6. Philip H. Abelson, 1971, "Continuing Increase in Use of Energy," *Science,* Vol. 172, p. 795. Copyright 1971 by the American Association for the Advancement of Science.

7. S. Dillon Ripley and Helmut K. Buechner, 1967, "Ecosystem Science as a Point of Synthesis," Reprinted by permission of *Daedalus,* Journal of the Academy of Arts and Sciences, Boston, Mass., Fall 1967, *America's Changing Environment.*

8. Managing the Environment, 1968, Publ. Committee on Scientific Astronautics, 90th Congress, 2nd Session, 1968, U.S. Government Printing Office, Washington, D.C., pp. 14-15.

9. Vincent E. McKelvey and S. Fred Singer, 1971, "Conservation and the Minerals Industry — a Public Dilemma, "*Geotimes,* Vol. 16, No. 12, p. 21.

Appendix
World maps
for physical
geography

MAP PROJECTIONS MUST be carefully selected to show geographical information on a global scale. In this book we use three basic types of projections; each has special properties best suited to present a category of information.

Figures A.1, A.2, and A.3 show the three projections we selected. First is the _Goode projection,_ named for its designer, Dr. J. Paul Goode. The second is the _Mercator projection,_ a classic navigator's map invented in 1569 by Gerhardus Mercator. The third is a _polar projection._ Some observations and comparisons concerning these three map projections follow.

Both Goode and Mercator projections have straight horizontal parallels of latitude. This feature is good for showing any subject matter relating to climate, because latitude is a strong control of many climate factors. Because the parallels are horizontal, we can scan across the map at a given level to compare regions most likely to be similar in climate. In contrast, the polar projection is centered on either of the poles; the parallels are concentric circles. The polar projection is best used to show geographical information in very high latitudes. It gives a good sense of true relationships among lands and seas grouped around the north pole. On the other hand, the polar projection has to be limited to areas near the pole; it is almost useless for a single map of the entire globe.

The Goode projection has a most important property with regard to the way in which it presents areas of the earth's surface. This network is an _equal-area projection,_ on which the size (areal extent) of any small area is correctly scaled, no matter where it falls on the map. If we were to draw a very small circle on a sheet of clear plastic and move this circle over all parts of the Goode world map, the circle would always show an area with a constant value in square miles or square kilometers. The equal-area projection is used to show any geographical entity that occupies

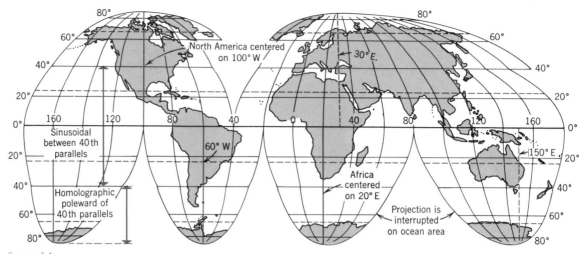

figure A.1

Goode's interrupted homolosine projection. The projection makes use of two well-known older projections. The sinusoidal projection is used between lat. 40° N and S; the homolographic projection is used poleward of lat. 40° N. (Copyright by the University of Chicago. Used by permission of the University of Chicago Press.)

surface area. For example, soil types and vegetation occupy surface area. It is important for the geographer to know at a glance how great an area is occupied by a given type of soil or vegetation.

The Goode projection suffers from a serious defect: it distorts the shapes of areas, particularly in high latitudes and at the far right and left of the network. To minimize this defect, Dr. Goode split his map apart on selected meridians. Each separate sector of the map is centered on a vertical meridian. This type of split map is called an *interrupted projection*. Although the interrupted projection reduces shape distortion a great deal, it has the drawback of separating parts of the earth's surface that are actually close together, particularly in the high latitudes.

The Mercator projection has its own unique quality. The shape of any small area — such as an island or lake — is preserved almost perfectly, no matter in what part of the earth it lies. In other words, a small square-shaped patch of ground is shown as a true square in any part of the grid. A map that preserves

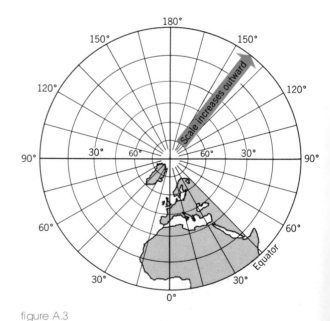

figure A.3

A polar projection. The map is centered on the north pole. All meridians are straight lines radiating from the center point, while all parallels are concentric circles. This particular network is of the stereographic type, which is conformal.

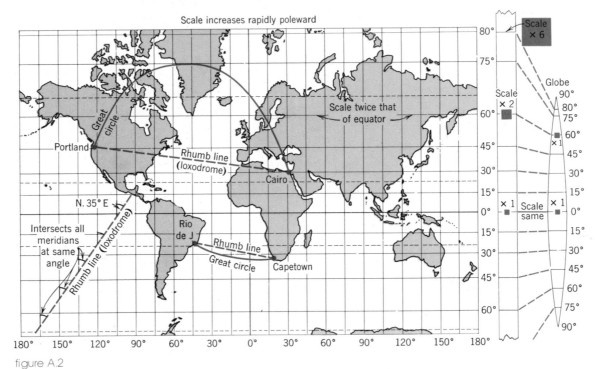

figure A.2

The Mercator projection. The rhumb line is always a straight line and holds a constant compass direction between any two points. The great circle is the shortest distance between any two points on the globe. The diagram at the right shows that map scale increases rapidly into higher latitudes. At lat. 60° the scale is double the equatorial scale; at lat. 80° the scale is six times greater than at the equator.

shapes correctly is called a *conformal projection*. Because shapes are correctly shown, it follows that the directions of lines on the earth's surface are also shown in their true cardinal directions. Notice that the meridians are all vertical straight lines, showing the north-south direction as being at right angles to the east-west direction. The lines we show on this map include shorelines and flow lines such as ocean currents or winds. Isopleths (lines of equal value of any selected property) are also shown in their correct directions and patterns. For this reason we use the Mercator projection to show all kinds of line features, among them lines of equal temperature (isotherms), lines of equal pressure (isobars), and crustal lines such as faults or chains of volcanoes.

The Mercator projection has a serious defect. It distorts the scale of global features very greatly in the higher latitudes. Areas become expanded enormously as we move into latitudes above 60°. In fact, the map stretches toward infinity both north and south, and it is impossible to depict either pole. For this reason the Mercator map is best terminated at about the 75th parallel. When the Mercator projection is used, it is often supplemented by two polar projections to show high latitudes.

Bear in mind that every map projection has both useful qualities and serious defects. Select a projection so that the useful qualities match the intended use as closely as possible, but remain aware at all times of the defects and of the possible misconceptions those defects can cause.

Index